Compressed Video Communications

Compressed Video Communications

Abdul H. Sadka
University of Surrey, Guildford, UK

JOHN WILEY & SONS, LTD

Baffins Lane, Chichester,
West Sussex PO19 1UD, England

National 01243 779777
International (+44) 1243 779777
e-mail (for orders and customer service enquiries): cs-books@wiley.co.uk
Visit our Home Page on: http://www.wiley.co.uk or http://www.wiley.com

Other Wiley Editorial Offices

John Wiley & Sons, Inc., 605 Third Avenue,
New York, NY 10158-0012, USA

Wiley-VCH Verlag GmbH, Pappelallee 3,
D-69469 Weinheim, Germany

Jacaranda Wiley Ltd, 33 Park Road, Milton,
Queensland 4064, Australia

John Wiley & Sons (Canada) Ltd, 22 Worcester Road,
Rexdale, Ontario M9W 1L1, Canada

John Wiley & Sons (Asia) Pte Ltd, Clementi Loop #02-01,
Jin Xing Distripark, Singapore 129809

Library of Congress Cataloguing in Publication Data

Sadka, Abdul H.
Compressed video communications / Abdul H. Sadka.
p. cm.
Includes bilbiographical references and index.
ISBN 0-470-84312-8
1. Digital video. 2. Video compression. I. Title.
TK6680.5.S24 2002
621.388—dc21 2001058144

British Library Cataloguing in Publication Data

A catalogue record for this book is available from the British Library

ISBN 0 470 84312 8

Typeset in 10.5/13pt Times by Vision Typesetting, Manchester

To both my parents,
my lovely wife and
our little daughter.

Contents

The content noted as on CD can now be found at www.wiley.com/go/sadkacompressed

Preface

This book examines the technologies underlying the compression and transmission of digital video sequences over networking platforms. The incorporated study covers a large spectrum of topics related to compressed video communications. It presents to readers a comprehensive and structured analysis of the issues encountered in the transmission of compressed video streams over networking environments. This analysis identifies the problems impeding the progress of compressed video communication technologies and impairing the quality of networked video services; it also presents a wide range of solutions that would help in the quality optimisation of video communication services. The book is a unique reference in which the author has combined the discussions of several topics related to digital video technology such as compression, error resilience, rate control, video transmission over mobile networks and transcoding. All the techniques, algorithms, tools and mechanisms connected to the provision of video communication services are explained and analysed from the application layer down to the transport and network layers. The description of these technologies is accompanied by a large number of video subjective and objective results in order to help graduate students, engineers and digital video researchers make the best understanding of the presented material and build a comprehensive view on the progress of this rather fertile and exciting field of technology. A number of video clips have also been prepared and put at the disposal of readers on the supplementary CD in order to back up the results depicted and conclusions reached within the book chapters.

Abdul H. Sadka

Acknowledgement

The author would like to thank his research assistants, namely Dr S. Dogan for his valuable contribution to Chapter 6, and both Dr S. Dogan and Dr S. Worrall for their assistance with preparing the video clips on the supplementary CD. Particular thanks also go to our computing staff for their understanding and cooperation in dealing with persistent requests for extra video storage space and processing power.

About the Author

Dr Sadka obtained a PhD in Electrical & Electronic Engineering from the University of Surrey in 1997, an MSc with distinction in Computer Engineering from the Middle East Technical University in 1993, and a BSc in Computer and Communications Engineering from the American University of Beirut in 1990. In October 1997, he became a lecturer in Multimedia Communication Systems in the Centre for Communication Systems Research at the University of Surrey, and has been a member of the academic staff of the university since then.

He and his researchers have pioneered work on various aspects of video coding and transcoding, error resilience techniques in video communications and video transmissions over networks. His work in this area has resulted in numerous novel techniques for error and flow control in digital video communications. He has contributed to several short courses delivered to UK industry on various aspects of media compression and multimedia communication systems. He served on the program, advisory and technical committees of various specialised international conferences and workshops. He acts as consultant on image/video compression and mobile video communications to potential companies in the UK Telecommunications industry. He is well supported by the industry in the form of research projects and consultancy contracts. He runs a private consultancy company VIDCOM.

Dr Sadka is the instigator of numerous funded projects covering a wide spectrum of multimedia communications. He has been serving as a referee to a number of *IEE Transactions* and *IEE Electronics Letters* since 1997. He has 2 patents filed in the area of video compression and mobile video systems and over 40 publications in peer refereed conferences and journals, in addition to several contributions to multi-authored books on mobile multimedia communications in the UK and abroad. He is a member of the IEE and a chartered electrical engineer.

1

Introduction

1.1 Background

Both the International Standardisation Organisation (ISO) and the International Telecommunications Union (ITU) standardisation bodies have been releasing recommendations for universal image and video coding algorithms since 1985. The first image coding standard, namely JPEG (Joint Picture Experts Group), was released by ISO in 1989 and later by ITU-T as a recommendation for still image compression. In December 1991, ISO released the first draft of a video coding standard, namely MPEG-1, for audiovisual storage on CD-ROM at 1.5–2 Mbit/s. In 1990, CCITT issued its first video coding standard which was then, in 1993, subsumed into an ITU-T published recommendation, namely ITU-T H.261, for low bit rate communications over ISDN networks at $p \times 64$ kbit/s. ITU-T H.262, alternatively known as MPEG-2, was then released in 1994 as a standard coding algorithm for HDTV applications at 4–9 Mbit/s. Then, in 1996, standardisation activities resulted in releasing the first version of a new video coding standard, namely ITU-T H.263, for very low bit rate communications over PSTN networks at less than 64 kbit/s. Further work on improving the standard has ended up with a number of annexes that have produced more recent and comprehensive versions of the standard, namely H.263+ and H.263++ in 1998 and 1999 respectively. In 1998, the ISO MPEG (Motion Picture Experts Group) AVT (Audio Video Transport) group put forward a new coding standard, namely MPEG-4, for mobile audiovisual communications. MPEG-4 was the first coding algorithm that used the object-based strategy in its layering structure as opposed to the block-based frame structure in its predecessors. In March 2000, the standardisation sector of ISO published the most recent version of a standard recommendation, namely JPEG-2000 for still picture compression. Most of the aforementioned video coding algorithms have been adopted as the standard video codecs used in contemporary multimedia communication standards such as ITU-T H.323 and H.324 for the provision of multimedia communications over packet-switched and circuit-switched networks respectively. This remarkable evolution of video coding technology has underlined the development of a multitude of novel signal compression techniques that aimed to optimise the compression efficiency and quality of service of standard video coders. In this book, we put at the disposal of readers a comprehensive but simple explanation of the basic principles of video coding techniques employed in this long series of standards. Emphasis is placed on the major building blocks that constitute the body of the standardised video coding algorithms. A large number of tests are carried out and included in the book to enable the readers to evaluate the performance of the video coding standards and establish comparisons between them where appropriate, in terms of their coding efficiency and error robustness.

From a network perspective, coded video streams are to be transmitted over a variety of networking platforms. In certain cases, these streams are required to travel across a number of asymmetric networks until they get to their final destination. For this reason, the coded video bit streams have to be transmitted in the form of packets whose structure and size depend on the underlying transport protocols. During transmission, these packets and the enclosed video payload are exposed to channel errors and excessive delays, hence to information loss. Lost packets impair the reconstructed picture quality if the video decoder does not take any action to remedy the resulting information loss. This book covers a whole range of error handling mechanisms employed in video communications and provides readers with a comprehensive analysis of error resilience techniques proposed for contemporary video coding algorithms. Moreover, this book provides the readers with a complete coverage of the quality of service issues associated with video transmissions over mobile networks. The book addresses the techniques employed to optimise the quality of service for the provision of real-time MPEG-4 transmissions over GPRS radio links with various network conditions and different error patterns.

On the other hand, to allow different video coding algorithms to interoperate, a heterogeneous video transcoder must be employed to modify the bit stream generated by the source video coder in accordance with the syntax of the destination coder. Some heterogeneous video transcoders are enabled to operate in two or more directions allowing incompatible streams to flow across 2 or more networks for inter-network video communications. If both sender and receiver are utilising the same video coding algorithm but are yet located on dissimilar networks of different bandwidth characteristics, then a homogeneous video transcoder is required to adapt the information rate of flowing coded streams to the available bandwidth of the destination network. Hybrid video transcoding algorithms have both heterogeneous and homogeneous transcoding capabilities to adapt the transmitted coded video streams to the destination network in terms of both the end-user video decoding syntax and the destination network capacity respectively. This book addresses the technologies underpinning both the homogeneous and heterogeneous transcoding algorithms and presents the solutions proposed for the improvement of quality of service for both transcoding scenarios in error-free and error-prone environments.

1.2 Source Material

ITU has specified a number of test video sequences for use in the performance evaluation process of its proposed video coding paradigms. In this book, we focus most of the conducted tests and experiments onto six different conventional ITU test head-and-shoulder sequences to verify the study made on the performance of the presented video coding schemes and the efficiency of the corresponding error control algorithms. These test sequences have been selected to reflect a wide range of video sequences with different properties and behaviour. Foreman, Miss America, Carphone, Grandma, Suzie and Claire are the six sequences used throughout the book to conduct a large number of experiments and produce subjective and objective results. Other video sequences, such as Stefan and Harry for instance, are used only sporadically throughout the book with minor emphasis placed on their use and corresponding test results. All of the six chosen sequences represent a head-and-shoulder type of scene with different contrast and activity. Foreman is the most active scene of all since it includes a shaky background, high noise and a fair amount of bi-directional motion of the foreground object. Claire and Grandma are both typical head-and-shoulder sequences with uniform and stationary background and minimal amount of activity confined to moving lips and flickering eyelids. Both Claire and Grandma are low motion video sequences with moderate contrast and noise and a uniform background. Miss America is rather more active than Claire and Grandma with the subject once moving her shoulders before a static camera. Suzie is another head-and-shoulder video sequence with high contrast and moderate noise. It contains a fast head motion with the subject, being the foreground, holding a telephone handset with a stationary and plain-textured background. The carphone sequence shows a moving background with fair details. Though not a typical head-and-shoulder sequence, Carphone shows a talking head in a moving vehicle with more motion in the foreground object and a non-uniform changing background. All these sequences are in discrete video YUV format with a decomposition ratio 4:2:0, a QCIF (Quadrature Common Intermediate Format) resolution of 176 pixels by 144 lines and a temporal resolution (frame rate) of 25 frames per second. Figure 1.1 depicts some original frames extracted from each one of the six sequences.

1.3 Video Quality Assessment and Performance Evaluation

Since compression at low bit rates results in inevitable quality degradation, the performance of video coding algorithms must be assessed with regard to the quality of the reconstructed video sequence. Both subjective and objective methods are usually adopted to evaluate the performance of video coding algorithms. The decoded video quality can be measured by simply comparing the

(a) Foreman (b) Claire
(c) Grandma (d) Miss America
(e) Suzie (f) Carphone

Figure 1.1 Original frames of used ITU test sequences

original and reconstructed video sequences. Although the subjective evaluation of decoded video quality is quite cumbersome compared to the calculation of numerical values for the objective quality evaluation, it is still preferable especially for low and very low bit rate compression because of the inconsistency between the existing numerical quality measurements and the Human Visual System (HVS). On the other hand, in error-prone environments, errors might corrupt the coded video stream in a way that causes a merge or split in the transmitted video frames. In this case, using the objective numerical methods to compare the original and reconstructed video sequences would incorporate some errors in associating the peer frames (corresponding frames between the two sequences) in both sequences with each other. This leads to an inaccurate evaluation of the coder performance. A subjective measurement in this case would certainly yield a fairer and more precise evaluation of the decoded video quality.

There are two broad types of subjective quality evaluation, namely rating scale methods and comparison methods (Netravali and Limb, 1980). In the first method, an overall quality rating is assigned to the image (usually the last frame of a video sequence) by using one of several given categories. In the second method, a quality impairment of a standard type is introduced to the original image until the viewer decides the impaired and reference images are of equal quality. However, throughout this book, pair comparison is used where the original sequence and decoded sequence frames are displayed side by side for subjective quality evaluation. Original sequence frames are used as reference to demonstrate the performance of a video coding algorithm in error-free environments. However, when the aim is to evaluate the performance of an error resilience technique, the original frames are then replaced by error-free decoded ones since the improvement is then intended to be shown on the error performance of the coder (decoded video quality in error-prone environments) and not on its error-free compression efficiency.

The quality of the video sequence can also be measured by using some mathematical criteria such as signal-to-noise ratio (SNR), peak-to-peak signal-to-noise ratio (PSNR) or mean-squared-error (MSE). These measurement criteria are considered to be objective due to the fact that they rely on the pixel luminance and chrominance values of the input and output video frames and do not include any subjective human intervention in the quality assessment process. For image and video, PSNR is preferred for objective measurements and is frequently used by the video coding research community, although the other two criteria are still occasionally used. PSNR and MSE are defined in Equations 1.1 and 1.2, respectively:

$$PSNR = 10\log_{10}\frac{255^2}{\frac{1}{M \times N}\sum_{i=0}^{M-1}\sum_{j=0}^{N-1}[x(i,j)-\hat{x}(i,j)]^2} \tag{1.1}$$

$$MSE = \frac{1}{M \times N}\sum_{i=0}^{M-1}\sum_{j=0}^{N-1}[x(i,j)-\hat{x}(i,j)]^2 \tag{1.2}$$

where M and N are the dimensions of the video frame in width and height respectively, and $x(i,j)$ and $\hat{x}(i,j)$ are the original and reconstructed pixel luminance or chrominance values at position (i,j).

Additionally, for a fair performance evaluation of a video coding algorithm, the bit rate must also be included. The output bit rate of a video coder is expressed in bits per second (bit/s). Since the bit rate is directly proportional to the number of pixels per frame and the number of frames coded per second, both the picture resolution and frame rate have to be indicated in the evaluation process as well. QCIF picture resolution and a frame rate of 25 frames per second have been adopted throughout the book unless otherwise specified.

1.4 Outline of the Book

The book is divided into six chapters covering the core aspects of video communication technologies. Chapter 1 presents a general historical background of the area and introduces to the reader the conventional ITU video sequences used for low bit rate video compression experiments. This chapter also discusses the conventional methods used for assessing the video quality and evaluating the performance of a video compression algorithm both subjectively and objectively.

Chapter 2 presents an overview of the core techniques employed in digital video compression algorithms with emphasis on standard techniques. The author highlights the major motivations for video compression and addresses the main issues of contemporary video coding techniques, such as model-based, segmentation-based and vector-based coders. The standardised block-transform video coders are then analysed and their performance is evaluated in terms of their quality/bit rate optimisation. A comprehensive comparison of ITU-T H.261 and H.263 is carried out in terms of their compression efficiency and robustness to errors. Emphasis is placed on the improvements brought by the latter by highlighting its performance in both the baseline and full-option modes. Then, the object-based video coding techniques are addressed in full details and particular attention is given to the ISO MPEG-4 video coding standard. The main techniques used in MPEG-4 for shape, motion and texture coding are covered, and the coder performance is evaluated in comparison to the predecessor H.263 standard. Finally, the concept of layered video coding is described and the performance of a layered video coder is analysed objectively with reference to a single layer coder for both quality and bit rates achieved.

Chapter 3 analyses the flow control mechanisms used in video communications. The factors that lead to bit rate variability in video coding algorithms are first described and alternatives to variable rate, fixed quality video coding are examined. Fixed rate video coding is then discussed by explaining several techniques used to achieve a regulated output bit rate. A variety of bit rate control algorithms are presented and their performance is evaluated using PSNR and bit rate values. Furthermore, particular attention is given to the feed-forward MB-based bit rate control algorithm which outperforms the standard-compliant rate control algorithm used in H.263 video coder. The performance of the feed-forward rate control technique is evaluated and comparison is established with the conventional TM5 rate controller. Furthermore, the concept of Region-of-Interest (ROI) coding is introduced with particular emphasis on its use for rate control purposes. The main benefit of using ROI in rate control algorithms is demonstrated by means of objective and subjective illustrations. The issue of prioritising compressed video information is then described by shedding light on its applicability for video rate control purposes. The prioritised information drop technique is analysed and its effectiveness is substantiated using objective and subjective methods. Methods used to prioritise video data in accordance to its sensitivity to errors, its contribution to quality and the reported channel conditions are presented. Additionally, the new concept of the internal feedback loop within the video encoder is explained and its usefulness for rate control is consolidated by subjective and objective evaluation methods. The effect of rate control on the perceptual video quality is illustrated by means of PSNR graphs and some video frames extracted from the rate controlled sequences. The reduced resolution rate control algorithm is presented and its ability to operate under very tight bit rate budget considerations is demonstrated. An extended version of the reduced resolution rate controller is then described with adaptive frame rate for improved rate control mechanism. The multi-layer video coding, described in Chapter 2, is then presented as a bit rate control algorithm commonly used in video communications today. The video scaleability techniques are also a point of focus in this chapter with particular attention given to the Fine Granularity Scaleability (FGS) technique recently recommended for operation under the auspices of the MPEG-4 video standard.

Chapter 4 is solely dedicated to all aspects of error control in video communications. Firstly, the effects of transmission errors on the decoded video quality are analysed in order to provide the reader with an understanding of the severity of the errors problem and a feeling of the importance of error resilience schemes, especially in mobile video communications. The sensitivity of different video parameters to error is then analysed to determine the immunity of video data to transmission errors and decide about the level of error protection required for each kind of parameter. Then, the description of error control mechanisms starts with the zero-redundancy concealment algorithms that are usually decoder-based techniques. Several techniques proposed for the recovery of lost or damaged motion data, DC coefficients and MB modes are presented. Then, the author presents a wide range of error resilience schemes, both proprietary and standards-compliant, used in video communications. Examples of these error resilience techniques are the robust INTRA frame, two-way decoding with reversible codewords, EREC (Error Resilient Entropy Coding), Reference Picture Selection (RPS), Video Redundancy Coding (VRC), etc. The performances of these schemes and their effectiveness in achieving error control are evaluated using an extensive illustration of subjective and objective results obtained from transmitting video over several environments and subjecting it to different error patterns. A comprehensive error-resilient video coding algorithm, namely H.263/M for mobile applications, is explained and its performance examined in comparison with the core H.263 standard. Thereafter, optimal combinations of these error-resilience tools are shown and analysed to further improve the performance of video compression techniques over error-prone environments.

In Chapter 5, the main issues associated with the provision of video services over the new generation mobile networks are investigated. The author describes the main characteristics and features of IP-based mobile networks from the service perspective to assess the feasibility of providing mobile video services from the point of view of quality of service and error performance. The multi-slotting feature underpinning the new radio interface technology and the channel coding schemes of radio protocols are highlighted and their implications on the video quality of service are pinpointed and analysed. The video quality is then put into perspective with a view to analyse the QoS issues in mobile video communications. The effects of some QoS elements such as packet structure and size (mainly for real-time video communications using RTP over IP), channel coding and throughput control using time-slot multiplexing, on the perceptual video quality are discussed with a comprehensive analysis of their implications on the received video quality. Quality control methods are further elaborated by describing the effect of combined error resilience tools on the perceptual quality of video in GPRS and UMTS radio access networks. The combination of error resilience tools used in the performance evaluation of video transmissions over these networks is selected in accordance with the profiles specified in Annex X of H.263 for wireless video applications.

Chapter 6 covers all aspects of transcoding in video communications. Two different kinds of transcoding algorithms, namely homogeneous and heterogenous, are presented. Several types of bit-rate reduction homogeneous transcoding schemes are analysed. The picture drift phenomenon resulting from open-loop transcoding is explained and methods to counteract its effects on the perceptual video quality are presented. This chapter also describes a number of techniques used to improve the quality of transcoded video data especially in the re-estimation and refinement of transcoded motion data. In addition to bit rate reduction algorithms, frame-rate and resolution reduction transcoding schemes are also elaborated. On the other hand, heterogeneous video transcoding algorithms are also described in this chapter with emphasis on inter-network communications. Video transcoding for error resilience purposes and inter-network traffic planning is also covered and associated technologies are highlighted with a view to the multi-transcoder video proxy which is highly desirable in packet-switched (H.323-based inter-network) multi-party video communication services. The description of the transcoding concepts throughout the whole chapter is supported by a vast number of illustrations and subjective/objective results, reflecting their operation and performance, respectively.

A list of useful references is appended to the end of each chapter in order to provide the reader with a rich bibliography for further reading on related topics. Appendix A addresses the layering syntax and semantics of ITU-T H.263 video coding standard for comparison with the modified H.263/M coder presented in Chapter 4. Finally Appendix B explains the content of the video clips on the supplementary CD.

1.5 References

Netravali, A. N., and Limb, J. O., Picture coding: a review, *Proc. IEEE*, **68**, 366–406, Mar. 1980.

2

Overview of Digital Video Compression Algorithms

2.1 Introduction

Since the digital representation of raw video signals requires a high capacity, low complexity video coding algorithms must be defined to efficiently compress video sequences for storage and transmission purposes. The proper selection of a video coding algorithm in multimedia applications is an important factor that normally depends on the bandwidth availability and the minimum quality required. For instance, a surveillance application may only require limited quality, raising alarms on identification of a human body shape, and a user of a video telephone may be content with only sufficient video quality that enables him to recognise the facial features of his counterpart speaker. However, a viewer of an entertainment video might require a DVD-like service quality to be satisfied with the service. Therefore, the required quality is an application-dependent factor that leads to a range of options in choosing the appropriate video compression scheme. Moreover, the bit and frame rates at which the selected video coder must be adaptively chosen in accordance with the available bandwidth of the communication medium. On the other hand, recent advances in technology have resulted in a high increase of the power of digital signal processors and a significant reduction in the cost of semiconductor devices. These developments have enabled the implementation of time-critical complex signal processing algorithms. In the area of audiovisual communications, such algorithms have been employed to compress video signals at high coding efficiency and maximum perceptual quality. In this chapter, an overview of the most popular video coding techniques is presented and some major details of contemporary video coding standards are explained. Emphasis is placed on the study and performance analysis of ITU-T H.261 and H.263 video coding standards and a comparison is established between the two coders in terms of their performance and error robustness. The basic principles of the ISO MPEG-4 standard video coder are also explained. Extensive subjective and objective test results are depicted and analysed where appropriate.

2.2 Why Video Compression?

Since video data is either to be saved on storage devices such as CD and DVD or transmitted over a communication network, the size of digital video data is an important issue in multimedia technology. Due to the huge bandwidth requirements of raw video signals, a video application running on any networking platform can swamp the bandwidth resources of the communication medium if video frames are transmitted in the uncompressed format. For example, let us assume that a video frame is digitised in the form of discrete grids of pixels with a resolution of 176 pixels per line and 144 lines per picture. If the picture colour is represented by two chrominance frames, each one of which has half the resolution of the luminance picture, then each video frame will need approximately 38 kbytes to represent its content when each luminance or chrominance component is represented with 8-bit precision. If the video frames are transmitted without compression at a rate of 25 frames per second, then the raw data rate for video sequence is about 7.6 Mbit/s and a 1-minute video clip will require 57 Mbytes of bandwidth. For a CIF (Common Intermediate Format) resolution of 352×288, with 8-bit precision for each luminance or chrominance component and a half resolution for each colour component, each picture will then need 152 kbytes of memory for digital content representation. With a similar frame rate as above, the raw video data rate for the sequence is almost 30 Mbit/s, and a 1-minute video clip will then require over 225 Mbytes of bandwidth. Consequently, digital video data must be compressed before transmission in order to optimise the required bandwidth for the provision of a multimedia service.

2.3 User Requirements from Video

In any communication environment, users are expected to pay for the services they receive. For any kind of video application, some requirements have to be fulfilled in order to satisfy the users with the service quality. In video communications, these requirements are conflicting and some compromise must be reached to provide the user with the required quality of service. The user requirements from digital video services can be defined as follows.

2.3.1 Video quality and bandwidth

These are frequently the two most important factors in the selection of an appropriate video coding algorithm for any application. Generally, for a given compression scheme, the higher the generated bit rate, the better the video quality. However, in most multimedia applications, the bit rate is confined by the scarcity of transmission bandwidth and/or power. Consequently, it is necessary to trade-off the network capacity against the perceptual video quality in order to come up with the optimal performance of a video service and an optimal use of the underlying network resources.

On the other hand, it is normally the type of application that controls the user requirement for video quality. For videophony applications for instance, the user would be satisfied with a quality standard that is sufficient for him to identify the facial features of his correspondent end-user. In surveillance applications, the quality can be acceptable when the user is able to detect the shape of a human body appearing in the scene. In telemedicine however, the quality of service must enable the remote end user to identify the finest details of a picture and detect its features with high precision. In addition to the type of application, other factors such as frame rate, number of intensity and colour levels, image size and spatial resolution, also influence the video quality and the bit rate provided by a particular video coding scheme. The perceptual quality in video communications is a design metric for multimedia communication networks and applications development (Damper, Hall and Richards, 1994). Moreover, in multimedia communications, coded video streams are transmitted over networks and are thus exposed to channel errors and information loss. Since these two factors act against the quality of service, it is a user requirement that video coding algorithms are robust to errors in order to mitigate the disastrous effects of errors and secure an acceptable quality of service at the receiving end.

2.3.2 Complexity

The complexity of a video coding algorithm is related to the number of computations carried out during the encoding and decoding processes. A common indication of complexity is the number of floating point operations (FLOPs) carried out during these processes. The algorithm complexity is essentially different from the hardware or software complexity of an implementation. The latter depends on the state and availability of technology while the former provides a benchmark for comparison purposes. For real-time communication applications, low cost real-time implementation of the video coder is desirable in order to attract a mass market. To minimise processing delay in complex coding algorithms, many fast and costly components have to be used, increasing the cost of the overall system. In order to improve the take up rate of new applications, many original complex algorithms have been simplified. However, recent advances in VLSI technology have resulted in faster and cheaper digital signal processors (DSPs). Another problem related to complexity is power consumption. For mobile applications, it is vital to minimise the power requirement of mobile terminals in order to prolong battery life. The increasing power of standard computer chips has enabled the implementation of some less complex video codecs in standard personal computer for real-time application. For instance, Microsoft's Media player supports the real-time decoding of Internet streaming MPEG-4 video at QCIF resolution and an average frame rate of 10 f/s in good network conditions.

2.3.3 Synchronisation

Most video communication services support other sources of information such as speech and data. As a result, synchronisation between various traffic streams must be maintained in order to ensure satisfactory performance. The best-known instance is lip reading whereby the motion of the lips must coincide with the uttered words. The simplest and most common technique to achieve synchronisation between two or more traffic streams is to buffer the received data and release it as a common playback point (Escobar, Deutsch and Partridge, 1991). Another possibility to maintain synchronisation between various flows is to assign a global timing relationship to all traffic generators in order to preserve their temporal consistency at the receiving end. This necessitates the presence of some network jitter control mechanism to prevent the variations of delay from spoiling the time relationship between various streams (Zhang and Keshav, 1991).

2.3.4 Delay

In real-time applications, the time delay between encoding of a frame and its decoding at the receiver must be kept to a minimum. The delay introduced by the codec processing and its data buffering is different from the latency caused by long queuing delays in the network. Time delay in video coding is content-based and tends to change with the amount of activity in the scene, growing longer as movement increases. Long coding delays lead to quality reduction in video communications, and therefore a compromise has to be made between picture quality, temporal resolution and coding delay. In video communications, time delays greater than 0.5 second are usually annoying and cause synchronisation problems with other session participants.

2.4 Contemporary Video Coding Schemes

Unlike speech signals, the digital representation of an image or sequence of images requires a very large number of bits. Fortunately however, video signals naturally contain a number of redundancies that could be exploited in the digital compression process. These redundancies are either statistical due to the likelihood of occurrence of intensity levels within the video sequence, spatial due to similarities of luminance and chrominance values within the same frame or even temporal due to similarities encountered amongst consecutive video frames. Video compression is the process of removing these redundancies from the video content for the purpose of reducing the size of its digital representation. Research has been extensively conducted since the mid-eighties to produce efficient and robust techniques for image and video data compression.

Image and video coding technology has witnessed an evolution, from the first-generation canonical pixel-based coders, to the second-generation segmentation-based, fractal-based and model-based coders to the most recent third-generation content-based coders (Torres and Kunt, 1996). Both ITU and ISO have released standards for still image and video coding algorithms that employ waveform-based compression techniques to trade-off the compression efficiency and the quality of the reconstructed signal. After the release of the first still-image coding standard, namely JPEG (alternatively known as ITU T.81) in 1991, ITU recommended the standardisation of its first video compression algorithm, namely ITU H.261 for low-bit rate communications over ISDN at $p \times 64$ kbit/s, in 1993. Intensive work has since been carried out to develop improved versions of this ITU standard, and this has culminated in a number of video coding standards, namely MPEG-1 (1991) for audiovisual data storage on CD-ROM, MPEG-2 (or ITU-T H.262, 1995) for HDTV applications, ITU H.263 (1998) for very low bit rate communications over PSTN networks; then the first content-based object-oriented audiovisual compression algorithm was developed, namely MPEG-4 (1999), for multimedia communications over mobile networks. Research on video technology also developed in the early 1990s from one-layer algorithms to scaleable coding techniques such as the two-layer H.261 (Ghanbari, 1992) two-layer MPEG-2 and the multi-layer MPEG-4 standard in December 1998. Over the last five years, switched-mode algorithms have been employed, whereby more than one coding algorithm have been combined in the same encoding process to result in the optimal compression of a given video signal. The culmination of research in this area resulted in joint source and channel coding techniques to adapt the generated bit rate and hence the compression ratio of the coder to the time-varying conditions of the communication medium.

On the other hand, a suite of error resilience and data recovery techniques, including zero-redundancy error concealment techniques, were developed and incorporated into various coding standards such as MPEG-4 and H.263+ (Cote *et al.*, 1998) to mitigate the effects of channel errors and enhance the video quality in error-prone environments. A proposal for ITU H.26L has been submitted (Heising *et al.*, 1999) for a new very low bit rate video coding algorithm which considers the combination of existing compression skills such as image warping prediction, OBMC (Overlapped Block Motion Compensation) and wavelet-based compression to claim an average improvement of 0.5–1.5 dB over the existing block-based techniques such as H.263++. Major novelties of H.26L lie in the use of integer transforms as opposed to conventional DCT transforms used in previous standards the use of $\frac{1}{6}$ pixel accuracy in the motion estimation process, and the adoption of 4×4 blocks as the picture coding unit as opposed to 8×8 blocks in the traditional block-based video coding algorithms. In March 2000, ISO has published the first draft of a recommendation for a new algorithm JPEG2000 for the coding of still pictures based on wavelet transforms. ISO is also in the process of drafting a new model-based image compression standard, namely JBIG2 (Howard, Kossentini and Martins, 1998), for the lossy and lossless compression of bilevel images. The design goal for JBIG2 is to enable a lossless compression performance which is better than that of the existing standards, and to enable lossy compression at much higher compression ratios than the lossless ratios of the existing standards, with almost no degradation of quality. It is intended for this image compression algorithm to allow compression ratios of up to three times those of existing standards for lossless compression and up to eight times those of existing standards for lossy compression. This remarkable evolution of digital video technology and the development of the associated algorithms have given rise to a suite of novel signal processing techniques. Most of the aforementioned coding standards have been adopted as standard video compression algorithms in recent multimedia communication standards such as H.323 (1993) and H.324 (1998) for packet-switched and circuit-switched multimedia communications, respectively. This chapter deals with the basic principles of video coding and sheds some light on the performance analysis of most popular video compression schemes employed in multimedia communication applications today. Figure 2.1 depicts a simplified block diagram of a typical video encoder and decoder.

Each input frame has to go through a number of stages before the compression process is completed. Firstly, the efficiency of the coder can be greatly enhanced if some undesired features of the input frames are primarily suppressed or enhanced. For instance, if noise filtering is applied on the input frames before encoding, the motion estimation process becomes more accurate and hence yields significantly improved results. Similarly, if the reconstructed pictures at the decoder side are subject to post-processing image enhancement techniques such as edge-enhancement, noise filtering (Tekalp, 1995) and de-blocking artefact suppression for block-based compressions schemes, then the decoded picture quality can be substantially improved. Secondly, the video frames are subject to a mathematical

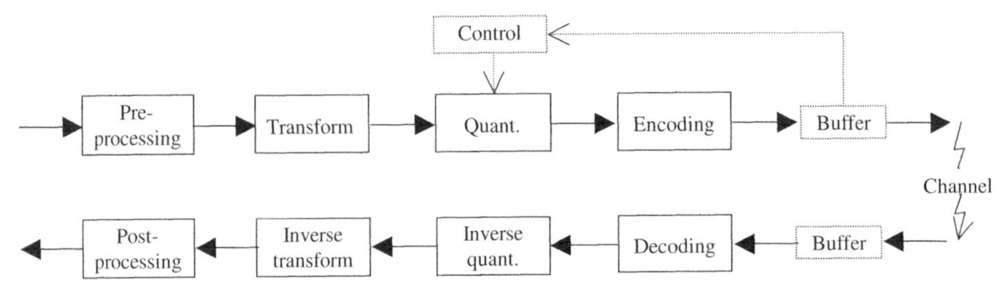

Figure 2.1 Block diagram of a basic video coding and decoding process

transformation that converts the pixels to a different space domain. The objective of a transformation such as the Discrete Cosine Transform (DCT) or Wavelet transforms (Goswami and Chan, 1999) is to eliminate the statistical redundancies presented in video sequences. The transformation is the heart of the video compression system. The third stage is quantisation in which each one of the transformed pixels is assigned a member of a finite set of output symbols. Therefore, the range of possible values of transformed pixels is reduced, introducing an irreversible degradation to quality. At the decoder side, the inverse quantisation process maps the symbols to the corresponding reconstructed values. In the following stage, the encoding process assigns code words to the quantised and transformed video data. Usually, lossless coding techniques, such as Huffman and arithmetic coding schemes, are used to take advantage of the different probability of occurrence of each symbol. Due to the temporal activity of video signals and the variable-length coding employed in video compression scenarios, the bit rate generated by video coders is highly variable. To regulate the output bit rate of a video coder for real-time transmissions, a smoothing buffer is finally used between the encoder and the recipient network for flow control. To avoid overflow and underflow of this buffer, a feedback control mechanism is required to regulate the encoding process in accordance with the buffer occupancy. Rate control mechanisms are extensively covered in the next chapter. In the following sections, the basic principles of contemporary video coding schemes are presented with emphasis placed on the most popular object-based video coding standard today, namely ISO MPEG-4, and the block-based ITU-T standards H.261 and H.263. A comparison is then established between the latter two coders in terms of their performance and error robustness.

2.4.1 Segmentation-based coding

Segmentation-based coders are categorised as a new class of image and video compression algorithms. They are very desirable as they are capable of producing very high compression ratios by exploiting the Human Visual System (Liu and Hayes, 1992; Soryani and Clarke, 1992). In segmentation-based techniques, the image is split into several regions of arbitrary shape. Then, the shape and texture parameters that represent each detected region are coded on a per-region basis. The decomposition of each frame to a number of homogeneous or uniform regions is normally achieved by the exploitation of the frame texture and motion data. In certain cases, the picture is passed through a nonlinear filter before splitting it into separate regions in order to suppress the impulsive noise contained in the picture while preserving the edges. The filtering process leads to a better segmentation result and a reduced number of regions per picture as it eliminates inherent noise without incurring any distortion onto the edges of the image. Pixel luminance values are normally used to initially segment the pictures based on their content.

Then, motion is analysed between successive frames in order to combine or split the segments with similar or different motion characteristics respectively. Since the segmented regions happen to be of arbitrary shape, coding the contour of each region is of primary importance for the reconstruction of frames at the decoder. Figure 2.2 shows the major steps of a segmentation-based video coding algorithm.

Therefore, in order to enhance the performance of segmentation-based coding schemes, motion estimation has to be incorporated in the encoding process. Similarities between the regions boundaries in successive video frames could then be exploited to maximise the compression ratio of shape data. Predictive differential coding is then applied to code the changes incurred on the boundaries of detected regions from one to another. However, for minimal complexity, image segmentation could only be utilised for each video frame with no consideration given to temporal redundancies of shape and texture information. The choice is a trade-off between coding efficiency and algorithmic complexity.

Contour information has critical importance in segmentation-based coding algorithms since the highest portion of output bits are specifically allocated to coding the shape. In video sequences, the shape of detected regions changes significantly from one frame to another. Therefore, it is very difficult to exploit the inter-frame temporal redundancy for coding the region boundaries. A new segmentation-based video coding algorithm (Eryurtlu, Kondoz and Evans, 1995) was proposed for very low bit rate communications at rates as low as 10 kbit/s. The proposed algorithm presented a novel representation of the contour information of detected regions using a number of control points. Figure 2.3 shows the contour representation using a number of control points.

These points define the contour shape and location with respect to the previous frame by using the corresponding motion information. Consequently, this coding scheme does not consider *a priori* knowledge of the content of a certain frame.

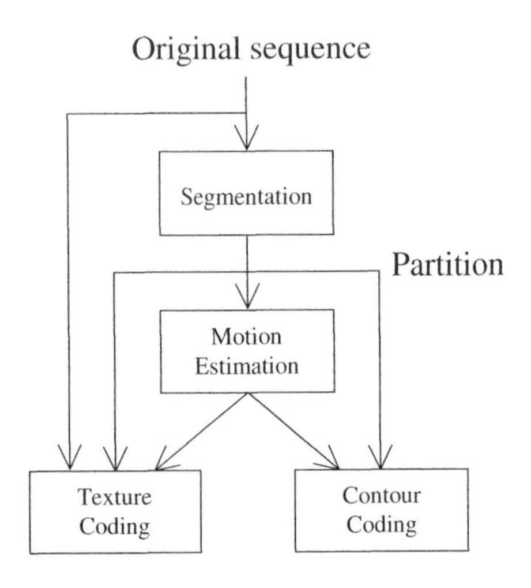

Figure 2.2 A segmentation-based coding scheme

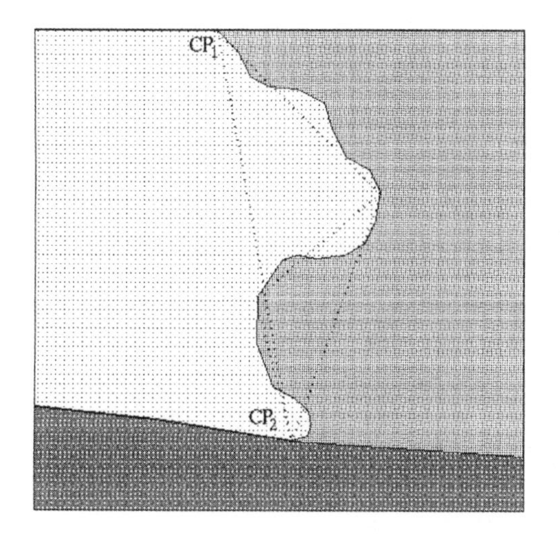

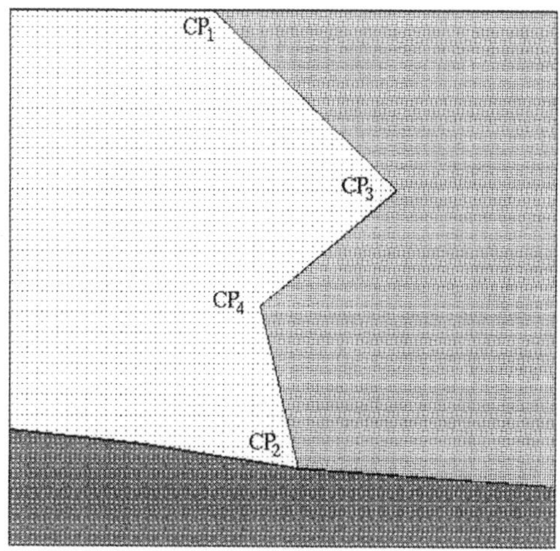

Figure 2.3 Region contour representation using control points

Alternatively, the previous frame is segmented and the regions shape data of the current frame is then estimated by using the previous frame segmentation information. The texture parameters are also predicted and residual values are coded with variable-length entropy coding. For still picture segmentation, each image is split into uniform square regions of similar luminance values. Each square region is successively divided into four square regions until it ends up with homogeneous enough regions. The homogeneity metric could then be used as a trade-off between bit rate and quality. Then, the neighbouring regions that have similar luminance properties are merged up.

ISO MPEG-4 is a recently standardised video coding algorithm that employs the object-based structure. Although the standard did not specify any video compression algorithm as part of the recommendation, the encoder operates in the object-based mode where each object is represented by a video segmentation mask, called the alpha file, that indicates to the encoder the shape and location of the object. The basic features and performance of this segmentation-based, or alternatively called object-based, coding technique will be covered later in this chapter (Section 2.5).

2.4.2 *Model-based coding*

Model-based coding has been an active area of research for a number of years (Eisert and Girod, 1998; Pearson, 1995). In this kind of video compression algorithms, a pre-defined model is generally used. During the encoding process, this model is adapted to detect objects in the scene. The model is then deformed to match the contour of the detected object and only model deformations are coded to represent the object boundaries. Both encoder and decoder must have the same pre-defined model prior to encoding the video sequence. Figure 2.4 depicts an example of a model used in coding facial details and animations.

As illustrated, the model consists of a large set of triangles, the size and orientation of which can define the features and animations of the human face. Each triangle is identified by its three vertices. The model-based encoder maps the texture and shape of the detected video object to the pre-defined model and only model deformations are coded. When the position of a vertex within the model changes due to object motion for instance, the size and orientation of the corresponding triangle(s) change, hence introducing a deformation to the pre-defined model. This deformation could imply either one or a combination of several changes in the mapped object such as zooming, camera pan, object motion, etc. The decoder uses the deformation parameters and applies them on the pre-defined model in order to restore the new positions of the vertices and reconstruct the video frame. This model-based coding system is illustrated in Figure 2.5.

The most prominent advantage of model-based coders is that they could yield very high compression ratios with reasonable reconstructed quality. Some good results were obtained by compressing a video sequence at low bit rates with a model-aided coder (Eisert, Wiegand and Girod, 2000). However, model-based coders have a major disadvantage in that they can only be used for sequences in which the foreground object closely matches the shape of the pre-defined reference model (Choi and Takebe, 1994). While current wire-frame coders allow for the position of the inner vertices of the model to change, the contour of the model must remain fixed making it impossible to adapt the static model to an arbitrary-shape object (Hsu and Harashima, 1994; Kampmann and Ostermann, 1997). For in-

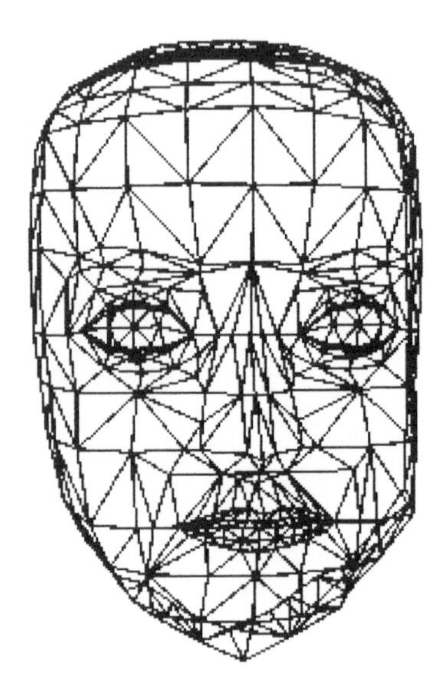

Figure 2.4 A generic facial prototype model

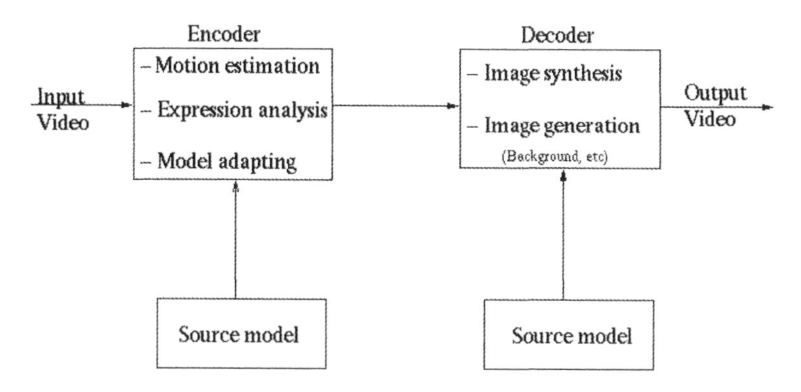

Figure 2.5 Description of a model-based coding system applied to a human face

stance, if the pre-defined reference model represents the shape of a human head-and-shoulder scene, then the video coder would not produce optimal results were it to be used to code sequences featuring for example, a car-racing scene, thereby limiting its versatility. In order to enhance the versatility of a model-based coder, the pre-defined model must be applicable to a wide range of video scenes. The first dynamic model generation technique was proposed (Siu and Chan, 1999) to build a model and dynamically modify it during the encoding process in accordance with new video frames scanned. Thus, the model generation is content-based, hence more flexible. This approach does not specify any prior orientation of a video object since the model is built according to the position and orientation of the object itself. Significant improvement was achieved on the flexibility and compression efficiency of the encoder when the generic model was dynamically adapted to the shape of the object of interest anytime new video information is available. Figure 2.6(b) shows frame 60 of sequence Claire coded using the 3-D pre-defined model depicted in (a). The average bit rate generated by the model-aided coder was almost 19 kbit/s for a frame rate of 25 f/s and CIF (352×288) picture resolution. For a 3-D model of 316 vertices (control points), the coder was able to compress the 60th frame with a luminance PSNR value of 35.05 dB.

2.4.3 Sub-band coding

Sub-band coding is one form of frequency decomposition. The video signal is decomposed into a number of frequency bands using a filter bank. The high-frequency signal components usually contribute to a low portion of the video quality so they can either be dropped out or coarsely quantised. Following the filtering process, the coefficients describing the resulting frequency bands are transformed and quantised according to their importance and contribution to reconstructed video quality. At the decoder, sub-band signals are up-sampled by zero insertion, filtered and de-multiplexed to restore the original video signal.

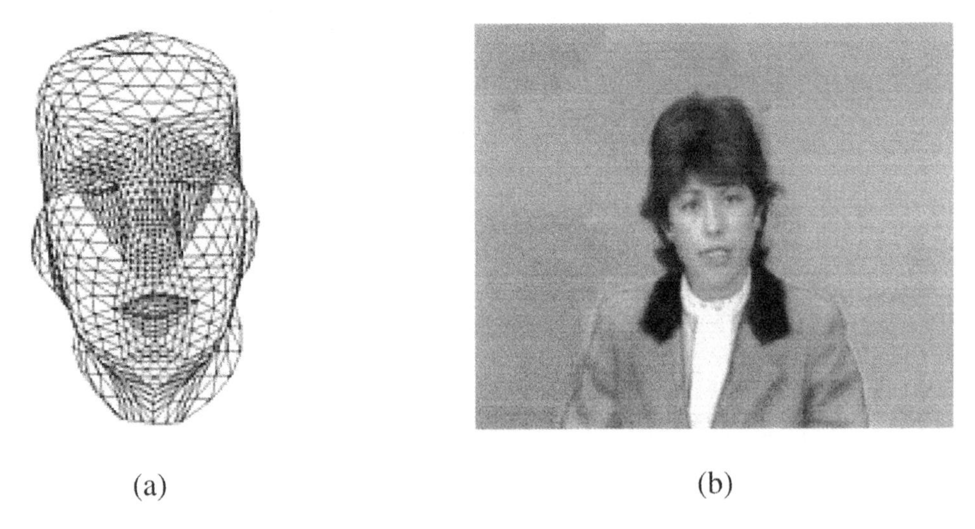

Figure 2.6 (a) 3-D model composed of 316 control points; (b) 60th frame of CIF-resolution Claire model-based coded using 716 bits

Figure 2.7 shows a basic two-channel filtering structure for sub-band coding.

Since each input video frame is a two-dimensional matrix of pixels, the sub-band coder processes it in two dimensions. Therefore, when the frame is split into two bands horizontally and vertically, respectively, four frequency bands are obtained: low-low, low-high, high-low and high-high. The DCT transform is then applied to the lowest sub-band, followed by quantisation and variable-length coding (entropy coding). The remaining sub-bands are coarsely quantised. This unequal decomposition was employed for High Definition TV (HDTV) coding (Fleisher, Lan and Lucas, 1991) as shown in Figure 2.8.

The lowest band is predictively coded and the remaining bands are coarsely quantised and run-length coded. Sub-band coding is naturally a scaleable compression algorithm due to the fact that different quantisation schemes could be used for various frequency bands. The use of the properties of HVS could also be incorporated into the sub-band compression algorithm to improve the coding efficiency. This could be achieved by taking into account the non-uniform sensitiv-

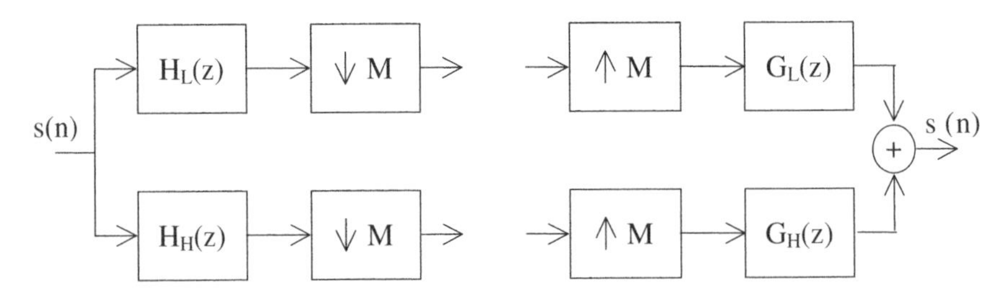

Figure 2.7 Basic two-channel filter structure for sub-band coding

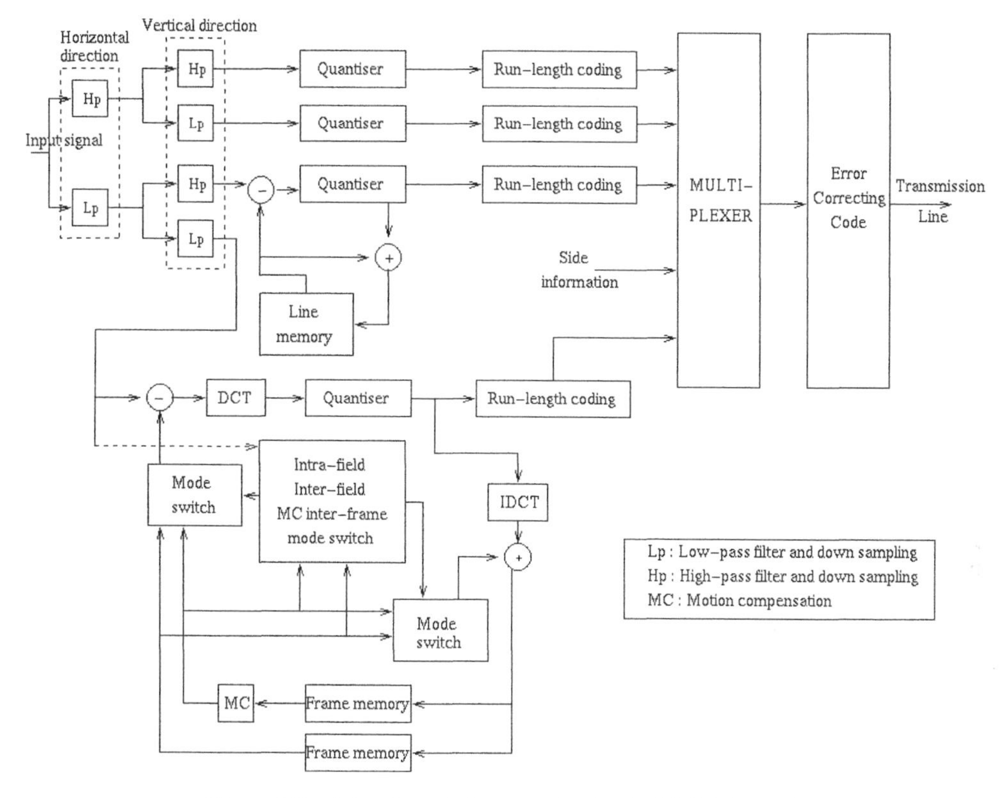

Figure 2.8 Adaptive sub-band predictive-DCT HDTV coding

ity of the human eye in the spatial frequency domain. On the other hand, improvement could be achieved during the filtering process through the use of a special filter structure (Lookabaugh and Perkins, 1990) or by allocating more bits to the eye-sensitive portion of the frame spatial frequency band.

2.4.4 *Codebook vector-based coding*

A vector in video can be composed of prediction errors, transform coefficients, or sub-band samples. The concept of vector coding consists of identifying a vector in a video frame and representing it by an element of a codebook based on some criteria such as minimum distance, minimum bit rate or minimum mean-squared error. When the best-match codebook entry is identified, its corresponding index is sent to decoder. Using this index, the decoder can restore the vector code from its own codebook which is similar to that used by the encoder. Therefore, the codebook design is the most important part of a vector-based video coding scheme. One popular procedure to design the codebook is to use the Linde–Buzo–Gray (LBG) algorithm (Linde, Buzo and Gray, 1980) which consists of an iterative search to achieve an optimum decomposition of the vector space into subspaces. One criterion for the optimality of the codebook design process is the smallest achieved distortion with respect to other codebooks of the same size. A replication of the optimally trained codebook must also exist in the decoder. The codebook is normally transmitted to the decoder out-of-band from the data transmission, i.e. using a separate segment of the available bandwidth. In dynamic codebook structures, updating the decoder codebook becomes a rather important factor of the coding system, hence leading to the necessity of making the update of codebooks a periodic process. In block-based video coders, each macroblock of a frame is mapped to a codebook vector that best represents it. If the objective is to achieve the highest coding efficiency then the vector selection must yield the lowest output bit rate. Alternatively, if the quality is the ultimate concern then the vector must be selected based on the lowest level of distortion. The decoder uses the received index to find the corresponding vector in the codebook and reconstruct the block. Figure 2.9 depicts the block diagram of a vector coding scheme

The output bit rate of a vector-based video encoder can be controlled by the design parameters of the codebook. The size M of the codebook (number of vectors) and the vector dimension K (number of bits per vector) are the major factors that affect the bit rate. However, increasing M would entail some quantisation complexities such as large storage requirements and added search complexity. For quality/rate optimisation purposes, the vectors in the codebook are variable-length coded.

2.4.5 *Block-based DCT transform video coding*

In block-based video coding schemes, each video frame is divided into a number of 16×16 matrices or blocks of pixels called macroblocks (MBs). In block-based transform video coders, two coding modes exist, namely INTRA and INTER

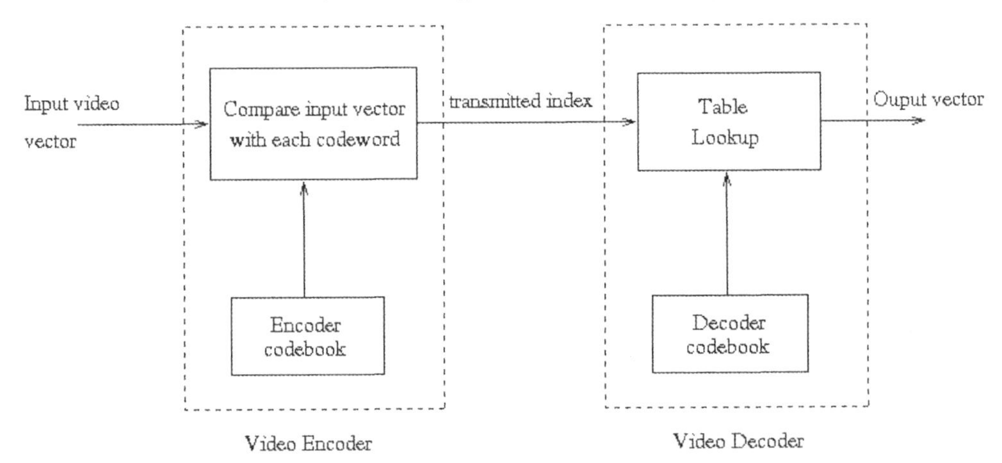

Figure 2.9 A block diagram of a vector-based video coding scheme

modes. In INTRA mode, a video frame is coded as an independent still image without any reference to precedent frames. Therefore, the DCT transform and the quantisation of transformed coefficients are applied to suppress only the spatial redundancies of a video frame. On the contrary, INTER mode exploits the temporal redundancies between successive frames. Therefore, INTER coding mode achieves higher compression efficiency by employing predictive coding. A motion search is first performed to determine the similarities between the current frame and the reference one. Then, the difference image, known as the residual error frame, is DCT-transformed and quantised. The resulting residual matrix is subsequently converted to a one-dimensional matrix of coefficients using the zigzag-pattern coding in order to exploit the long runs of zeros that appear in the picture after quantisation. A run-length coder, which is essentially a Huffman coder, assigns variable-length codes to the non-zero levels and the runs of zeros in the resulting one-dimensional matrix. Table 2.1 lists the ITU and ISO block-based DCT video coding standards and their corresponding target bit rates.

2.4.5.1 Why block-based video coding?

Given the variety of video coding schemes available today, the selection of an appropriate coding algorithm for a particular multimedia service becomes a crucial issue. By referring to the brief presentation of video coding techniques in previous sections, it is straightforward to conclude that the choice of a suitable video coder depends on the associated application and available resources. For instance, although model-based coders provide high coding efficiency, they do not give enough flexibility when detected objects do not properly match the predefined model. Segmentation-based techniques are not optimal for real-time applications since segmenting a video frame prior to compression introduces a considerable amount of delay especially when the segmentation process relies on the temporal dependencies between frames. On the other hand, block-based video coders seem to be more popular in multimedia services available today. Moreover, both ISO and ITU-T video coding standards are based on the DCT transformation of 16×16 blocks of pixels, hence their block-based structure. Although MPEG-4 is considered an exception, for it is an object-based video compression algorithm, encoding each object in MPEG-4 is a MB-based process similar to

Table 2.1 List of block-based DCT video coding standards and their applications

Standard	Application	Bit rates
MPEG-1	Audio/video storage on CD-Rom	1.5–2 Mbit/s
MPEG-2	HDTV/DVB	4–9 Mbit/s
H.261	Video over ISDN	$p \times 64$ kbit/s
H.263	Video over PSTN	<64 kbit/s

other block-based standards. The popularity of this coding technique must then have its justifications. In this section, some of the reasons that have led to the success and the widespread deployment of block-based coding algorithms are discussed.

The primary reason for the success achieved by block-based video coders is the quality of service they are designed to achieve. For instance, ITU-T H.261 demonstrated a user-acceptable perceptual quality when used in videoconferencing applications over the Internet. With frame rates of 5 to 10 f/s, H.261 provided a decent perceptual quality to end-users involved in a multicast videoconference session over Internet. This quality level was achievable using a software-based implementation of the ITU standard (Turletti, 1993). With the standardisation of ITU-T H.263, which is an evolution of H.261, the video quality can be remarkably enhanced even at lower bit rates. With the novelties brought forward by H.263, a remarkable improvement on both the objective and subjective performance of the video coding algorithm can be achieved as shall be discussed later in this chapter. H.263 is one of the video coding schemes supported by ITU-T H.323 standard for packet-switched multimedia communications. Microsoft is currently employing the MPEG-4 standard for streaming video over the Internet. In good network conditions, the streamed video could be received with minimum jitter at a bit rate of around 20 kbit/s and a frame rate of 10 f/s on average for a QCIF resolution picture. In addition to the quality of service, block-based coders achieve fairly high compression ratios in real-time scenarios. The motion estimation and compensation process in these coders employs block matching and predictive motion coding to suppress the temporal redundancies of video frames. This process yields high compression efficiency without compromising the reconstructed quality of video sequences. For all the ITU conventional QCIF test sequences illustrated in Chapter 1, H.263 can provide an output of less than 64 kbit/s with 25 f/s and an average PSNR of 30 dB. The coding efficiency of block-based algorithms makes them particularly suitable for services running over bandwidth- restricted networks at user-acceptable quality of service. Another feature of block-based coding is the scaleability of their output bit rates. Due to the quantisation process, the variable-length coding and the motion prediction, the output bit rate of a block-based video scheme can be tuned to meet bandwidth limitations. Although it is very preferable to provide a constant level of service quality in video communications, it is sometimes required to scale the quantisation parameter of a video coder to achieve a scaleable output that can comply with the bandwidth requirements of the output channel. The implications of the bit rate control on the quality of service in video communications will be examined in more details in the next chapter. In addition to that, block-based video coders are suitable for real-time operation and their source code is available on anonymous FTP sites. ANSI C code for H.261 was developed by the Portable Video Research Group at Stanford University (1995) and was placed on their website for public access and download. Telenor R&D (1995) in Norway has developed C code for the H.263 test model and made it available for free download. In 1998, the first version of the MPEG-4 verification model software was released within the framework of the European funded project ACTS MoMuSys (1998) on Mobile Multimedia Systems.

2.4.5.2 Video frame format

A video sequence is a set of continuous still images captured at a certain frame rate using a frame grabber. In order to comply with the CCIR-601 (1990) recommendation for the digital representation of television signals, the picture format adopted in block-based video coders is based on the Common Intermediate Format (CIF). Each frame is composed of one luminance component (Y) that defines the intensity level of pixels and two chrominance components (Cb and Cr) that indicate the corresponding colour (chroma) difference information within the frame. The five standard picture formats used today are shown in Table 2.2, where the number of lines per picture and number of pixels per line are shown for both the luminance and chrominance components of a video frame.

As shown in Table 2.2, the resolution of each chrominance component is equal to half its value for the luminance component in each dimension. This is justified by the fact that the human eye is less sensitive to the details of the colour information. For each of the standard picture formats, the position of the colour difference samples within the frames is such that their block boundaries coincide with the block boundaries of the corresponding luminance blocks, as shown in Figure 2.10.

2.4.5.3 Layering structure

Each video frame consists of $k \times 16$ lines of pixels, where k is an integer that depends on the video frame format ($k = 1$ for sub-QCIF, 9 for QCIF, 18 for CIF, 4CIF and 16CIF. In block-based video coders, each video frame is divided into

Table 2.2 Picture resolution of different picture formats

Picture format	Number of pixels for luminance (dx)	Number of lines for luminance (dy)	Number of pixels for chrominance (dx/2)	Number of lines for chrominance (dy/2)
sub-QCIF	128	96	64	48
QCIF	176	144	88	72
CIF	352	288	176	144
4CIF	704	576	352	288
16CIF	1408	1152	704	576

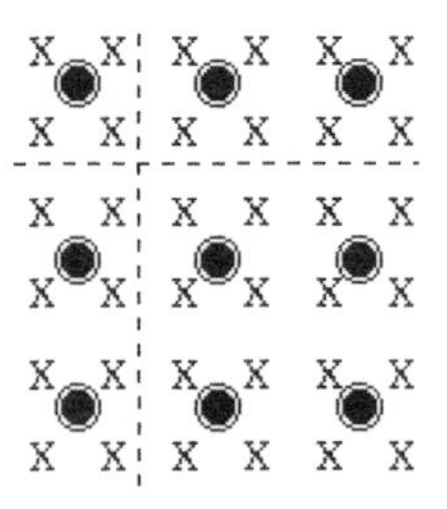

Figure 2.10 Position of luminance and chrominance samples in a video frame

groups of blocks (GOB). The number of GOBs per frame is 6 for sub-QCIF, 9 for QCIF and 18 for CIF, 4CIF and 16CIF. Each GOB is assigned a sequential number starting with the top GOB in a frame. Each GOB is divided into a number of macroblocks (MB). A MB corresponds to 16 pixels by 16 lines of luminance Y and the spatially corresponding 8 pixels by 8 lines of Cb (U) and Cr (V). An MB consists of four luminance blocks and two spatially corresponding colour difference blocks. Each luminance or chrominance block consists of 8 pixels by 8 lines of Y, U or V. Each MB is assigned a sequence number starting with the top left MB and ending with the bottom right one. The block-based video coder processes MBs in ascending order of MB numbers. The blocks within an MB are also encoded in sequence. Figure 2.11 depicts the hierarchical layering structure of a video frame in block-based video coding schemes for QCIF picture format.

2.4.5.4 INTER and INTRA coding

Two different types of coding exist in a block-transform video coder, namely INTER and INTRA coding modes. In a video sequence, adjacent frames could be strongly correlated. This temporal correlation could be exploited to achieve higher compression efficiency. Exploiting the correlation could be accomplished by coding only the difference between a frame and its reference. In most cases, the reference frame used for prediction is the previous frame in the sequence. The resulting difference image is called the residual image or the prediction error. This coding mode is called INTER frame or predicted frame (P-frame) coding. However, if successive frames are not strongly correlated due to changing scenes or fast camera pans, INTER coding would not achieve acceptable reconstructed quality.

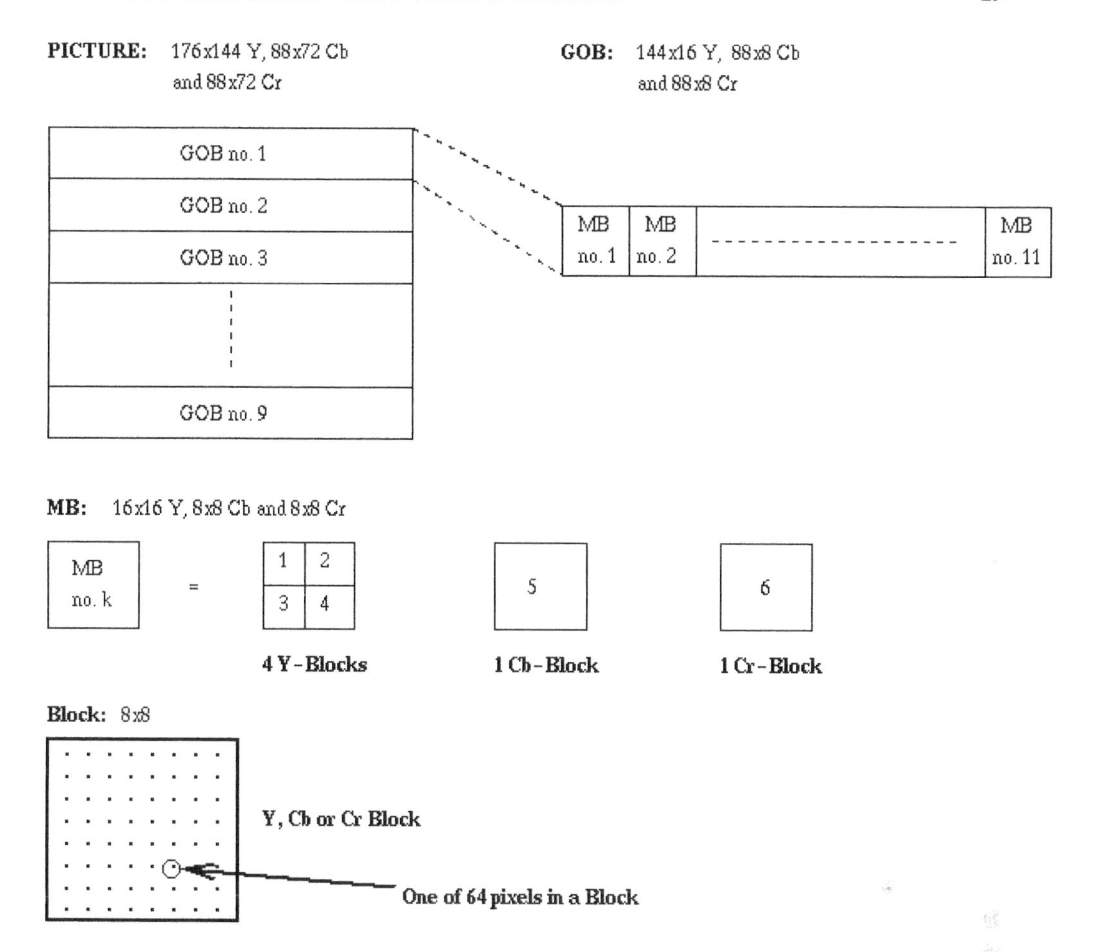

Figure 2.11 Hierarchical layering structure of a QCIF frame in block-based video coders

In this case, the quality would be much better if prediction was not employed. Alternatively, the frame is coded without any reference to video information in previous frames. This coding mode is referred to as INTRA frame (I-Frame) coding. INTRA treats a video frame as a still image without any temporal prediction employed. In INTRA frame coding mode, all MBs of a frame are INTRA coded. However, in INTER frame coding, some MBs could still be INTRA coded if a motion activity threshold has not been attained. For this reason, it is essential in this case that each MB codes a mode flag to indicate whether it is INTRA or INTER coded. Although INTER frames achieve high compression ratios, an accumulation of INTER coded frames could lead to fuzzy picture quality due to the effect of repeated quantisation. Therefore, an INTRA frame could be used to refresh the picture quality after a certain number of frames have been INTER coded. Moreover, INTRA frames could be used as a trade-off between the bit rate and the error robustness as will be discussed in Chapter 4.

2.4.5.5 Motion estimation

INTER coding mode uses the block matching (BM) motion estimation process where each MB in the currently processed frame is compared to MBs that lie in the previous reconstructed frame within a search window of user-defined size. The search window size is restricted such that all referenced pixels are within the reference picture area. The principle of block matching is depicted in Figure 2.12.

The matching criterion may be any error measure such as mean square error (MSE) or sum of absolute difference (SAD) and only luminance is used in the motion estimation process. The 16×16 matrix in the previous reconstructed frame which results in the least SAD is considered to best match the current MB. The displacement vector between the current MB and its best match 16×16 matrix in the previous reconstructed frame is called the motion vector (MV) and is represented by a vertical and horizontal components. Both the horizontal and vertical components of the MV have to be sent to the decoder for the correct reconstruction of the corresponding MB. The MVs are coded differentially using the coordinates of a MV predictor, as discussed in the MV prediction subsection. The motion estimation process in a P-frame of a block-transform video coder is illustrated in Figure 2.13.

If all SADs corresponding to 16×16 matrices within the search window fall below a certain motion activity threshold then the current MB is INTRA coded within the P-frame. A positive value of the horizontal or vertical component of a MV signifies that the prediction is formed from pixels in the reference picture that are spatially to the right or below the pixels being referenced respectively. Due to the lower resolution of chrominance data in the picture format, the MVs of

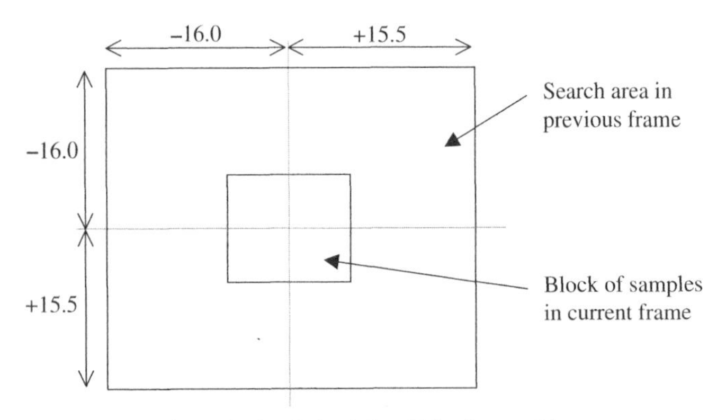

Figure 2.12 Principle of block matching

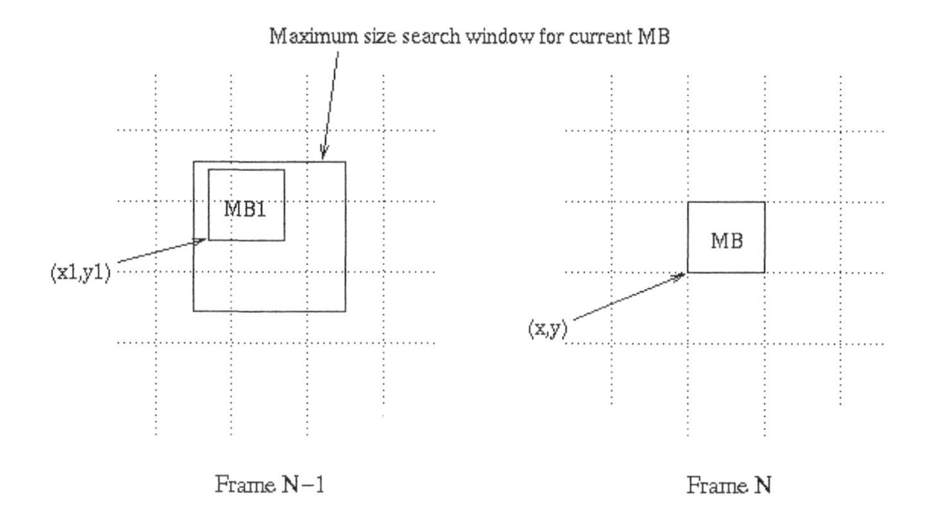

Figure 2.13 Motion estimation in block-transform video coders

chrominance blocks are derived by dividing the horizontal and vertical component values of corresponding luminance MVs by a factor of two in each dimension. If the search window size is set to zero, the best-match 16×16 matrix in the previous reconstructed frame would forcibly then be the coinciding MB, i.e. the MB with zero displacement. This scenario is called the no-motion compensation case of P-frame coding mode. In the no-motion compensation scenario, no MV is coded since the decoder would automatically figure out that each coded MB is assigned an MV of (0,0). Conversely, full motion compensation is the case where the search window size is set at its maximum value.

2.4.5.6 Half-pixel motion prediction

For better motion prediction, half-pixel accuracy is used in the motion estimation of ITU-T H.263 video coding standard. Half-pixel prediction implies that a half-pixel search is carried out after the MVs of full-pixel accuracy have been estimated. To enable half-pixel precision, H.263 encoder employs linear interpolation of pixel values, as shown in Figure 2.14, in order to determine the coordinates of MVs in half-pixel accuracy.

Half-pixel accuracy adds some computational load on the motion estimation process of a video coder. In H.263 Telenor (1995) software, an exhaustive full-pixel search is first performed for blocks within the search window. Then, another search is conducted in half-pixel accuracy within ± 1 pixel of the best match block. This implies that the displacement could either be an integer pixel value meaning that no filtering applies or half-pixel value as if a prediction filter was used.

2.4.5.7 Motion vector prediction

In order to improve the compression efficiency of block-transform video coding algorithms, MVs are differentially encoded. H.261 and H.263 have a different MV predictor selection mechanism. In H.261, the predictor is the MV of the left-hand side MB. In H.263, the predictors are calculated separately for the horizontal and vertical components. For each component, the predictor is the median value of three different candidate predictors. Once the MV predictor has been determined, only the difference between the actual MV components and those of the predictor is encoded using variable-length codewords. At the decoder, the MV components are recovered by adding the predictor MV to the received vector differences. A positive value of the horizontal or vertical component of the MV signifies that the prediction is formed from pixels in the previous picture which are spatially to the right or below the pixels being predicted, respectively. The MV prediction process in both ITU-T H.261 and H.263 video coding algorithms is illustrated in Figure 2.15. This MV predictor selection process has an impact on the error performance analysis of each video coding algorithm. This will be examined in more details later in this chapter.

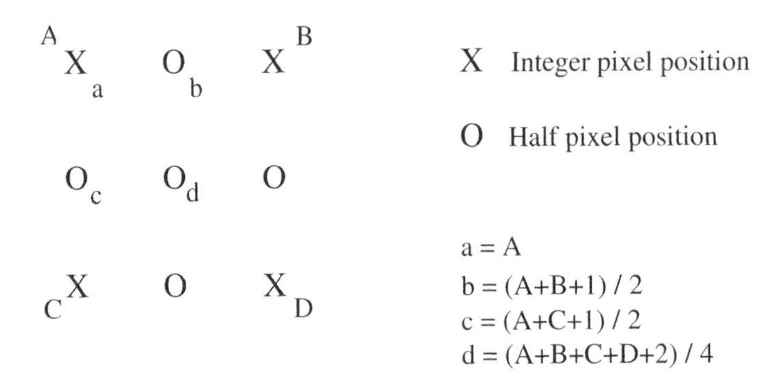

Figure 2.14 Half-pixel prediction by linear interpolation in ITU-T H.263 video coder

In H.263, if the current MB happens to be on the border of a GOB or video frame, the following rules are then applied with reference to Figure 2.16.

1. When the corresponding MB is INTRA coded or not coded at all, the candidate predictor is set to zero.
2. The candidate predictor MV1 is set to zero if the corresponding MB is outside the picture (at the left).
3. The candidate predictors MV2 and MV3 are set to MV1 if the corresponding MBs are outside the picture (at the top) or outside the GOB (at the top) if the current GOB is non-empty.
4. The candidate predictor MV3 is set to zero if the corresponding MB is outside the picture (at the right side).

2.4.5.8 Fundamentals of block-based DCT video coding

The architecture of a typical block-based DCT transform video coder, namely ITU-T H.263, is shown in Figure 2.17.

For each MB in a predicted frame, the SADs are compared to a motion activity threshold to decide whether INTRA or INTER mode is to be used for a specific MB. If INTRA mode is decided, the coefficients of the six 8×8 blocks of this

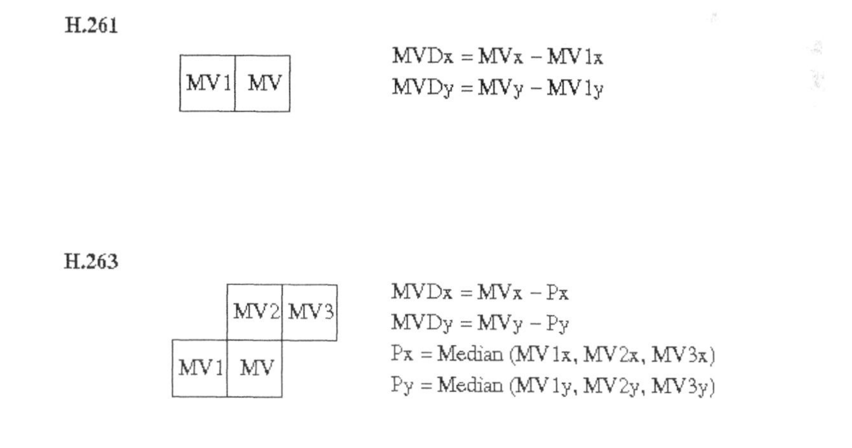

Figure 2.15 MV prediction in H.261 and H.263 video coding standards

	MV2	MV3
MV1	MV	

MV : Current motion vector
MV1 : Previous motion vector
MV2 : Above motion vector
MV3 : Above right motion vector

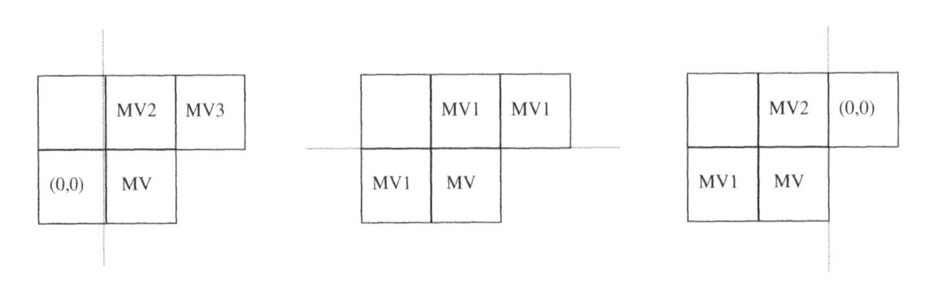

: Picture or GOB border

Figure 2.16 MV prediction at picture or GOB border in H.263 video coder

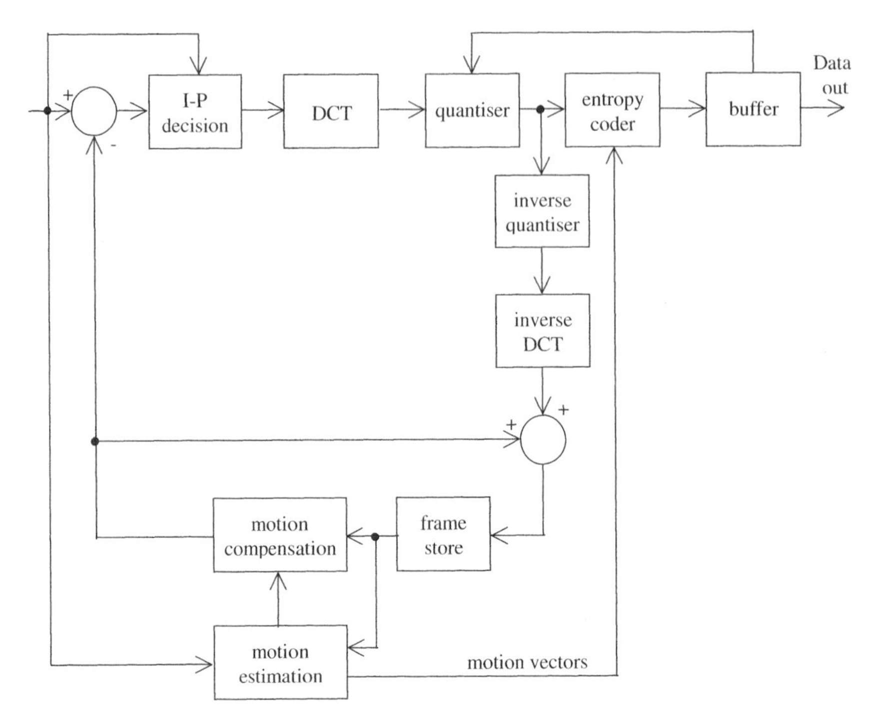

Figure 2.17 Architecture of ITU-T H.263 video coding algorithm

MB are DCT transformed, quantised, zigzag coded, run-length coded, and then variable-length coded using a Huffman encoder. However, if INTER mode is chosen, the resulting MV is differentially coded and the same encoding procedure as in INTRA mode above is applied on the residual matrix. INTRA coded MBs in a P-frame are processed as INTRA MBs in I-frames except that a MB-type flag must be sent in the former case to notify the decoder of the MB mode. The block-transform encoder contains a local decoder that internally reconstructs the video frames to employ them in the motion prediction process. The locally reconstructed frame is a replication of the decoded video frame, assuming error-free video transmission. Therefore, using previous reconstructed frames in the motion prediction process, as opposed to previous original frames, assures an accurate match between encoder and decoder reference pictures and hence a better decoded video quality. The block diagram of ITU-T H.261 is very similar to that of H.263 depicted in Figure 2.17, with only one major difference: the presence of a spatial filter in the prediction loop. Since H.263 introduces more accuracy to motion prediction by using half-pixel coordinates, the spatial filter is removed from the prediction loop. The buffer is used to regulate the output bit rate of the encoder, as will be discussed in the next chapter. The building blocks of a typical block-transform video coding algorithm are explained in the following subsections.

2.4.5.9 DCT transformation

To reduce the correlations between the coefficients of a MB, the pixels are transformed into a different domain space by means of a mathematical transform. There are a number of transforms that may be used for this purpose, such as the Discrete Cosine Transform (DCT) (Ahmed, Natarajan and Rao, 1974), the Hadamard Transform (Frederick, 1994) used in the NetVideo (NV) Internet videoconferencing tool, and the Karhunen Loeve Transform (Pearson, 1991). The latter requires *a priori* knowledge of the stochastic properties of the frame and is therefore inappropriate for real-time applications. However, the DCT transform is relatively fast to perform, and hence is adopted in most block-based video coding standards today, such as MPEG-1, MPEG-2, ITU-T H.261 and H.263. DCT is also used in object-based MPEG-4 to reduce the spatial correlations between the coefficients of MBs located in the tightest rectangle that embodies a detected arbitrary-shape object.

In block-based video coding algorithms, the 64 coefficients of every 8×8 block in a video frame are passed through a two-dimensional DCT transform stage. DCT converts the pixels in a block to vertical and horizontal spatial frequency coefficients. The 2-D 8×8 DCT employed in block-transform video coders is given in Equation 2.1.

$$F(u,v) = \frac{1}{4}C(u)C(v)\sum_{x=0}^{7}\sum_{y=0}^{7} f(x,y)\cos\left[\frac{\pi(2x+1)u}{16}\right]\cos\left[\frac{\pi(2y+1)v}{16}\right] \tag{2.1}$$

$F(u,v)$ represents the transformed coefficient at position (u,v) and $f(x,y)$ is the original pixel value (either luminance or chrominance) at position (x,y).

$$C(u) = \frac{1}{\sqrt{2}} \text{ for } u = 0; \text{ 1 otherwise}$$

$$C(v) = \frac{1}{\sqrt{2}} \text{ for } v = 0; \text{ 1 otherwise}$$

DCT produces real outputs for real inputs and is computationally a fast operation. For $u = v = 0$, Equation 2.1 yields the average of the block pixels, which is referred to as the DC value. If the corresponding block is INTRA coded, this (0,0) coefficient is referred to as the INTRADC coefficient and the remaining 63 coefficients are called the AC coefficients. The inverse DCT transform is given by Equation 2.2.

$$f(x,y) = \frac{1}{4}C(u)C(v)\sum_{u=0}^{7}\sum_{v=0}^{7} F(u,v)\cos\left[\frac{\pi(2x+1)u}{16}\right]\cos\left[\frac{\pi(2y+1)v}{16}\right] \tag{2.2}$$

A straightforward example of a 2-D DCT transformation process applied on an 8×8 block of video data is depicted in Figure 2.18.

It is obvious from Figure 2.18 that the distribution of coefficients in the transformed block is far from uniform, with a few large coefficients positioned in the upper left-hand corner of the block (the largest coefficient amplitude is that of the INTRADC coefficient) and small coefficient values elsewhere. Therefore, the DCT transform has considerably reduced the spatial redundancies of the block and suppressed the correlations of original pixels. The energy of the block was concentrated in the top left-hand section where lower frequency coefficients of the original sample block are located. Since the human visual system is more sensitive to low-order DCT coefficients, the block-transform video coding algorithms exploit this sensitivity by coding the perceptually important DC coefficient of the block more accurately than the remaining 63 AC coefficients. Each of the DC coefficients is assigned an 8-bit length codeword, while the AC coefficients are run-length coded.

2.4.5.10 Quantisation

The compression process in block-based video coders is mainly attributed to the

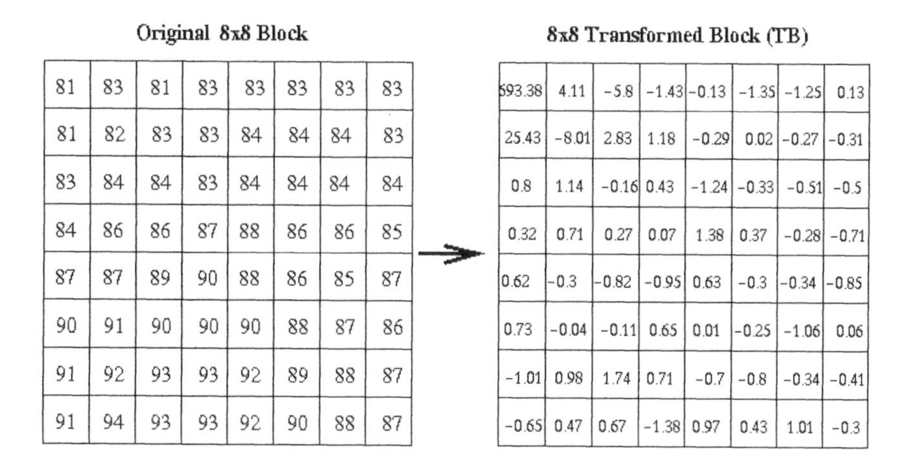

Original 8x8 Block

81	83	81	83	83	83	83	83
81	82	83	83	84	84	84	83
83	84	84	83	84	84	84	84
84	86	86	87	88	86	86	85
87	87	89	90	88	86	85	87
90	91	90	90	90	88	87	86
91	92	93	93	92	89	88	87
91	94	93	93	92	90	88	87

8x8 Transformed Block (TB)

593.38	4.11	-5.8	-1.43	-0.13	-1.35	-1.25	0.13
25.43	-8.01	2.83	1.18	-0.29	0.02	-0.27	-0.31
0.8	1.14	-0.16	0.43	-1.24	-0.33	-0.51	-0.5
0.32	0.71	0.27	0.07	1.38	0.37	-0.28	-0.71
0.62	-0.3	-0.82	-0.95	0.63	-0.3	-0.34	-0.85
0.73	-0.04	-0.11	0.65	0.01	-0.25	-1.06	0.06
-1.01	0.98	1.74	0.71	-0.7	-0.8	-0.34	-0.41
-0.65	0.47	0.67	-1.38	0.97	0.43	1.01	-0.3

Figure 2.18 An example of a DCT transform of a block of pixels

quantisation of the transformed coefficients. The quantiser is regarded as the most important component of the video encoder since it controls both the coding efficiency and the quality of the reconstructed video sequence. The coding efficiency and decoded video quality could be considerably improved if the quantisation operation is based on the human visual sensitivity to both luminance and chrominance. It has been observed experimentally that it is not necessary to convey to the decoder the full numerical precision of digital video information to achieve excellent quality reproduction. Therefore, the range of possible signal values which must be accommodated by the encoder can be reduced by means of quantisation. In video coding, there exist several techniques to quantise a frame. If each sample is quantised independently then the process is known as scalar quantisation (Max, 1960). Conversely, the quantisation of a group of samples or vectors is referred to as vector quantisation (Gray, 1984).

The quantiser maps the values of the DCT transformed coefficients to a smaller range of values in order to reduce the number of bits required to encode the block. The quantisation is a lossy process since the exact original pixel value cannot be restored using inverse quantisation. Therefore, quantisation introduces quality degradation with the benefit of improved coding efficiency. As will be described in the next chapter, adjusting the quantiser is one method to regulate the output bit rate of a block-based video coder. The following equations show the quantisation and inverse quantisation processes performed by the H.263 encoder and decoder, respectively. COF is the transformed coefficient to be quantised, LEVEL is the absolute value of the quantised coefficient and COF′ is the reconstructed transformed coefficient after inverse quantisation. Qp is called the quantiser level or quantisation parameter, and $2 \times$ Qp is the quantisation step size.

Quantisation

- INTRADC coefficient:
  LEVEL = COF/8
- INTRA AC coefficients:
  LEVEL = |COF|/(2 × Qp)
- INTER coefficients:
  LEVEL = (|COF| − Qp/2)/(2 × Qp)

Inverse quantisation

- INTRADC coefficient:
  COF′ = LEVEL × 8
- INTRA or INTER coefficients:
  |COF′| = 0 if LEVEL = 0
  |COF′| = 2Qp × LEVEL + Qp if LEVEL ≠ 0, Qp is odd
  |COF′| = 2Qp × LEVEL × Qp − 1 if LEVEL ≠ 0, Qp is even

The sign of COF is then added to obtain COF′ = sign (COF) × |COF′|.

Figure 2.19 shows the quantisation, inverse quantisation and inverse DCT of the block of transformed coefficients depicted in Figure 2.18.

2.4.5.11 Zigzag pattern coding

The two-dimentional quantised block of DCT coefficients consists of a small number of non-zero coefficients in the top left part of the block and a large number of zero coefficients elsewhere. The concentration of the non-zero high-energy coefficients in the upper left-hand corner of the block can be exploited by performing a zigzag scan of the 2-D block. The order of this zigzag scan adopted in block-based video coding standards such as H.261 and H.263 is depicted in Figure 2.20.

As a result of the zigzag-pattern coding, the non-zero low frequency coefficients will be concentrated sequentially in a one-dimensional stream of coefficients with a number of successive zeros and then followed by a long string of zeros at the end.

2.4.5.12 Run-length coding

As the name implies, the run-length coder generates output codewords for the runs of zeros and the non-zero levels rather than coding each coefficient in the block

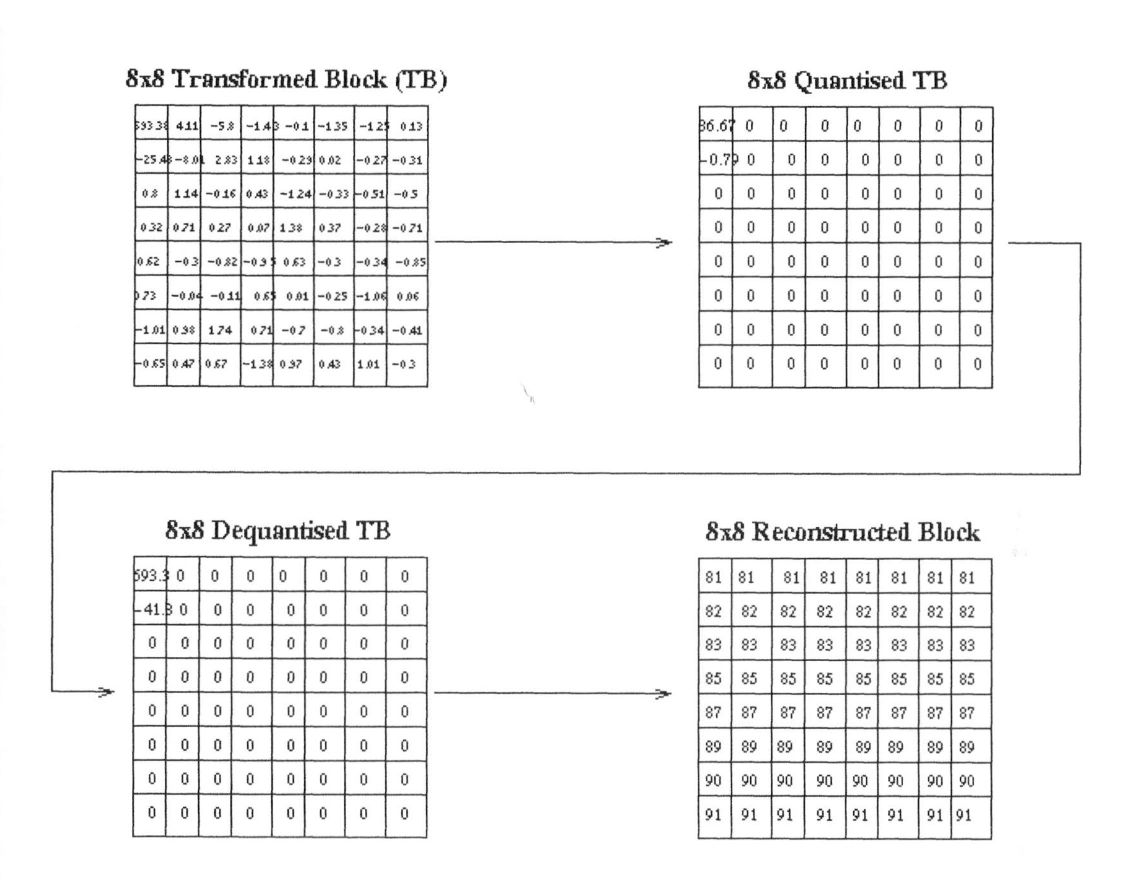

8x8 Transformed Block (TB)

593.38	4.11	-5.8	-1.43	-0.1	-1.35	-1.25	0.13
-25.48	-8.01	2.83	1.18	-0.29	0.02	-0.27	-0.31
0.8	1.14	-0.16	0.43	-1.24	-0.33	-0.51	-0.5
0.32	0.71	0.27	0.07	1.38	0.37	-0.28	-0.71
0.62	-0.3	-0.82	-0.35	0.63	-0.3	-0.34	-0.85
0.73	-0.04	-0.11	0.65	0.01	-0.25	-1.06	0.06
-1.01	0.38	1.74	0.71	-0.7	-0.8	-0.34	-0.41
-0.65	0.47	0.67	-1.38	0.37	0.43	1.01	-0.3

8x8 Quantised TB

36.67	0	0	0	0	0	0	0
-0.79	0	0	0	0	0	0	0
0	0	0	0	0	0	0	0
0	0	0	0	0	0	0	0
0	0	0	0	0	0	0	0
0	0	0	0	0	0	0	0
0	0	0	0	0	0	0	0
0	0	0	0	0	0	0	0

8x8 Dequantised TB

593.3	0	0	0	0	0	0	0
-41.3	0	0	0	0	0	0	0
0	0	0	0	0	0	0	0
0	0	0	0	0	0	0	0
0	0	0	0	0	0	0	0
0	0	0	0	0	0	0	0
0	0	0	0	0	0	0	0
0	0	0	0	0	0	0	0

8x8 Reconstructed Block

81	81	81	81	81	81	81	81
82	82	82	82	82	82	82	82
83	83	83	83	83	83	83	83
85	85	85	85	85	85	85	85
87	87	87	87	87	87	87	87
89	89	89	89	89	89	89	89
90	90	90	90	90	90	90	90
91	91	91	91	91	91	91	91

81	83	81	83	83	83	83	83
81	82	83	83	84	84	84	83
83	84	84	83	84	84	84	84
84	86	86	87	88	86	86	85
87	87	89	90	88	86	85	87
90	91	90	90	90	88	87	86
91	92	93	93	92	89	88	87
91	94	93	93	92	90	88	87

Original Block

Figure 2.19 An example of quantisation, inverse quantisation and inverse DCT of an INTRA coded 8×8 block of Suzie sequence with Qp = 10

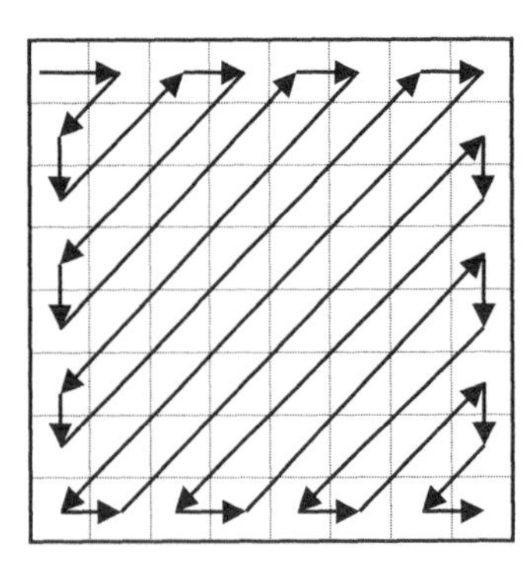

Figure 2.20 Zigzag scanning of quantised transform coefficients

separately. The length of each run of zeros and the preceding non-zero level are coded. A further 1-bit flag (LAST) is coded for each run in order to indicate whether the corresponding run is the last one in the block. Therefore, an EVENT is a combination of three parameters:

LAST 0 there are no more non-zero coefficients in the block
1 this is the last non-zero coefficient in the block

RUN number of zero coefficients preceding the current non-zero coefficient

LEVEL magnitude of the coefficient

The most frequently occurring combinations of the above three parameters (LAST, RUN, LEVEL) are coded with VLC codes. The remaining combinations are coded with a 22-bit word consisting of: ESCAPE, 7 bits; RUN, 6 bits; LAST, 1 bit; LEVEL, 8 bits.

2.4.6 *Novelties of ITU-T H.263 video coding standard*

In addition to half-pixel accuracy and the new MV prediction scheme with three candidate predictors described earlier in this chapter, ITU-T H.263 introduced four new negotiable options. They are called 'negotiable' because the decoder signals to the encoder which option it has the ability to decode before the encoder switches on the option. These negotiable options are as follows.

2.4.6.1 Unrestricted MV (Annex D)

In the default mode of H.263, MVs are only allowed to point to pixels in the reference frame that are within the coded picture area. If this mode is switched on, this restriction is released and MVs are allowed to point outside the picture, hence the expression unrestricted MV. If a pixel referenced by an MV lies outside the coded picture area, then it is replaced by the nearest edge pixel. Unrestricted MVs are particularly useful to achieve better prediction for coding objects moving into the scene. Conversely, if an object is moving out of the picture, then it will be predicted using MVs pointing inside the picture, which is a feature that is enabled by both H.261 and H.263. The former scenario arises when there is a camera pan involving new objects joining in the video scene or when an object impinges on a boundary.

On the other hand, this option provides an extension of the overall range of an MV so that larger MVs can be used. In the default prediction mode, the values of an MV in half-pixel accuracy are restricted to the range $[-16.0, +15.5]$. In the unrestricted MV mode, the range of pixels is extended to $[-31.5, +31.5]$. However, this vector range is not fully used in one vector prediction. In fact, if the predictor is outside $[-15.5, +16]$, then the vector values that are within the extended range on only one side of the zero vector could be reached. For instance, if the prediction of a vector component is -17.5, then only vectors in the range $[-31.5, 0]$ could be reached. Obviously, this option achieves negligible gain for a low-activity picture or a stationary camera, but is particularly useful in the cases of fast camera pans or new objects going into the coded picture area.

2.4.6.2 Syntax arithmetic coding (Annex E)

Syntax arithmetic coding is a variant of arithmetic coding and is occasionally used for the lossless compression of video data instead of traditional Huffman coding. The limitation of Huffman coding is that each code must be assigned an integer number of bits. If the optimum length of the codes that is calculated from the entropy of data is non-integer, then the length must be rounded to the nearest integer. This process introduces inefficiency in the compression scheme. Arithmetic coding largely eliminates this inefficiency by effectively allowing fractional bits per symbol. Arithmetic coding is used in conjunction with a modeller which estimates the probability of a particular symbol in the stream. In H.263, the used models are switched in accordance with the type of information being coded, hence the name syntax arithmetic coding. The resulting PSNR values of reconstructed video frames remain the same with the use of this option but a bit rate reduction is achieved due to the optimised bit representation of each individual symbol. The reduction of bit rate is certainly video sequence-dependent and could vary between an INTRA and an INTER frame. For INTRA frames, the reduction in bit rate is more noticeable. When syntax arithmetic coding is employed, an average reduction of 3–4 per cent of overall bit rate could be achieved for INTER frames and 10 per cent for INTRA frames.

2.4.6.3 Advanced prediction mode (Annex F)

This option allows the possibility of using four MVs instead of just one per MB to compensate for the motion of an MB in a P-frame. It also employs a scheme called overlapped block motion compensation (OBMC) to produce a smoother prediction image by mitigating the effects of block artefacts caused by block coding. In OBMC, each pixel in an 8×8 luminance prediction block is a weighted sum of three prediction values, divided by eight (with rounding). In order to obtain the three prediction values, three motion vectors are used: The motion vector of the current luminance block and two out of four 'remote' MVs: the MV of the block at the left or right side of the current luminance block; the MV of the block above or below the current luminance block.

If this option is switched on, then the unrestricted motion vector mode (Annex D) is automatically enabled. However, the extended MV range feature of Annex D is not automatically allowed. The advanced prediction mode leads to a significant subjective improvement especially when small moving objects are found in the video scene. This more accurate motion prediction is compromised by an additional bit overhead for coding the four MVs. A trade-off between bit rate and quality is then established on an MB basis to decide whether one or four MVs are to be used for each MB. The four MVs are differentially coded. The MV predictors are calculated separately for each of the vertical and horizontal components, as shown in Figure 2.21.

2.4.6.4 PB-frame mode (Annex G)

This mode introduces a new type of predicted frame which is particularly useful for low bit rate compression. This frame consists of a bi-directional motion prediction as used in ISO MPEG-2 standard. A PB-frame is composed of two frames, a predicted or INTER frame (P) that is predicted from the previous reconstructed P or I frame, and a bi-directional (B) frame that is predicted bi-directionally from the previous reconstructed frame (I or P) and the P-frame that is currently being coded as shown in Figure 2. 22. The B and P frames are coded as a single unit, hence the name PB-frame; this is a picture type used in the MPEG-2 standard.

The motion vectors of the B-frame are obtained from scaling down vectors from the corresponding P-frame, but additional 'delta' vectors may also be transmitted. The PB-frame mode is very bit-efficient since B-frames achieve very high compression ratios. This mode is particularly useful in cases when slow motion is found between adjacent video frames but very inefficient with highly active scenes and low frame rates. This could be justified by the inaccuracy of the interpolation scheme used to predict B-frames. Moreover, because of the high compression efficiency of B-frames, this mode allows for doubling the frame rate of a video sequence due to the introduction of efficiently coded B-frames with only a slight

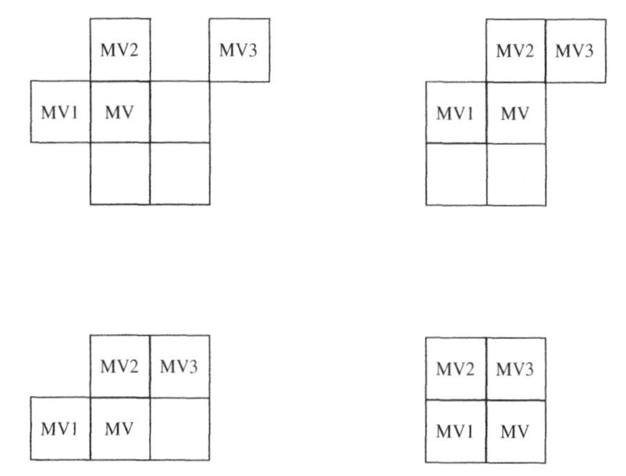

Figure 2.21 Candidate predictors MV1, MV2 and MV3 for the advanced prediction mode

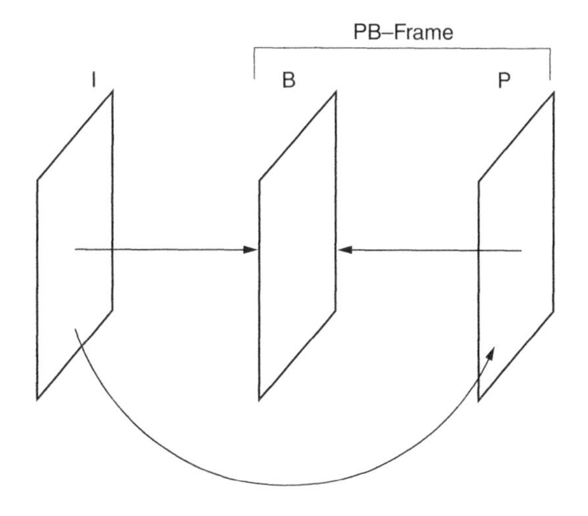

Figure 2.22 Prediction of a PB-frame

increase in bit rate. It also allows a constant frame rate (by skipping one original frame during the encoding process and replacing the skipped frame by a B-frame) with a considerable reduction in the output bit rate of the encoder.

2.4.7 Performance evaluation of ITU-T H.263 video coding standard

The order of transmission of video stream parameters is defined by the syntax of the video coding standard. The layering structure of H.263 is similar to that of its predecessor H.261, since the upper layer is reserved for a video frame and the lower layer is for a block of pixels, as described earlier. The syntax of ITU-T H.263 coding algorithm is depicted in Appendix A and the semantics of the video parameters are all defined there. The order of transmission of MBs is from left to right and top to bottom. In other words, the top-left MB is the first one to be coded in a frame and the right-bottom one is placed at the end of the frame bit stream. Therefore, the order of transmission of video data is not based on the video content or its contribution to perceptual quality even when bandwidth and channel error effects are two major limitations. The main and only objective of the standard coding algorithm was to achieve an optimised coding efficiency with improved perceptual video quality. The error performance of the coding scheme was not a matter of priority throughout the development process of the standard. However, ITU has released new versions of the standard, namely H.263+ and H.263++, with additional annexes, where a suite of error resilience techniques have been proposed to enhance the robustness of the coded video stream to channel errors and information loss. The absence of any error resilience mechanism in the original H.263 standard has left ample room to improve the performance of the source coding algorithm in error-prone conditions. In Chapter 4, we will examine a variety of these techniques proposed for improving the subjective as well the objective quality of decoded images in cases where video streams are subject to channel errors (such as in mobile radio links) and to information loss (mainly in congested networks). In the following subsections, we evaluate the performance of ITU-T H.263 and present the average objective results taken for both the luminance PSNR values and bit rates achieved by the standard with and without its negotiable options. The purpose is to assess the usefulness of each of the four negotiable options of the standard in comparison with the performance of the standard coder in default mode (with all negotiable options switched off). These results were obtained by encoding five different ITU test video sequences (150 frames per sequence captured at 25 f/s) at QCIF resolution, a coding frame rate of 12.5 f/s and fixed quantiser parameters. The average luminance PSNR and bit rate of the 150 frames were recorded in each of the cases.

2.4.7.1 No negotiable options

The results in Table 2.3 show the quantiser parameter used for each of the sequences, the average luminance PSNR (*Y*-PSNR) values and the bit rates obtained using the default mode of H.263. As expected, Foreman tends to be the most demanding sequence in terms of throughput, since it involves a high amount of activity that is due to persistent camera motion. For a quantiser value of 20, the Foreman sequence generated a bit rate of 28 kbit/s for a PSNR value of 29 dB. The high quantisation parameter value results in a coarse quantisation process, leading to distortion and hence degraded picture quality. The degradation of the quality could be observed in the blurriness of the picture that is a consequence of the inaccurate reconstruction of high-frequency components of the video frames. The Miss America sequence has produced the lowest output bit rate and highest

Table 2.3 Objective results obtained using the default mode of H.263

Sequence name	Quantiser, QP	PSNR Y(dB)	Bit rate (kbit/s)
Claire	8	37.27	20.08
Miss America	8	38.05	19.27
Suzie	15	32.68	20.94
Carphone	18	30.51	20.26
Foreman	20	29.03	27.99

objective quality for a quantiser parameter of 8. Different quantiser parameters were set for different sequences in order to customise the output bit rates to about 20 kbit/s on average. The same quantiser parameter was used for each sequence when the negotiable options were switched on.

2.4.7.2 *Unrestricted MV mode: Annex D*

Annex D allows the MVs to point outside the coded picture area, resulting in a more precise motion prediction mainly in cases where new objects are moving in the scene or an object is lingering on the edge of video frames. The objective results of Table 2.4 indicate an improvement in compression efficiency in comparison with the default mode. This could be observed in the percentage savings in it rate achieved by using this coding option with respect to the default mode. In particular, the reduction in bit rate was significant in the case of Foreman which is a highly active video sequence. This is an indication of the usefulness of this mode in coding such sequences where a finer motion prediction yields an enhanced reconstructed quality with a smaller number of bits. If this mode was disabled, it would not be possible to predict the motion of pixels on the edge of a frame and therefore they would have to be INTRA coded, hence the increase in bit rate.

2.4.7.3 *Syntax-based arithmetic coding mode: Annex E*

Annex E allows for a more efficient variable length coding than ordinary Huffman

Table 2.4 Comparison of core H.263 with the use of Annex D

Sequence name	PSNR Y(dB)	Bit rate (kbit/s)	% change in bit rate
Claire	37.30	19.87	−1.05
Miss America	38.11	19.06	−1.09
Suzie	32.76	20.33	−2.91
Carphone	30.57	20.20	−0.30
Foreman	29.18	25.68	−8.25

coding. The results shown in Table 2.5 indicate that additional compression can be obtained using this mode with almost the same average PSNR values. The percentage of bit rate reduction remains sequence-dependent, but generally a reduction of 4–10 per cent can be expected for head-and-shoulder sequences. The more motion-active the sequence is, the higher the reduction of bit rate, as clearly shown in the results of Table 2.5.

2.4.7.4 Advanced prediction mode: Annex F

When Annex F is used, four MVs are coded per MB. Theoretically, coding four MVs, as opposed to just one in default mode, should lead to an increase in bit rate. However, the results obtained in Table 2.6 show that a reduction in bit rate is achieved in all cases. The justification for this reduction is that Annex F uses smaller blocks for motion estimation, and therefore the motion prediction tends to be smoother, resulting in a less number of bits to code the residuals. This saving in coding residuals compensates for the extra bits needed to code the four MVs of the MB, resulting in a finer motion prediction at a lower bit rate. This justification is also confirmed when we see that the bit rate reduction increases with better quantisation (lower quantiser value). When a lower quantiser parameter is set, encoding the residual images will generate a higher bit rate, and therefore the bit rate reduction achieved by this mode in comparison with the less accurately predicted picture (default mode is one MV per MB) becomes more noticeable. The considerable reduction in bit rate for Foreman is also attributed to the use of

Table 2.5 Compression improvement using Annex E

Sequence name	PSNR Y(dB)	Bit rate (kbit/s)	% chang in bit rate
Claire	37.27	19.31	–3.83
Miss America	38.05	18.53	–3.84
Suzie	32.68	19.48	–6.97
Carphone	30.51	18.93	–6.56
Foreman	29.03	26.03	–7.00

Table 2.6 Use of Annex F for more accurate motion prediction

Sequence name	PSNR Y(dB)	Bit rate (kbit/s)	% change in bit rate
Claire	37.35	19.19	–4.43
Miss America	38.14	18.58	–3.58
Suzie	32.88	20.64	–1.43
Carphone	30.67	19.98	–1.38
Foreman	29.38	26.14	–6.61

Annex D which is automatically switched on when this mode is active. On the other hand, the most obvious contribution of Annex F is the significant subjective quality improvement. This is achieved by the use of the overlapped block motion compensation which alleviates the blocking artefacts that are usually caused by the block-based video coding algorithms. Despite the reduction in bit rates, the results of Table 2.6 show a slight objective quality improvement in the range 0.1–0.3 dB.

2.4.7.5 PB-frame mode: Annex G

As described earlier, this mode could be used in two different ways. If the coding objective is to provide a smoother motion reconstruction and minimal frame jerkiness then the B-frames could be used to double the frame rate of the sequence for a relatively modest increase in bit rate. Alternatively, annex G could be used to achieve higher compression at a given frame rate through the more efficient coding of B pictures. Tables 2.7 and 2.8 show the result of using annex G for each purpose. For sequences containing a high amount of motion, the PB-frame mode becomes inefficient as the interpolation process employed to code B-frames becomes inaccurate. However, for low-activity scenes, this mode results in much higher compression efficiencies compared to the default mode, as shown in the bit rate savings achieved with the Claire and Miss America sequences.

2.4.8 Performance comparison between ITU-T H.261 and H.263

In order to establish a comparison between two video coding schemes, we ought to examine their performance from two different angles. The reconstructed video quality is analysed for each coding scheme for the same target bit rates and frame rates. The comparison of the compression efficiency of the coders is performed both subjectively and objectively. Alternatively, the same quantisation parameter may be used to compare the output bit rates generated by both coders for almost the same output quality of service. Another angle of comparison is the error robustness of the coders. It is not sufficient to compare the performance of the coders in error-free conditions. Video streams are sometimes transmitted over networks where bit errors and information loss are likely to happen. Therefore, the robustness of the coding schemes to these error conditions is an important metric of comparison since it reflects the performance of the video coding algorithms in communication services. In the following two subsections, a performance comparison is established between two ITU-T video coding standards, namely H.261 and H.263, for the compression efficiency and error robustness of the coders, respectively.

Table 2.7 Use of Annex G for achieving higher compression ratios

Sequence name	PSNR *Y*(dB) (No. of P-frames)	PSNR *Y*(dB) (No. of B-frames)	Bit rate (kbit/s)	% change in bit rate
Claire	37.22 (39)	36.89 (34)	16.36	−18.53
Miss America	38.04 (39)	37.65 (34)	14.53	−24.60
Suzie	32.67 (47)	32.37 (26)	19.02	−9.17
Carphone	30.48 (62)	30.15 (11)	19.45	−4.00
Foreman	29.07 (63)	28.54 (10)	26.41	−5.64

Table 2.8 Use of Annex G for doubling the frame rate

Sequence name	PSNR *Y*(dB) (No. of P-frames)	PSNR *Y*(dB) (No. of B-frames)	Bit rate (kbit/s)	% change in bit rate
Claire	37.28 (76)	37.20 (73)	21.27	+5.93
Miss America	38.05 (75)	38.00 (74)	20.39	+5.81
Suzie	32.77 (78)	32.53 (71)	24.37	+16.38
Carphone	30.41 (100)	30.20 (49)	26.30	+29.81
Foreman	29.08 (77)	28.88 (72)	32.98	+17.83

2.4.8.1 Compression efficiency

H.261 and H.263 are both block-based DCT transform video coding algorithms which were standardised by ITU-T for low and very low bit rate video communications, respectively. They have the same layering structure, which starts with a frame and ends with a block. However, H.263 achieves a finer motion prediction due to the half-pixel accuracy employed in coding MVs. This helps smooth out the effects of blocking artefacts that normally result from block-based video compression. Moreover, H.263 introduces a refinement to the motion prediction used in H.261 by adopting three different MV predictor candidates as opposed to just one predictor. Furthermore, H.263 employs more optimised Huffman tables for the variable-length coding of runs of zeros and non-zero levels. This enhances the compression efficiency of the coder and optimises its performance for all kinds of video sequences. In addition to the default mode differences, H.263 introduces four negotiable options that improve the performance of the coding scheme compared to its default mode, as discussed in the previous section. Figure 2.23 shows frame 150 of the Suzie sequence encoded with both coders at 64 kbit/s and 25 f/s. As expected, the H.263 coded sequence shows a smoother quality with fewer blocking artefacts. When the negotiable options are used, the subjective quality is further improved and fewer artefacts appear in the video frames. Figure 2.24 depicts the objective results showing the PSNR value of each video frame in the sequence. When negotiable options are used, some fluctuations in the PSNR levels of video frames can be noticed. These fluctuations are mainly due to the inclusion of

Figure 2.23 (a) 150th frame of original Suzie sequence, its compressed version at 64 kbit/s and 25 f/s; (b) H.261; (c) H.263 default mode; (d) H.263 with four negotiable options

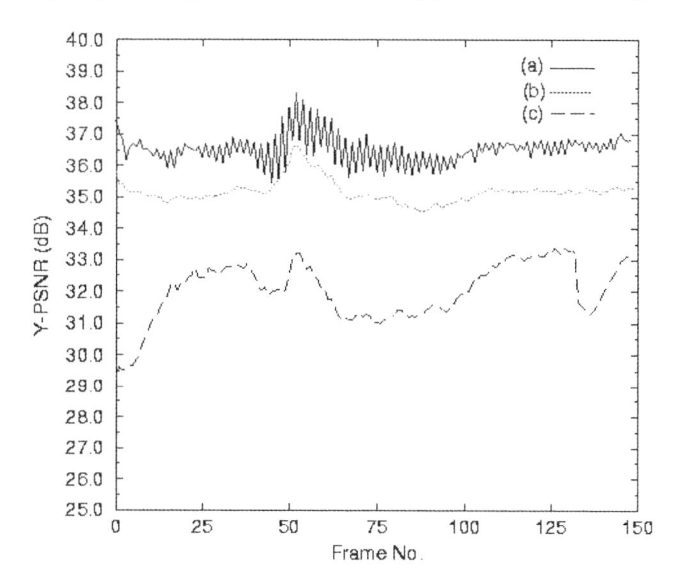

Figure 2.24 PSNR values of Suzie sequence compressed at 64 kbit/s: (a) H.263 default mode; (b) H.263 with negotiable options; (c) H.261

B-frames in the sequence. B-frames have a high compression ratio and thus present a higher-distortion compression than P-frames. The frame increment (frameskip) of the PB-frame mode is set to one to yield the same coding frame rate as the original sequence (25 f/s). For bit rates less than 64 kbit/s, H.261 fails to deliver an acceptable video quality, especially for sequences which involve a high amount of motion.

Table 2.9 shows the performance comparison between the baseline H.263, H.263

Table 2.9 Average Luminance PSNR (dB) and total bit rate values (kbit/s) for three ITU QCIF sequences encoded at 12.5 f/s, with H.261, H.263 with and without options

	Baseline H.263		H.263 + all options		H.261	
Sequence	*Y*-PSNR	kbit/s	*Y*-PSNR	kbit/s	*Y*-PSNR	kbit/s
Claire	36.12	13.84	37.23	13.49	33	14.05
Suzie	30	14.63	29.77	14.52	29.33	15.83
Foreman	28.58	14.57	28.97	14.37	28.08	15.67

with negotiable options and H.261 for three different video sequences, namely Claire, Suzie and Foreman, at a frame rate of 12.5 f/s.

This table shows that H.263 achieves a higher PSNR than H.261 for all sequences and bit rates tested. For all types of video sequence, H.263 shows a PSNR advantage over H.261 of the order of 2 to 3 dB, coupled with a significant subjective improvement. It is agreed that H.263 provides equivalent subjective quality to H.261 for less than half the bit rate.

2.4.8.2 Error robustness

Errors have a detrimental effect on the decoded video quality due to spatial and temporal inter-dependencies of video data. The motion prediction process in block-based video coding algorithms makes the compressed bit stream more sensitive to errors and information loss. For instance, a bit error in the differential motion vector (MVD) of a MB or one of its candidate MV predictors could lead to the incorrect reconstruction of the MB. The reason is that the decoder becomes unable to compensate for the motion of the currently processed MB with respect to its best-match matrix in the reference frame. An erroneous MV leads to the false reconstruction of its corresponding MB and other MB(s) whose MV(s) depend on the erroneous MV as a candidate predictor. Therefore, this MV dependency is the main reason for this video quality degradation in both the spatial and temporal domains. On the other hand, if a bit error corrupts the texture data of a MB (not necessarily the MV), then the decoder will fail to correctly reconstruct the currently processed MB that is predicted from pixels in the corrupted texture segment. Similarly, the erroneous decoding of the current MB prevents the decoder from correctly reconstructing upcoming MB(s) which are predicted from pixels in the current MB. The accumulative damage due to this temporal correlation might be caused by a single bit error in the video bit stream. The effects of errors on the video bit stream and their implications for the perceptual quality will be examined in detail in Chapter 4.

In H.261 a MV is predicted using only one MV, as illustrated in Figure 2.15. However, in H.263 the prediction of a MV engages three candidate MV predictors which belong to three neighbouring MBs. Therefore, it is expected that the loss of an MV in H.263 may lead to the false prediction of three other MVs. This can be considered as a drawback of the error robustness of H.263. Although the three-candidate predictors technique ensures a smoother quality and reduced bit rate due to better motion prediction, it helps speed up the propagation of errors in both the spatial and temporal domains. Consequently, a bit error would certainly have a more widespread effect in H.263 than H.261 and the fast accumulation of errors in H.263 has a more damaging effect on the reconstructed video quality. Figure 2.25 shows that for the same percentage of information loss, H.261 outperforms H.263 for both the Foreman and Suzie sequences coded at 64 kbit/s when full motion compensation is used (motion search window = ± 15). The PSNR graphs depicted in Figure 2.26 prove objectively that H.261 is more robust to errors than H.263 when full motion compensation is applied.

Apart from the motion prediction, another factor that has a direct influence on the error robustness of a block-based video coder is the size of the motion estimation search window. The larger the window size, the wider the range of MVs is. In the 'no motion compensation' case, the motion estimation process is put on hold and the best-match matrix for a predicted MB is always the coinciding MB in the previous frame. In this case, the MV of any INTER coded MB is implicitly always (0,0) and not encoded. The residual matrix is composed of the difference between the MB and the overlapping MB in the previous frame. When no motion compensation is performed, a MB loss results in little damage to decoded picture quality. This is mainly explained by the fact that a MB corruption does not lead to MV prediction errors since the MBs are always assumed to have undertaken a

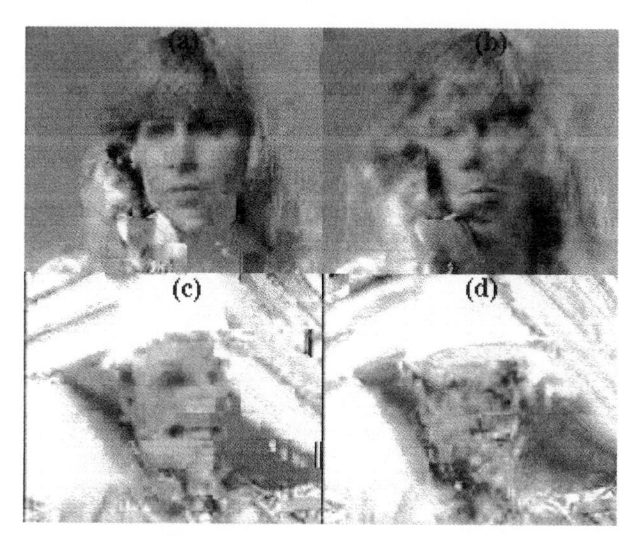

Figure 2.25 Subjective performance of the Suzie and Foreman sequences coded at 64 kbit/s for 6 per cent information loss with full motion compensation: (a) and (c) H.261, (b) and (d) H.263

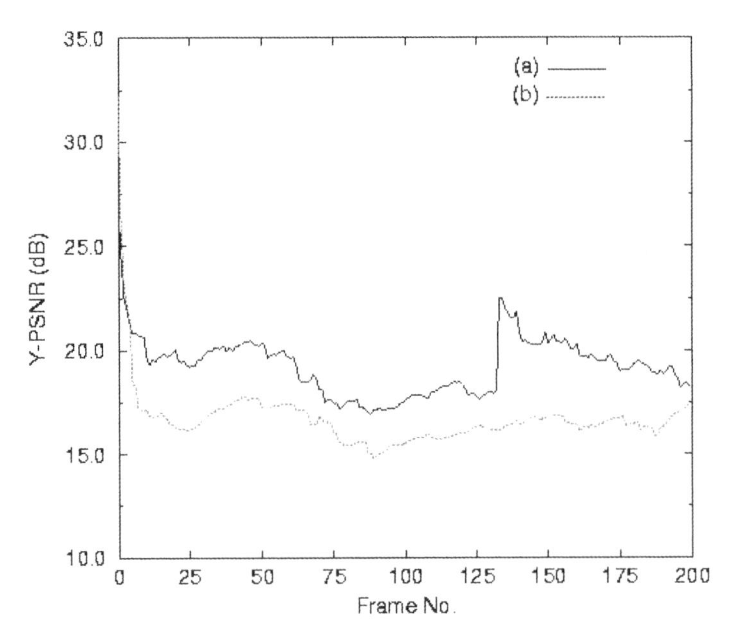

Figure 2.26 PSNR values for the Foreman sequence coded at 64 kbit/s with full motion compensation and subjected to 6 per cent information loss: (a) H.261, (b) H.263

zero displacement. Although a wider search window could lead to a better motion prediction and hence reduced bit rate, increasing the window size reduces the robustness of video coders to data loss. For this reason, the most disastrous effect of errors appears at full motion compensation; that is, when the window size is set to its maximum value.

In the no-motion compensation case, the window size is set to zero and the encoder is forced to predict the coded MB from the overlapping MB in the previous frame. If the residual error falls below a certain threshold, the encoder assumes that no prediction could be performed and thus forces the MB coding mode to INTRA. Therefore, the cost of reducing the search window size for improved error robustness is a considerable bit rate overhead. Figure 2.27 shows that H.263 presents a better performance than H.261 for a 6 per cent information loss and when no motion compensation is applied. Since a target bit rate of 64 kbit/s is set, the damage to video quality is more due to the quantisation distortion than to information loss. Since H.263 has a better compression efficiency than H.261, mainly here due to its optimised Huffman tables, it produces a better image quality than H.261 when it is made more robust to information loss after the cancellation of the motion prediction process. Figure 2.28 explains this conclusion using the objective analysis. In Figures 2.27 and 2.28, the results indicate that both coders are less sensitive to errors, since no motion prediction is used, while H.263 provides a better quality of service in this case due to its half-pixel precision and optimised Huffman tables. On the other hand, H.263

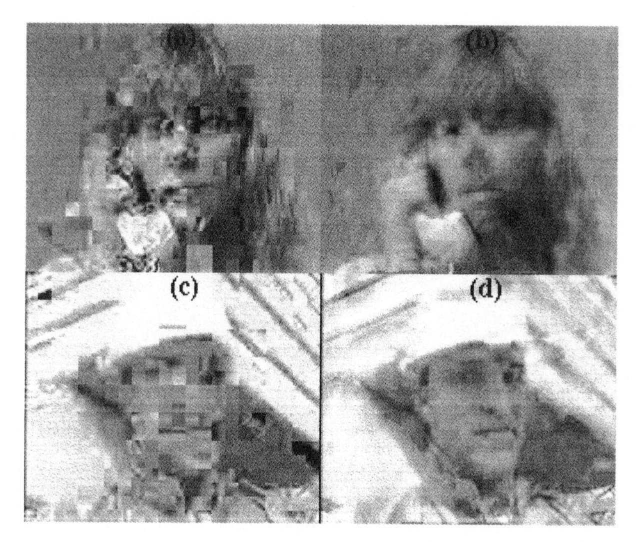

Figure 2.27 Subjective performance of the Suzie and Foreman sequences coded at 64 kbit/s for a 6 per cent information loss with no motion compensation: (a) and (c) H.261, (b) and (d) H.263

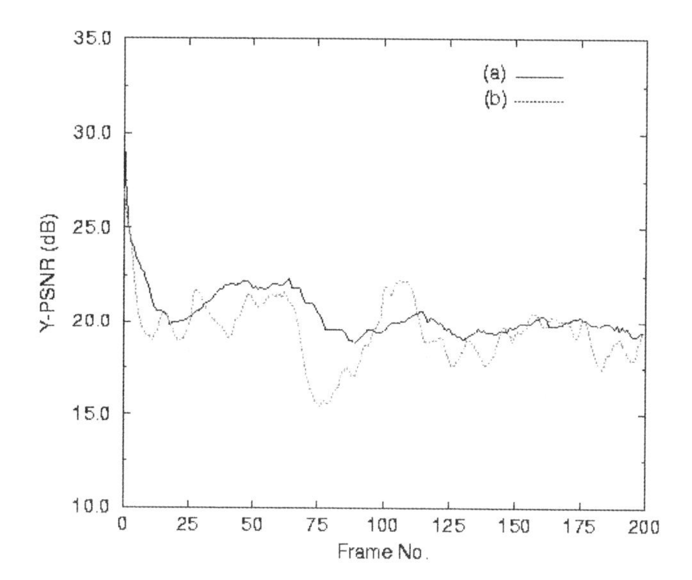

Figure 2.28 PSNR values of Foreman sequence frames compressed at 64 kbit/s with no motion compensation and subject to 6 per cent information loss: (a) H.263, (b) H.261

streams can be more fragile and sensitive to errors when the negotiable options are switched on, especially the PB-frame mode. A B-frame helps broaden the effect of errors in both time and space. Since a B-frame prediction depends on two other predicted frames, a MV corruption in a P-frame leads to the erroneous prediction of three other MVs in three separate frames (2 B-frames and 1 P-frame), in addition to the corruption of MVs within the same P-frame due to motion prediction. Consequently, a single bit error could result in a fast propagation of errors and considerable quality degradation.

2.5 Object-based Video Coding

As the name implies, object-based video coding is a concept of video compression that relies on the detection of arbitrary-shape objects in a video scene. The most widely known object-based video coding scheme today is the ISO standard MPEG-4, which is an algorithm for coding of moving pictures and associated audio information. Unlike block-based video coders such as MPEG-1, MPEG-2, H.261 and H.263, MPEG-4 detects entities in the video frame that the user can access and manipulate (crop, cut, paste, re-scale, move, rotate, etc.), hence providing the user with content-based functionalities for the processing and compression of any video scene. These entities are called Video Object Planes (VOPs). Therefore, in addition to coding the texture and motion information traditionally encountered in block-based video coders, MPEG-4 codes the shape of each VOP, as illustrated in Figure 2.29, so that the composition of objects could be done at the decoder end.

A VOP can be a semantic object in the video scene; it consists of luminance and chrominance components (Y,U,V) in addition to shape information. The VOP can be of arbitrary shape and can be detected by semi-automatic segmentation. Figure 2.30 shows a number of arbitrary shape objects that compose a video frame. Each detected object in the frame is a VOP with different contour, texture and perhaps motion specifications.

When the video sequence has only one VOP of rectangular shape that is

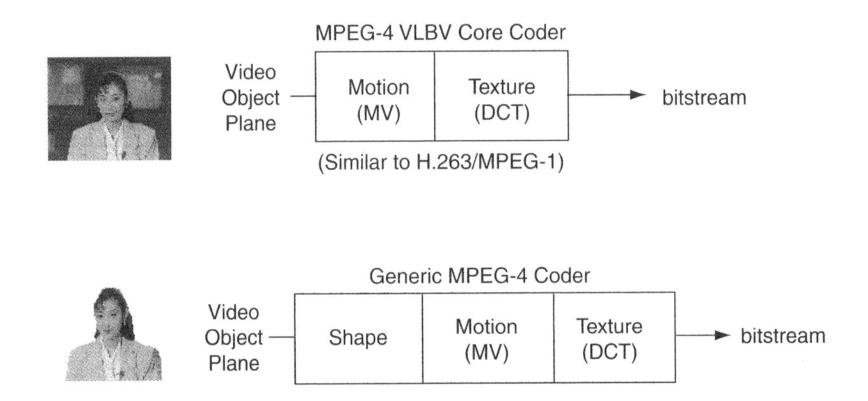

Figure 2.29 Content-based video coding in MPEG-4

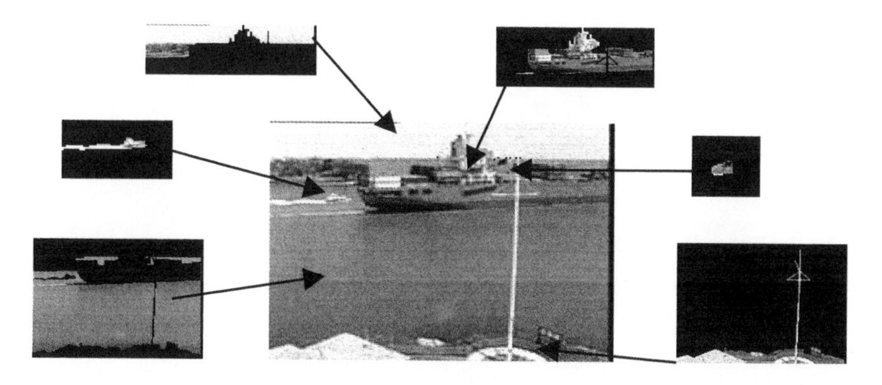

Figure 2.30 Creation of content-based VOPs for object-based MPEG-4 video coding

displayed at fixed intervals, it corresponds to the frame-based coding technique employed by block-based video compression algorithms. In order to code VOPs whose width and height are not integer multiples of 16 (MB size), the width and height of these VOPs are extended to be the smallest integer multiples of 16. The extended areas of the VOPs are then padded using a repetitive padding technique described in Section 2.5.4. MPEG-4 supports two ranges of output bit rates. The first is tailored to low bit rate video communications below 64 kbit/s for low-detail video scenes, and the second is focussed on high bit rate video applications (mobile TV for instance) up to 2 Mbit/s for high-activity fine-detail sequences. In addition to that, MPEG-4 defines a suite of error resilience algorithms to improve the robustness of coded bit streams to channel errors. Error handling mechanisms employed in video codecs are examined in Chapter 4.

2.5.1 VOP encoder

The VOP encoder structure is composed of two parts: the shape coder, and the traditional motion and texture coder that is found in block-based video coders such as H.263. Figure 2.31 depicts the block diagram of the VOP encoder structure.

2.5.2 Shape coding

Two methods are employed by MPEG-4 to code the shape information of detected VOPs, namely binary shape coding and grey scale shape coding. The shape information is referred to as alpha planes. An alpha plane is bounded by the tightest rectangle that includes the shape of a VOP. The bounding rectangle of the VOP is then extended on the right-bottom side to multiples of 16 × 16 blocks. The extended alpha samples are set to zero. The extended alpha plane is partitioned

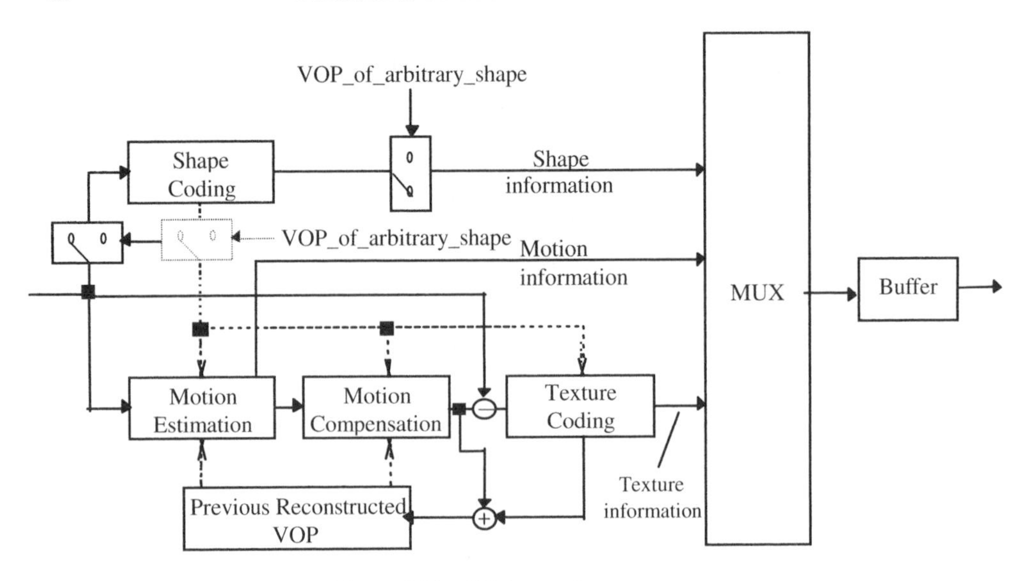

Figure 2.31 VOP encoder structure

into blocks of 16 × 16 samples referred to as alpha blocks, and the encoding/decoding process is done per alpha block. If the pixels in an MB are all transparent (all zero), the MB is skipped before motion and/or texture coding. No overhead is required to indicate this mode since this transparency information can be obtained from shape coding. Binary alpha planes are encoded by quadtree without vector quantisation and grey scale alpha planes are encoded by quadtree with vector quantisation.

2.5.2.1 Binary shape coding: quadtree without vector quantisation

An alpha plane is divided into 16 × 16 alpha blocks and coding is done on an alpha-block basis. A quadtree is associated with each alpha block. The quadtree is composed of four levels, as shown in Figure 2.32. At level 3 (the lowest level), there are 64 sub-blocks of 2 × 2 samples. A higher level of the quadtree is formed by grouping four sub-blocks at the lower level.

An index is assigned to each 2 × 2 sub-block and used in calculating indices for the sub-blocks of the upper level. Figure 2.33 shows the 2 × 2 sub-block at level 3. Each index is assigned to the 2 × 2 sub-block in the following order.

1. If upper_b[0] is less than upper_b[1], then swap b[0] with b[1] and b[2] with b[3] except for the sub-blocks of the top line of level 3.
2. If left_b[0] is less than left_b[1], then swap b[0] with b[2] and b[1] with b[3] except for the sub-blocks of the first column to the left of level 3.

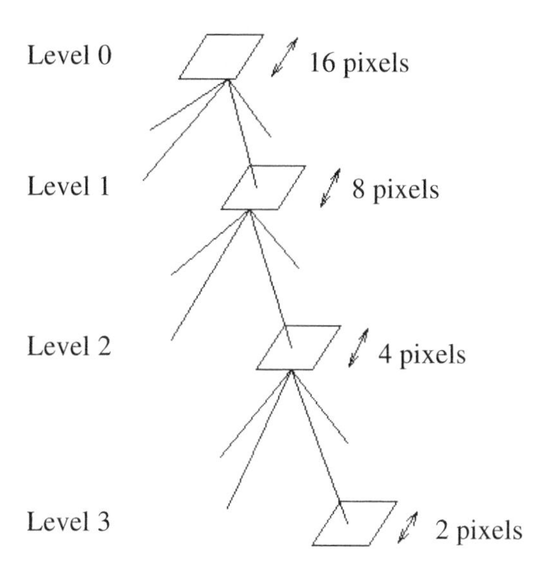

Figure 2.32 Quadtree structure for binary shape coding

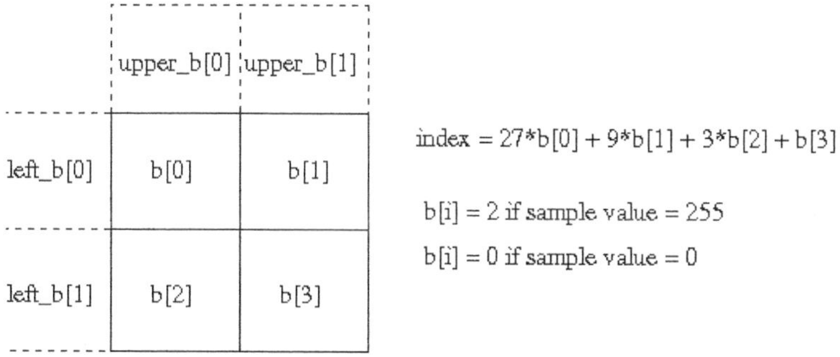

	upper_b[0]	upper_b[1]
left_b[0]	b[0]	b[1]
left_b[1]	b[2]	b[3]

index = 27*b[0] + 9*b[1] + 3*b[2] + b[3]

b[i] = 2 if sample value = 255

b[i] = 0 if sample value = 0

Figure 2.33 Indexing of a 2 × 2 sub-block at level 3

3. If upper_b[0]+ upper_b[1] is less than left_b[0]+ left_b[1], then swap b[1] with b[2] except for sub-blocks of both the top line and the leftmost column of level 3.
4. The index of the 2 × 2 sub-block is computed from b[0], b[1], b[2] and b[3], as shown in Figure 2.33.

The swapping operation adopted in assigning an index to the 2 × 2 sub-block helps gather large values to the upper left of each sub-block so that more efficient variable length codes are generated, leading to a reduced bit rate. Figure 2.34

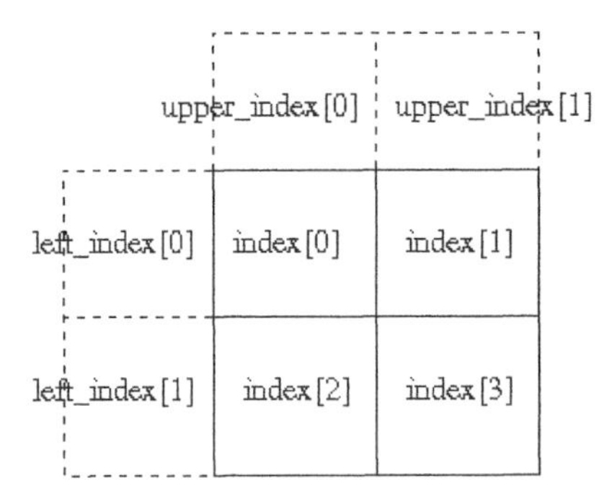

Index at higher level = 27*f(index [0]) + 9*f(index [1]) + 3*f(index [2]) + f(index [3])

$$f(x) = \begin{cases} 0, & \text{if } x=0 \\ 2, & \text{if } x=80 \\ 1, & \text{otherwise} \end{cases}$$

Figure 2.34 Assigning an index to a sub-block of a quadtree level based on the indices of the corresponding sub-blocks of the lower level

shows how indices of four sub-blocks belonging to the same level are used to calculate another index for a sub-block at a higher level. Again the same order used in assigning an index to sub-blocks of level 3 is followed at the other levels. This technique produces a total of 85 ($= 1+4+16+64$) indices for each 16×16 alpha block.

2.5.2.2 Grey scale shape coding: quadtree with vector quantisation

An alpha plane is divided into 16×16 alpha blocks and the encoding process is done per alpha block by producing its quadtree and applying vector quantisation. The quadtree structure for the grey scale shape coding is composed of three levels, as shown in Figure 2.35. At the lowest level (level 2), the alpha block is partitioned into 16 sub-blocks of 4×4 samples. At level 1, the quadtree consists of four sub-blocks of 8×8 samples. Each sub-block of the lowest level is encoded by vector quantisation using a codebook containing 256 vectors. For the upper two levels, indices are computed for each sub-block using vectors of sub-blocks belonging to the lower level.

An index is assigned to each 4×4 sub-block in level 2 in the following manner (Figure 2.36).

1. If g(x(0)) is less than g(x(1)), then swap vector components horizontally (first column with fourth and second with third) except for sub-blocks of the first column and first row.

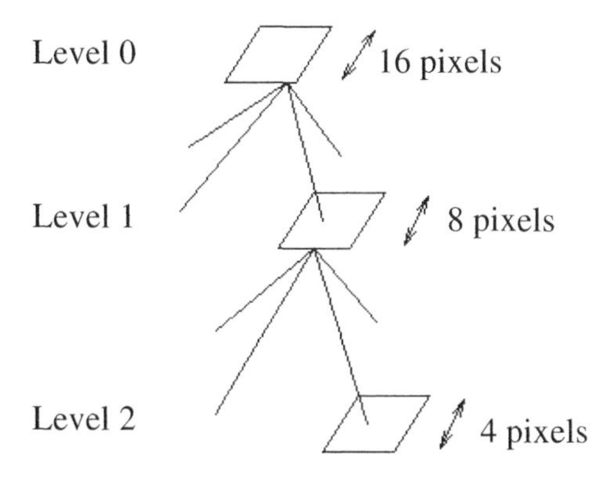

Figure 2.35 Quadtree structure for grey scale shape coding

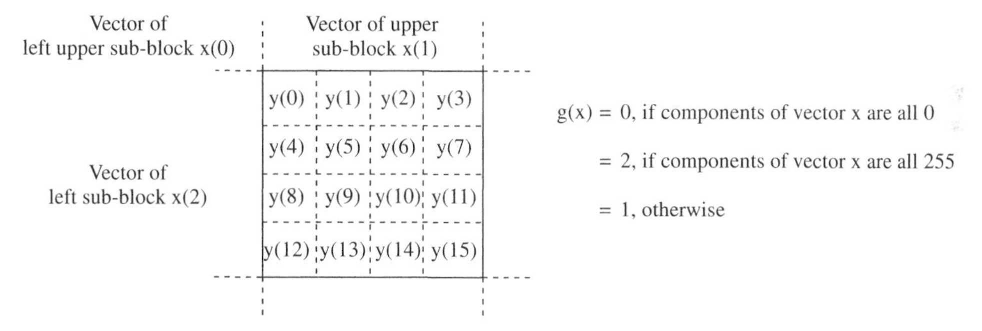

Figure 2.36 Indexing of a 4 × 4 sub-block at level 2 in grey shape coding

2. If g(x(0)) is less than g(x(2)), then swap vector components vertically (first row with fourth and second with third) except for sub-blocks of the first column and first row.
3. If g(x(1)) is less than g(x(2)), then swap vector components diagonally, as shown in Figure 2.37, except for sub-blocks of the first column and first row.
4. Each vector is encoded using vector quantisation. The vector indices are taken from codebooks stored in both shape encoder and decoder.
5. Reconstruct the current sub-block to use as a reference for the subsequent sub-block, and continue processing until the last sub-block.

For levels 0 and 1, the computation of indices is done in a way similar to that used in binary shape coding where each index is a weighted sum of the four corresponding reconstructed vectors of the lower level:

$$\begin{aligned}\text{index of upper level} = {} & 27 \times g(\text{reconstructed vector}[0]) + 9 \times g(\text{reconstructed vector}[1]) \\ & + 3 \times g(\text{reconstructed vector}[2]) + g(\text{reconstructed vector}[3])\end{aligned}$$

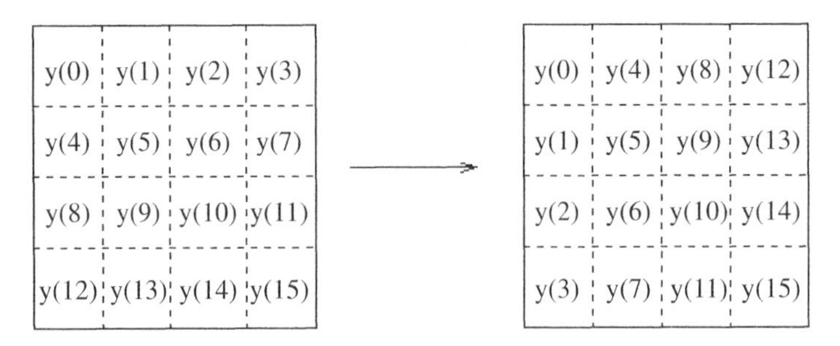

Figure 2.37 Diagonal swapping of vector components

In this shape coding process, a total of 21 ($= 1 + 4 + 16$) indices for each 16×16 alpha block are generated. The order of transmission of the indices does not change with the swapping process. An index at level 0 or level 1 is variable-length coded and an index at level 2 is encoded with an 8-bit fixed length code, except vectors of all 0 and all 255.

2.5.3 *Motion estimation and compensation*

Since a VOP has an arbitrary shape, the blocks on the boundaries of the reference VOP need to be extended to multiples of 16 samples in both height and width. The motion estimation process of the blocks at the boundaries of the VOP has to be achieved with polygon matching rather than block matching. At each particular time instance a bounding box, the tightest rectangle that includes the shape of that VOP is defined as shown in Figure 2.38. The top and left corners of the bounding box (in their absolute coordinates) are encoded in the VOP spatial reference. To estimate the motion of the blocks located on the borders of the VOP (which has an arbitrary shape rather than a rectangular one), a special padding technique is applied.

2.5.4 *Padding technique*

An image repetitive padding technique is applied on the reference VOP to perform the motion estimation/compensation. It consists of five steps.

1. Every undefined pixel in the extended region of the VOP area (outside the VOP boundaries) is set to zero.
2. Each line of the original image region is scanned horizontally. The scan lines are of two kinds: zero segments that consist of zero pixels only and non-zero segments that have all non-zero pixels. If there are no non-zero segments in the

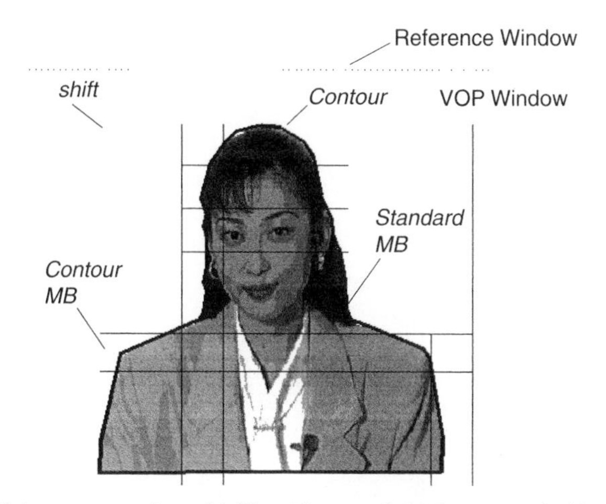

Figure 2.38 The tightest rectangle and MB-grid around the foreground object of a video frame

current scan line, then skip to the next line. Otherwise, either one of two cases arises for a particular zero segment: The zero segment is located between an end point of the scan line and the end point of a non-zero segment or between the end points of two different non-zero segments. In the first case, the pixels of the zero segment are assigned the value of the end point of the non-zero segment. In the second case, the pixels of the zero segment are filled with the average pixel value of the two end points.

3. Each line of the original image is scanned vertically and the same procedure in step 2 is applied to each vertical scan line.
4. If a zero pixel is filled up in steps 2 and 3, the final value takes the average of the two possible values.
5. For any remaining zero pixel, scan horizontally to find the closest non-zero pixel on the same horizontal scan (if there is a tie, the non-zero pixel to the left of the current pixel is selected), and scan vertically to find the closest non-zero pixel on the same vertical scan (if there is a tie, the non-zero pixel on the top of the current pixel is selected). The zero pixel is then replaced by the average of these two horizontally and vertically closest non-zero pixels. Figure 2.39 illustrates the steps described above in the algorithm, and Figure 2.40 shows a frame of the Akiyo sequence after the padding technique has been applied to it.

2.5.5 Basic motion techniques

MPEG-4 employs a number of techniques similar to those used in ITU-T .263 for the coding of motion data. In the following subsections, the basic concepts of these techniques are explained.

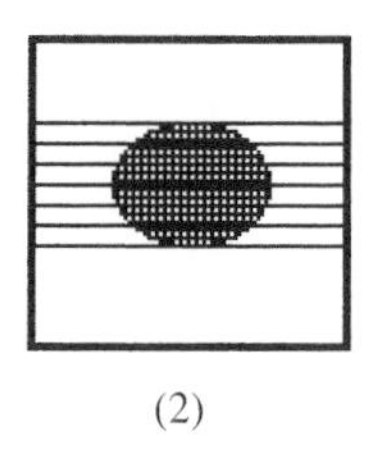
(2)

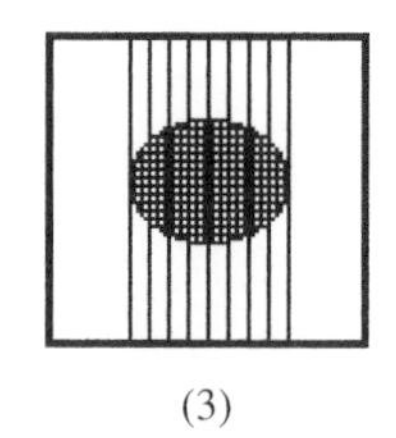
(3)

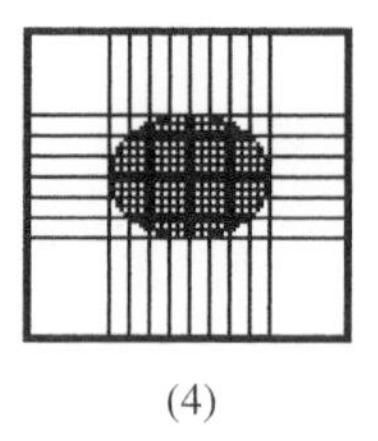
(4)

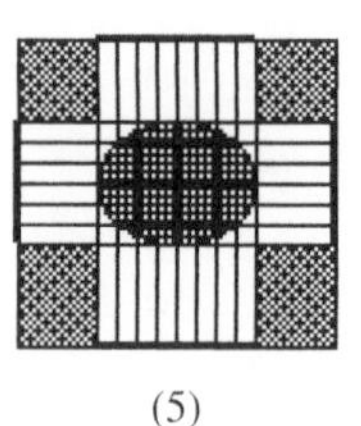
(5)

Figure 2.39 Steps of the padding technique

(a)

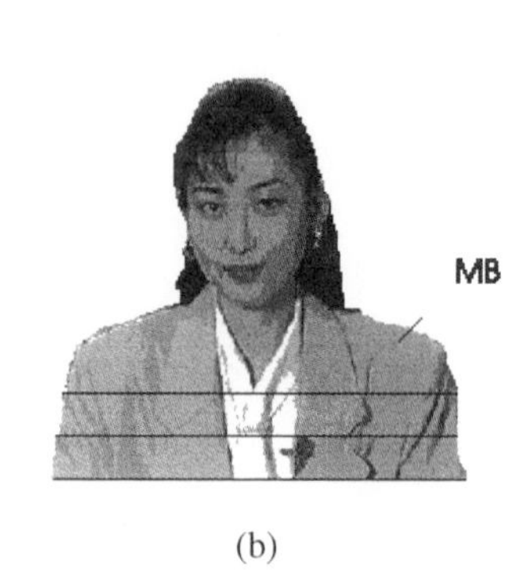

(b)

Figure 2.40 Sequence Akiyo: (a) reference VOP with padded background; (b) current VOP

2.5.5.1 Modified block matching (polygon matching)

The tightest bounding rectangle containing the VOP is extended on the right-bottom side to multiples of MB size. The size of the bounding rectangle for the luminance VOP is multiples of 16×16, while the size of the chrominance plane is multiples of 8×8. The alpha values of the extended pixels (i.e. outside the boundaries of the VOP) are set to zero. The MBs are formed by partitioning the extended bounding rectangles into 16×16 blocks. In the motion estimation process, SAD (Sum of Absolute Differences) is used as error measure. The reconstructed alpha plane for the VOP is used to exclude the pixels of the MB that are outside the VOP. SAD is computed only for the pixels with non-zero alpha value. This forms a polygon for the MB that includes the VOP boundary. Figure 2.41 shows an example of this technique.

2.5.5.2 Integer pixel motion estimation

The search for the best match MB is made with integer pixel displacement and for

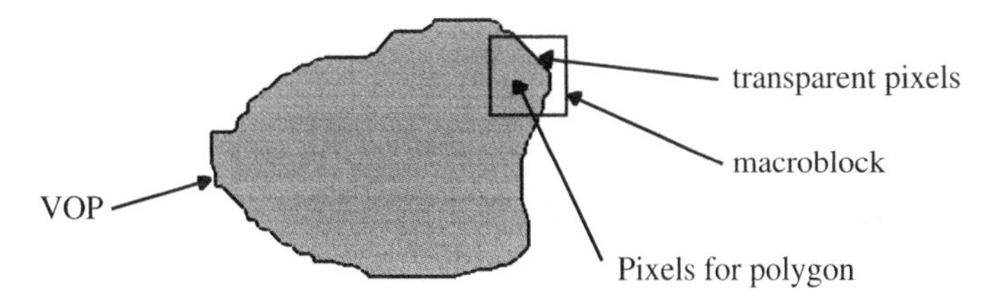

Figure 2.41 Polygon matching for an arbitrary shape VOP

the Y component. Comparison is made between the current MB and the displaced MB in the previous original VOP. A full search is carried out within a search window of up to ± 31.5 pixels in horizontal and vertical direction around the original MB position.

$$SAD_N(x, y) = \sum_{i=1,j=1}^{N,N} |original - previous| \times (!(Alpha_{original} = 0))$$

where x, y is up to ± 31.5 and $N = 16$ or 8. To favour the zero vector when there is no significant difference, the zero vector $SAD_{16}(0, 0)$ is used:

$$SAD_{16}(0, 0) = SAD_{16}(0, 0) - \left(\frac{N_B}{2} + 1\right)$$

where N_B is the number of MB pixels inside the VOP. The (x, y) pair resulting in the lowest SAD_{16} is chosen as the 16×16 integer pixel MV (motion vector), V0. The corresponding SAD is $SAD_{16}(x, y)$. In the advanced prediction mode, four 8×8 MVs represent each 16×16 MB. The 8×8 based SAD for the MB is then calculated as follows:

$$SAD_{Kx8} = \sum_{1}^{k} SAD_8(x, y)$$

where $0 < K \leq 4$ is the number of 8×8 blocks (per MB) that lie inside the VOP shape.

2.5.5.3 *INTRA/INTER mode decision*

When the integer pixel motion estimation has been completed, the coder makes a decision on whether to use INTRA or INTER coding mode.

$$\text{MB mean} = \frac{\sum_{i=1,j=1}^{N_B} |original|}{N_B}$$

$$A = \sum_{i=1,j=1}^{16,16} |original - MB\_mean| \times (!Alpha_{original} = 0))$$

INTRA mode is chosen if $A < (SAD_{\text{inter}} - 2N_B)$

If INTRA mode is chosen, no further operation is required for the motion search. However, if INTER mode is selected, the motion search continues with a half-pixel search around the V0 position.

2.5.5.4 *Half-pixel search*

This is performed using the previous reconstructed VOP for 16 × 16 vectors as well as 8 × 8 vectors. The search is done on the luminance component of the MB within a search area of ± 1 pixel around the target matrix whose points are V0, V1, V2, V3 or V4. The half-pixel values are found using the bilinear interpolation described in Figure 2.14. The vector which results from the half-pixel search is named MV and consists of horizontal and vertical components (MVx, MVy), both measured in half-pixel units.

2.5.5.5 *Decision on 16 × 16 or 8 × 8 prediction mode*

The SAD for the best-match half-pixel 16 × 16 vector, including subtraction of 100 if the vector is (0, 0), is $SAD_{16}(x, y)$, and the SAD for the whole MB for the best-match half-pixel 8 × 8 vectors is:

$$SAD_{Kx8} = \sum_{1}^{K} SAD_8(x, y)$$

where $0 < K \leq 4$ is the number of 8 × 8 blocks that do not lie outside of the VOP shape.

If $SAD_{Kx8} < SAD_{16} - (N_B/2 + 1)$, then choose 8 × 8 prediction, where N_B is the number of pixels inside the VOP multiplied by $2^{(\text{bits per pixel} - 8)}$; otherwise choose the 16 × 16 prediction.

2.5.5.6 *MV prediction*

When INTER mode is chosen, MVs are to be transmitted. The horizontal and vertical components of the MV are coded differentially by using a spatial neighbourhood of three MV candidate predictors, as is the case for H.263 MV prediction (refer to Figure 2.15).

At the borders of the current VOP, some rules are applied.

1. If the MB of one and only one candidate predictor is outside of the VOP, it is set to 0.
2. If the MBs of two and only two candidate predictors are outside of the VOP, they are set to the third candidate predictor.
3. If the MBs of all three candidate predictors are outside of the VOP, they are set to zero.

For each component (horizontal and vertical), the median value of the three candidates for the same component is computed, and the motion vector differences MVDx and MVDy are then transmitted, as shown in Figure 2.15.

2.5.5.7 Unrestricted motion vector mode

This is similar to Annex D of H.263, since it allows MVs to point outside the VOP for improved motion prediction. In this technique, the reference VOP is extended 32 pixels in all four directions and padded using the repetitive padding technique described in Section 2.5.4. Padding is performed only on the reference VOP, while the target VOP remains the same except for extending it to multiples of 16×16 blocks. This procedure is performed using the following three steps.

1. The bounding rectangle of the arbitrary shape VOP is extended by 32 pixels in all four directions and the alpha value of the extended pixels is set to zero.
2. The extended regions are padded using the repetitive padding technique and the padded VOP is used as the new reference VOP.
3. The modified block matching (polygon matching) described in Section 2.5.5.1 is applied to compute the motion vectors.

2.5.5.8 Advanced prediction mode

This is analogous to Annex F of H.263, as it allows four MVs to be coded for each INTER MB in a VOP. When the four MV decision is made, a vector for each 8×8 block of luminance is transmitted. The candidate predictors for each MV are indicated in Figure 2.21. Half-pixel MV values are found using bilinear interpolation, as described in Section 2.4.5.6. The prediction for luminance is obtained by the overlapped block motion compensation discussed in Section 2.4.6.3.

2.5.6 *Texture coding*

INTRA VOPs and residual data resulting from motion prediction are coded using the DCT scheme of H.263 (refer to Section 2.4.5.9). When the VOP has an arbitrary shape, three kinds of MB appear (Figure 2.42): MBs that lie completely inside the VOP shape, MBs that lie on the boundary of the arbitrary shape VOP, and MBs that lie completely outside the VOP but inside the tightest rectangle. MBs that lie completely inside the VOP are coded with a technique identical to that of H.263 (refer to Sections 2.4.5.10–2.4.5.13). MBs that lie on the boundary of the VOP are first padded with their transparent pixels (outside the VOP but inside the bounding box) all set to 0s, and then coded in a way identical to the internal MBs that lie inside the VOP. MBs that lie inside the bounding box and do not belong to the arbitrary shape VOP are coded as skipped MBs (COD = 1).

2.5.7 *MPEG-4 VOP decoder*

A general block diagram of the VOP decoder is shown in Figure 2.43. The decoder consists mainly of two major parts: the shape decoder, and the traditional motion and texture decoder. The reconstructed VOP is obtained by the right combination of the shape, texture and motion information. The same decoding scheme is applied when decoding all the VOPs of a given session. The reconstructed VOPs are then blended together in the compositor according to the order specified by the VOP composition order to reconstruct the video frames of the synthetic sequence.

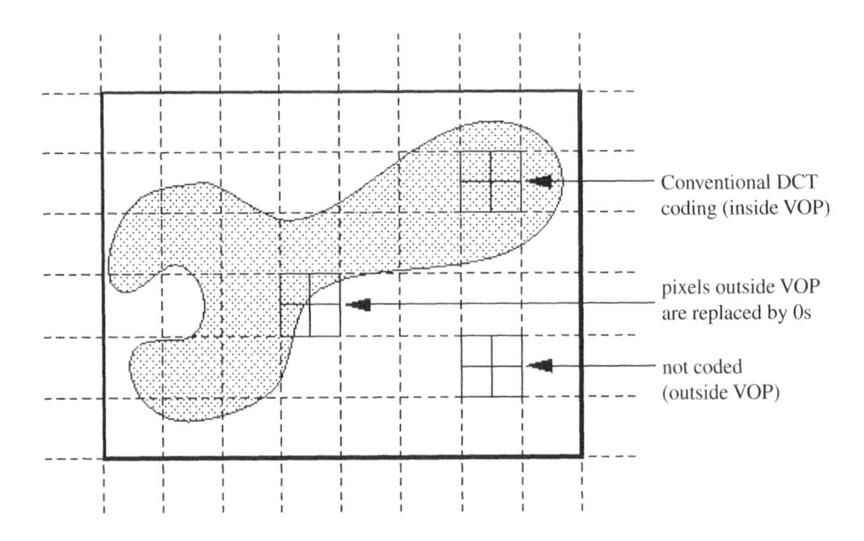

Figure 2.42 Texture coding in MPEG-4 standard

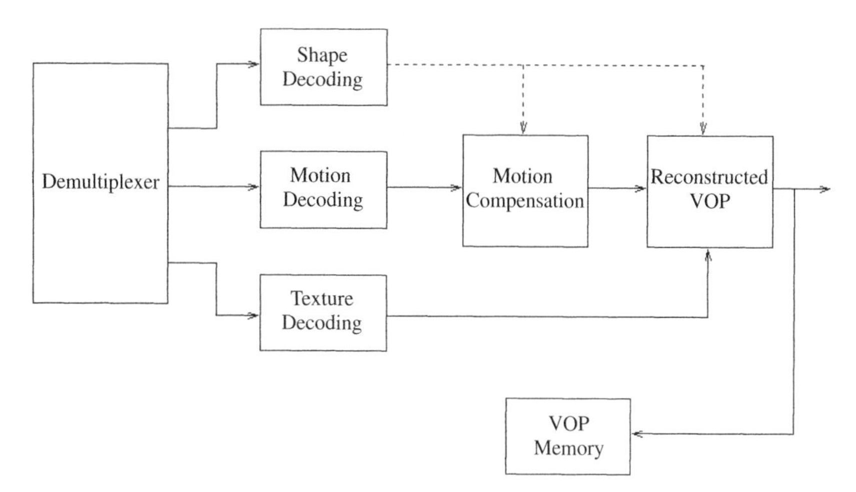

Figure 2.43 VOP decoder structure

2.5.8 Performance evaluation

As already described, MPEG-4 is identical to H.263 in its core structure. However, when object-oriented coding is enabled, MPEG-4 becomes a content-based coder, treating each detected VOP in a frame as a separate coding entity. MPEG-4 employs the same quality enhancement techniques used in H.263 such as the half-pixel motion prediction, the advanced prediction mode and the unrestricted motion vector. Therefore, it is expected that we will conclude that the baseline MPEG-4 video coder achieves a similar perceptual quality to that provided by H.263 when Annexes F and D are used. The quantiser level is normally kept constant throughout the encoding process to produce a constant quality video. However, if the encoding process is required to achieve a fixed bit rate, the coder applies its bit regulation algorithm to meet the pre-defined target rate. A higher quantisation parameter (Qp), frame skipping and coarser motion prediction are only some techniques used to code a video sequence at a fixed output rate. The bit rate control schemes used in video compression algorithms will be described in detail in the next chapter. Figure 2.44 shows the subjective quality achieved by the baseline MPEG-4 video coder when coding the Suzie sequence at three different bit rates. The advanced prediction and the unrestricted motion vector modes are both switched on. Since the core compression scheme is used, no alpha plane is required and therefore no shape coding is then performed. The perceptual quality is very comparable to that achieved by ITU-T H.263 with Annexes F and D employed. Figure 2.45 depicts the subjective quality achieved by both H.263 and MPEG-4 to code the Suzie sequence at 64 kbit/s. Figure 2.46 shows the luma PSNR values of the first 150 frames of the Suzie sequence encoded by either MPEG-4 or H.263. A quantisation parameter of 10 is used for both I and P frames,

(a) (b) (c)

Figure 2.44 Frame 100 of the Suzie sequence encoded at: (a) 128 kbit/s, (b) 64 kbit/s, (c) 32 kbit/s

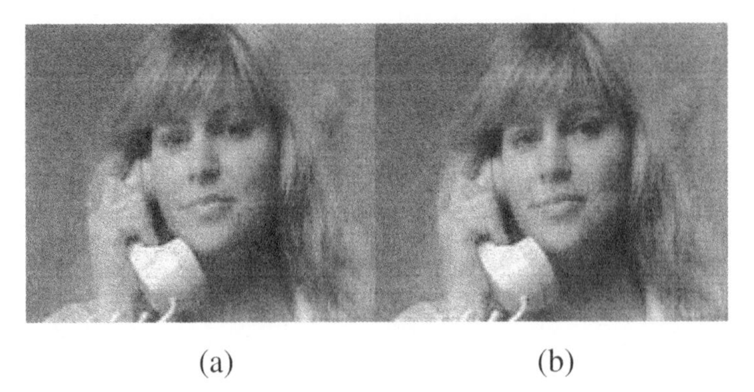

(a) (b)

Figure 2.45 Frame 100 of the Suzie sequence encoded at 128 kbit/s: (a) baseline MPEG-4, (b) H.263

and Annexes F and D are used for a frame rate of 25 f/s. Both results indicate that the core MPEG-4 coder is similar in performance to H.263.

On the other hand, using the object-oriented capability of MPEG-4 requires coding the contour of each detected VOP. Figure 2.47 shows the foreground arbitrary-shape object (ship) of the container ship sequence. Using the alpha plane of this VOP, the ship is segmented out of the sequence and its shape is coded using the binary shape coding technique described in Section 2.5.2.1. The background is coded as a separate VOP using a different alpha plane (segmentation file) and both objects are then decoded and composed to produce the reconstructed sequence at the decoder.

2.5.9 *Layered video coding*

For scaleability and error robustness purposes, MPEG-4 allows multi-layer video coding. The compressed bit stream of each VOP in the video sequence consists of a number of layers, namely the base layer and a number (1 or more) of enhancement

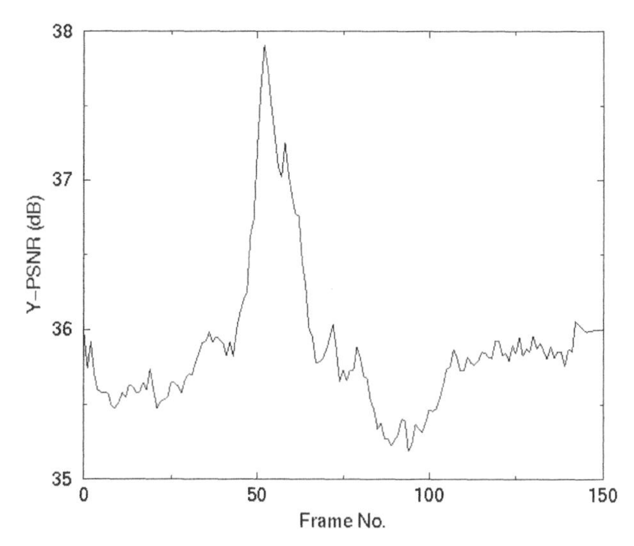

Figure 2.46 Luma PSNR values for 150 frames of the Suzie sequence encoded at 128 kbit/s with either H.263 or baseline MPEG-4

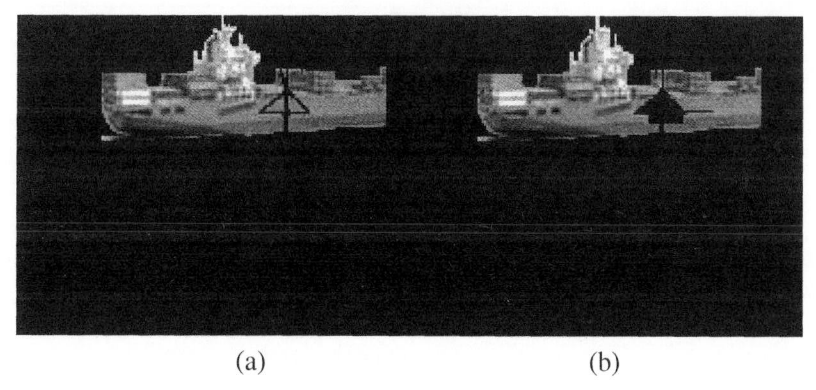

(a) (b)

Figure 2.47 Foreground object of a sequence at frame 200: (a) original, (b) MPEG-4 coded at 20 kbit/s and 25 VOP/s

layers. The base layer is essential for the reconstruction of output video, while the enhancement layers contribute to improving the perceptual quality at the expense of additional bit overhead. The compression ratio of enhancement layers is a compromise between coding efficiency and video quality. The usefulness of multi-layer coding is in producing a temporally scaleable video output in unreliable and highly varying network links. If the channel can handle high rates, more enhancement layers can then be accommodated for improved quality of service. Conversely, in situations such as congestion of network links, only the base layer is transmitted to avoid traffic explosion and guarantee the maximum possible video quality. MPEG-4 produces the base layer by coding a number of original sequence

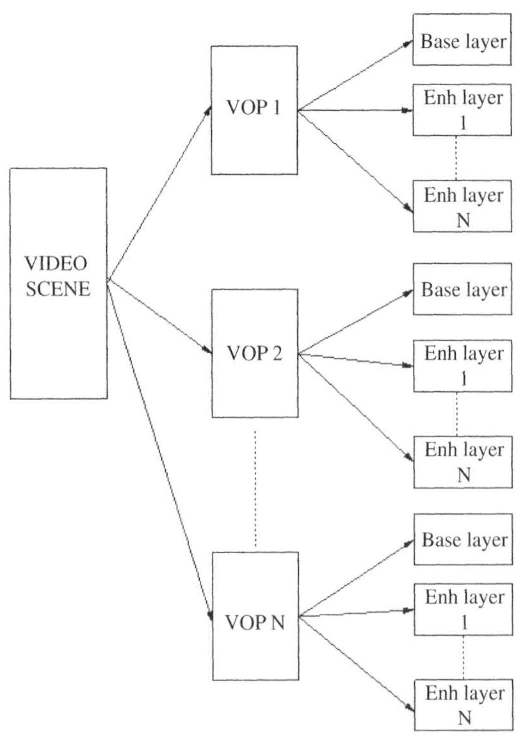

Figure 2.48 Multi-layer video coding in MPEG-4 algorithm

frames using a predefined VOP rate and quantisation parameter. Some of the original sequence frames are skipped to reduce the frame rate and hence allow more bits per frame for a target output bit rate. For example, the base layer could have a VOP rate that is the third of the VOP rate of the original sequence, thereby skipping two frames for each coded base layer frame. The enhancement layers are formed by predicting the skipped frames of the sequence from the coded base layer ones, normally using a coarser quantiser. For each VOP in a video scene, a base layer and a number of enhancement layers could be defined as shown in Figure 2.48. MPEG-4 allows coding of the video layers of various VOPs at different rates based on their importance and contribution to quality, hence providing a content-based scaleable output. This layered video coding is very useful in assigning levels of priority to video layers in accordance with their contribution to overall quality, their sensitivity to errors and the reported channel conditions at any instant of time. Table 2.10 shows the results obtained by coding the Foreman sequence using the temporal multi-layer feature of MPEG-4 video coding scheme where a frame is considered as one visual object and only one enhancement layer is coded per object. A fixed quantisation parameter value of 13 is used during the encoding process. The results indicate that the multi-layer coding produces a much higher bit rate than that generated by the single-layer MPEG-4 encoder for almost the

Table 2.10 Bit rates and Y-PSNR values for single and combined base and enhancement layers obtained by coding the Foreman sequence using MPEG-4 with the multi-layer capability

BL/(BL&EL) Frame rate (f/s)	BL&EL Y PSNR (bit rate)	BL Y PSNR (bit rate)	SL Y PSNR (bit rate)
7.5/15	31.04 (57)	30.97 (32)	31.01 (47)
5/10	31.00 (44)	30.94 (26)	31.00 (37)
3/6	30.92 (33)	30.89 (19)	30.96 (28)

same decoded video quality and aggregate frame rate. However, coding only the base layer results in a considerable reduction of bit rate with only a negligible, perceptually invisible, degradation to picture quality at the expense of a slower frame rate. This implies that for bandwidth-limited channels or congested networks where only low bit rates could be accommodated, the enhancement layer could be dropped without causing serious damage to the reconstructed picture quality. The techniques and effects of scaleability will be covered more extensively in the following chapter.

2.6 Conclusions

In this chapter, the basic coding principles of contemporary digital video compression algorithms have been presented. The concept of video compression implies the removal of spatial and temporal redundancies from the content of digital video sequences. The quality of service and storage are the basic user requirements from digital video content. In order to meet these requirements, the compression algorithms have to optimise their coding efficiency without compromising the perceptual quality of decoded video data. Therefore, video compression is a trade-off between coding efficiency and perceptual quality.

In addition to model-based, vector-based, sub-band and segmentation-based video coding techniques, standard-compliant video compression algorithms are mainly block-based coding schemes. ITU-T released two video compression schemes, namely H.261 and H.263 in the last decade of the twentieth century, and followed the standard with a number of annexes that address the error resilience and the coding efficiency aspects of H.263, in what has later been referred to as H.263+ and H.263++. H.263 has a number of attributes which make it a more efficient video coder. The half-pixel accuracy, the improved Huffman tables, the more accurate motion vector prediction and the four negotiable options are all novel techniques that were not introduced in H.261. For these four main reasons, H.263 outperforms H.261 both objectively and subjectively and for all types of video sequences. However, due to the three motion vector predictors used in its motion prediction, H.263 is more sensitive to information loss than H.261, which employs only one motion vector predictor. In the full motion compensation scenario, the subjective quality achieved by H.263 was perceptually worse than that of H.261 in error-prone cases, especially for active sequences.

MPEG-4 is an ISO object-based video coding standard that exploits the object segmentation of a video scene for flexible processing and compression of objects. The performance of the MPEG-4 baseline coder is similar to that of H.263 when no object segmentation is employed. Using the object-oriented structure of MPEG-4, detected objects can be rescaled, relocated and even omitted from a video frame. The object-based coding scheme of MPEG-4 enables different objects to be coded at different rates, thereby allowing for the content-based scaleability of the output bit stream. On the other hand, MPEG-4 supports the multi-layer coding of a video sequence in a way that allows each video object to be coded as a set of a base layer and a number of one or more enhancement layers. This mechanism enables the prioritisation of different video layers in accordance with their error sensitivity and contribution to overall video quality, especially when video streams are to be transmitted over packet radio networks.

2.7 References

Ahmed, N., Natarajan, T., and Rao, K. R., Discrete cosine transform, *IEEE Trans. Computers*, **23**, 90–93, Jan. 1974.

CCIR Rec. 601-2, Encoding parameters of digital television for studios, 1990.

Choi, C. S., and Takebe, T., Analysis and synthesis of facial image sequences in model-based image coding, *IEEE Trans. Video Technology*, **4**, No. 3, 257–275, June 1994.

Cote, G., Erol, B., Gallant, M., and Kossentini, F., H.263+: Video coding at low bit rates, *IEEE Trans. on Circuits and Systems for Video Technology*, **8**, No. 7, 849–866, Nov. 1998.

Damper, R. I., Hall, W., and Richards, J. W., *Multimedia technologies and future applications*, Pentech press, 1994.

Eisert, P., and Girod, B., Model-based coding of facial image sequences at varying illumination conditions, *Proc. 10th Image and Multidimensional Digital Signal Processing, IMDSP '98*, Alpbach, Austria, July 1998, 119–122.

Eisert, P., Wiegand, T., and Girod, B., Model-aided coding: a new approach to incorporate facial animation into motion-compensated video coding, *IEEE Trans. on Circuits and Systems for Video Technology*, **10**, No. 3, 344–358, Apr. 2000.

Eryurtlu, F., Kondoz, A. M., and Evans, B. G., Very low bit-rate segmentation-based video coding using contour and texture prediction, *IEE Proc. Vision, Image and Signal Processing*, **142**, No. 5, 253–261, Oct. 1995.

Escobar, J., Deutsch, D., and Partridge, C., A multi-service flow synchronisation protocol, BBN STC Tech. report, Cambridge, Mass, Mar. 1991.

Fleisher, P. E., Lan, C., and Lucas, M., Digital transport of HDTV on optical fiber, *IECE Communication Magazine*, **29**, 36–41, Aug. 1991.

Frederick, R., Experiences with real-time software video compression, *Sixth International Packet Video Workshop*, July 1994.

Frederick, R., Experiences with real-time software video compression, Xerox PARC, online paper about NV, ftp://parcftp.xerox.com/pub/net-research/nv-paper.ps, July 1994.

Ghanbari, M., An adapted H.261 adapted two-layer video codec for ATM networks, *IEEE Trans. Communications*, **40**, 1481–1490, Sept. 1992.

Goswami, J. C., and Chan, A. K., *Fundamentals of wavelets: theory, algorithms and applications*, J. Wiley & Sons, ISBN 0 471 19748 3, 1999.

Gray, M., Vector quantisation, *IEEE ASSP Magazine*, **1**, No. 2, 4–29, Apr. 1984.

Heising, G., Marpe, D., Cycon, H. L., and Petukhov, A., Proposal for ITU-T H.26L, A wavelet-based video coding scheme using OBMC and image warping prediction, http://invinet.hhi.de/itu/, (03/99).

Howard, P., Kossentini, F., and Martins, B., The emerging JBIG2 standard, *IEEE Trans. on Circuits and Systems for Video Technology*, **8**, No. 7, 838–848, Nov. 1998.

Hsu, R., and Harashima, H., Detecting scene changes and activities in video databases, *ICASSP 94*, **33–36**, Adelaide, 1994.

ISO/IEC CD 11172: Coding of moving pictures and associated audio for digital storage media at 1.5 Mbit/s, (12/91).

ISO/IEC JTC1/SC29/WG11/N2802: Information technology – Generic coding of audiovisual objects – Part 2: Visual, ISO/IEC 14496-2, MPEG Vancouver meeting, (07/99).

ITU-T H.261: Video codec for audiovisual services at $p \times 64$ kbit/s (03/93).

ITU-T H.262: Information technology – Generic coding of moving pictures and associated audio information: video (07/95).

ITU-T H.263: Video coding for low bit rate communications (02/98).

ITU-T recommendation H.263 (draft) Version 2 (H.263+): Video coding for low bit rate communications, (01/98).

ITU-T H.323: Packet-based multimedia communication systems (09/99).

ITU-T H.324: Terminal for low bit rate multimedia communications (02/98).

ITU T.81: Information technology – Digital compression and coding of continuous-tone still images – Requirements and guidelines – common text with ISO/IEC (09/92).

JPEG 2000, Coding of still pictures, Final committee draft version 1.0 ISO/IEC JTC 1/SC 29/WG N1646 R, (03/00).

Kampmann M., and Ostermann, J., Automatic adaptation of a face model in a layered coder with an object-based analysis-synthesis layer and a knowledge-based layer, *Signal Processing: Image Communication*, **9**, No. 3, 201–220, Mar. 1997.

Linde, Y., Buzo, A., and Gray, R. M., An algorithm for vector quantiser design, *IEEE Trans. on Communications*, **28**, 84–95, Jan. 1980.

Liu, S., and Hayes, M., Segmentation-based coding of motion difference and motion field images for low bit rate video compression, *Proc. ICASSP*, **3**, 525–528, 1992.

Lookabaugh , T. D., and Perkins, M. G., Application of the Princen–Balley filter bank to speech and image compression, *IEEE Trans. Acoustics, Speech and Signal Processing*, **39**, 1914–1926, 1990.

Max, J., Quantising for minimum distortion, *IEEE Trans. on Information Theory*, **6**, No. 1, 7–12, Mar. 1960.

MoMuSys MPEG-4 VM software, 1998.

Padgett, E., Gunther, C. G., and Hattori, T., Overview of wireless personal communications, *IEEE Communications Magazine*, **3**, No. 1, 28–41, Jan. 1995.

Pearson, D., Developments in model-based video coding, *Proc. IEEE*, **83**, No. 6, 892–906, June 1995.

Pearson, D. E., *Image processing*, McGraw-Hill, 1991.

Portable Video Research Group, Stanford University, PVRG-P64 Codec, 1995.

Siu, M., and Chan, Y. H., A robust universal texture extraction technique for model-based coding, *Proc. of SPIE Conference*, **3811**, 19–20 July 1999.

Soryani, M., and Clarke, R. J., Segmented coding of digital image sequences, *IEE Proc. I*, **139**,

No. 2, 212–218, Apr. 1992.

Tekalp, M., *Noise filtering, digital video processing*, Prentice Hall PTR, Chap. 14, 262–282, 1995.

Telenor R&D, H.263 video codec test model, Nov. 1995.

Torres, L., and Kunt, M., *Video coding: the second generation approach*, Kluwer Academic Publishers, ISBN 0-7923-9680-4, 1996.

Turletti, T., H.261 software codec for videoconferencing over the Internet, INRIA rapport de recherché N. 1834, Jan. 1993.

Zhang, H., and Keshav, S., Comparison of rate based service disciplines, *Proceedings of ACM SIGCOMM*, Zurich, Sept. 1991.

3

Flow Control in Compressed Video Communications

3.1 Introduction

In multimedia communications, compressed video streams need to be transmitted over networks that have inconsistent and time-varying bandwidth requirements. To make the best use of available network resources at any time and guarantee a maximum level of perceptual video quality from the end-user's perspective, a certain flow control mechanism must be introduced into the video communication system (Cote *et al.*, 1998; Wang, 2000). Over-rating the output of a video coder can cause an undesirable traffic explosion and lead to congested networks. On the other hand, uncontrolled reduction of the output bit rate of a video coder leads to unnecessary quality degradation and inefficient use of available bandwidth resources. Flow control techniques must then be employed to regulate and control the output bit rates of video sources in the network to achieve the best trade-off between quality and bandwidth utilisation (Girod, 1993).

One of the main challenges of video communications is to provide a guaranteed quality of service when the network is swamped with excessive delays and information loss rates (Kurose, 1993). Network congestion could be avoided by using preventive instead of reactive remedies. Congestion avoidance techniques in video communications must consist of an efficient flow control mechanism that regulates the rates of active video sources (Jacobson, 1988). In a bit rate regulation scheme, the video source might sometimes be required to decrease its output flow due to high traffic load across the network. This reduction in bit rate could certainly lead to quality degradation since the quantisation distortion becomes more noticeable at lower bit rates. However, the quality degradation resulting from a coarser quantisation process is far less detrimental to the video quality than the effect of intolerable time delays and high data loss rates caused by a state of network congestion. Network congestion effects could also be more disastrous in real-time video services where the decoded video quality is much less tolerant to delay and data loss. Therefore, some policy must be adopted to prevent the occurrence of congestion or reduce its effect in high traffic load conditions. A lot of research efforts have been exerted to establish efficient techniques for resolving congestion. Bolot and Turletti (1994) have developed a feedback control mechanism for flow control of video sources over the multicast backbone (Kumar, 1996) of the Internet. In this preventive rate control scheme, the rate control of a video encoder is regulated by modifying some encoding parameters, as indicated by some feedback messages sent by network receivers. Each receiver sends a feedback message that includes some statistics data such as average packet transit time, average loss rate for multicast traffic, average packet delay, etc. The sender collates this data and adjusts its output flow accordingly. Another feedback mechanism (Bolot, Turletti and Wakeman, 1994) employs a probing technique to solicit information and estimate the number of receivers in a multicast tree. A number of video scaleability paradigms (Radha *et al.*, 1999; Stuhlmuller, Link and Girod, 1999; Horn and Girod, 1997) have been proposed for Internet streaming applications. Other research efforts produced reactive approaches such as error concealment and video data recovery schemes, which we will elaborate on in the next chapter. In this chapter, we present a variety of rate control algorithms that can be used in compressed video communications today. These algorithms can perform dynamically in accordance with the varying channel conditions. The status of the channel is reported back to the video source by a number of receivers that have special traffic data compilation capabilities. These feedback reports make the video source more network-aware and thus contribute to efficiently adapting the flow control algorithms to the reported channel conditions at any instant of time.

3.2 Bit Rate Variability of Video Coders

All the standard video coding algorithms described in the previous chapter produce a variable bit rate per frame for a constant quantisation parameter. To guarantee a constant perceptual quality of the decoded sequence, it is necessary to keep a constant quantiser value Qp during the encoding process. Alternatively, varying the quantiser value on a frame or MB basis could achieve a constant output bit rate but at the expense of an undesirable variation in the decoded video quality. A new variable quantiser rate control algorithm has been proposed (Perra, Pinna and Giusto, 2000) to produce a minimal output bit rate for a fixed objective quality. The relationship between the temporal activity and quality of service in video communications is shown in Figure 3.1 for both fixed and variable bit rate encoding. In addition to the constant quality justification of variable rate video, the fluctuation of bit rates is also useful for the dynamic allocation of available bandwidth. As described in Chapter 2, a video source produces a higher output rate with a more active scene or more detailed texture. The drop in the output rate of a video source could be exploited to allocate a larger portion of bandwidth to a

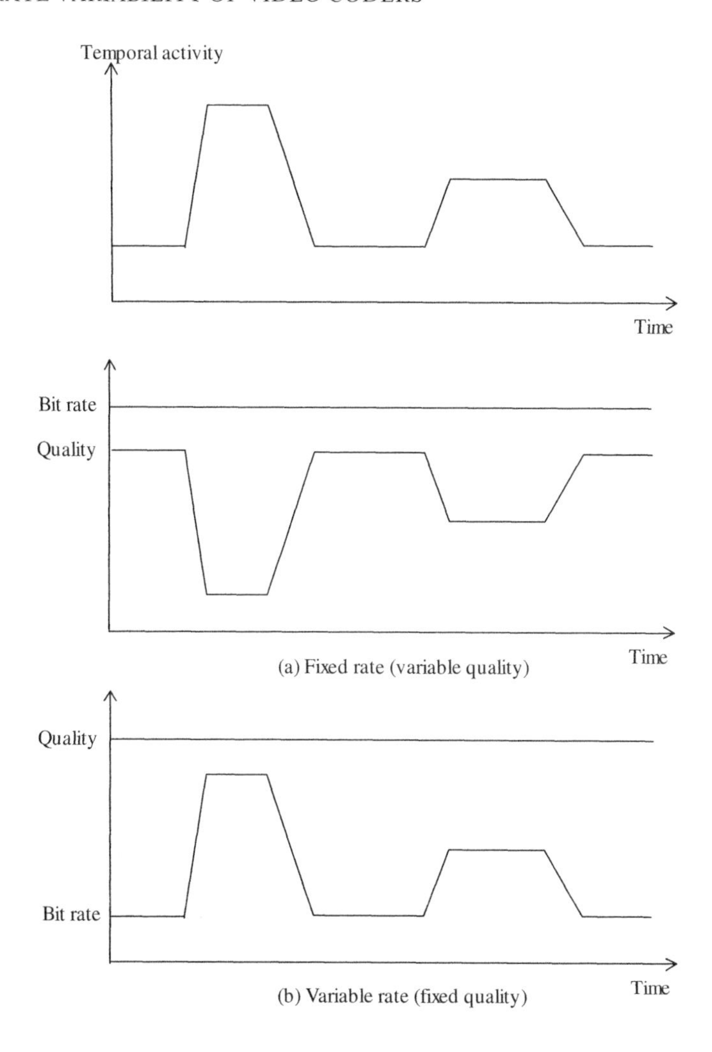

Figure 3.1 Relationship between quality and bit rate

more active source in the network, thereby ensuring a more efficient bandwidth sharing than for the fixed bandwidth scenario. However, this dynamic bandwidth allocation requires a flow control mechanism which can police and dictate the output traffic of each video source on the network in accordance with the time-varying network conditions and requirements. In general, there are two main reasons why a block-transform video coder has this variable bit rate characteristic.

A digital video signal incorporates a huge amount of sequence-dependent redundancies in both time and space. The compression efficiency of a video encoder is determined by the amount of redundancy that is detected and suppressed from the video sequence in both the spatial and temporal domains. It is the proportional removal of these spatial and temporal redundancies which make the output bit rate a variable function of time. For instance, an MB in a predicted frame could represent an unchanged picture area between two successive frames. Therefore, this MB remains stationary as compared to the corresponding MB in the preceding frame. In this case, the block-transform video encoder does not code the MB for improved coding efficiency but sets a single bit flag (COD = 1) indicating to the decoder that this MB has been skipped in the encoding process. The number of uncoded MBs in predicted frames is certainly a function of the temporal correlations in the video content. This number also depends on the temporal similarities criteria used by the encoder as to whether a certain MB in a predicted frame is to be coded or skipped. The variability of the number of coded MBs in predicted frames certainly leads to a variable output bit rate. On the other hand, the spatial correlations between pixels of the same video frame dictate the number of bits required to encode the 64 transform coefficients of each 8×8 block of data. This is in addition to the chosen quantisation parameter that controls the number of zero coefficients and non-zero levels that are fed into the run-length encoder. Obviously, since the quantised coefficients (TCOEFF) of the video blocks result in different levels and zero-run lengths, the run-length encoder produces a different number of VLC words (RUN, LEVEL) per block even when the quantisation parameter remains constant throughout the encoding process. Moreover, the temporal scaleability feature enabled by multi-layer coding, such as in MPEG-4 for instance, contributes towards the variable output bit rate. Different VOP rates, frame skipping, different quantisation parameters per video layer, are all factors that contribute to this highly time-varying output bit rate.

The second factor that leads to the bit rate variability in video coding algorithms is the presence of Huffman coding. Variable-length coding is used to optimise the compression efficiency by achieving an optimal average bit length per codeword. As opposed to fixed-length coding, Huffman coding attempts to assign a code to a certain event, such as a run of zeros, based on the likelihood of its occurrence. The more likely the event, the shorter the code and vice versa. For some video parameters defined by the syntax of a video coding algorithm, such as ITU-T H.263 (Refer to Appendix A), specific Huffman tables are defined. These tables are used to guarantee an optimal average number of bits per coded video parameter. However, due to spatial correlations of video data, different areas of a video frame could be coded at different compression ratios, hence with different number of bits, even if they happen to have an equal number of MBs and/or pixels. This could be best demonstrated by assigning variable-length codes to the different runs of zeros and non-zero levels produced by the run-length encoder. Table 3.1 lists the fixed and variable-length video parameters of the H.263 compression algorithm. Although the table shows more parameters that are fixed-length coded, the contribution of variable-length parameters to the overall output bit rate is much higher than that of fixed-length parameters. Therefore, the percentage of the bits corresponding to variable-length parameters is much higher than that of their fixed-length counterparts. This conclusion is better illustrated in Table 3.2 which

Table 3.1 Fixed and variable length video parameters in H.263 coding algorithm

Layers	Codes			
	Variable length		Fixed length	
Picture	*Bit Suffing*	ESTUF, PSTUF	*Synchronisation* *Addressing* *Quantisation step size* *Administrative* *Spare*	PSC(22), ECS (22) TR (8), TRB (3) PQUANT (5), DBQUANT (2) PTYPE (13), CPM (1), PSBI (2) PEI (1), PSPARE (8)
Group of Blocks	*Bit Suffing*	GSTUF	*Synchronisation* *Addressing* *Administrative* *Quantisation step size*	GBSC (17) GN (5) GSBI (2), GFID (2) GQUANT (5)
Macroblock	*Administrative*	MCBPC, MODB, CBPY	*Administrative*	COD (1), CBPB (6)
	Motion	MVD, MVD2-4, MVDB	*Quantisation step size*	DQUANT (2)
Block	*DCT Coefficients (except Intra DC terms)*	TCOEFF	*DC terms of Intra DCT Coefficients*	INTRADC (8)

shows that most of the bits of an H.263 stream, for the Foreman sequence coded at 30 kbit/s, are due to the variable-length codes. More precisely, the statistics show that the DCT coefficients (excluding the fixed-length INTRADC codes) and the differential MV components contribute to 75 per cent of the overall output flow of the encoder.

3.3 Fixed Rate Coding

Although a variable bit rate is sometimes desirable for dynamic bandwidth allocation, constant bit rate transmissions are useful for fixed bandwidth channels such as PSTN. To achieve fixed rate video transmissions, a buffer between the video encoder and the channel is used to smooth out the bit rate fluctuations. Obviously, buffering the compressed video streams before transmission entails a certain amount of delay, which must be avoided or at least minimised in real-time video services. This buffer could only regulate the output bit rate for short-term variations. In some video sequences, bit rate fluctuations could last for several frames and thus a large buffer would then be required to absorb long-term

Table 3.2 Contribution of video parameters to overall bit rate for Foreman coded by H.263 at 30 kbit/s

		Fixed	Variable	Total
Synchronisation	PSC	0.73		5.25
	GBSC	4.53		
Addressing	TR	0.27		1.60
	GN	1.33		
Quantisation	PQUANT	0.17		1.50
	GQUANT	1.33		
Administrative	PTYPE	0.43		16.67
	CPM	0.03		
	GFID	0.53		
	COD	3.27		
	CBPY		8.86	
	MCBPC		3.55	
DCT coefficients	INTRADC	5.07		46.42
	TCOEF		41.35	
Motion vectors	MVD		28.52	28.52
Spare	PEI	0.03		0.03
Total		17.72	82.28	100.00

fluctuations. This long-term buffering introduces intolerable details and makes the provision of real-time video services impossible. Therefore, in addition to buffering the video data, other measures need to be taken in order to reduce the burstiness of the output flow of video coders.

The most commonly used technique is to adjust some video encoding parameters as a function of the buffer fullness, i.e. by feedback control. On the other hand, the use of current picture activity, i.e. feed-forward control, provides an alternative means of indicating to the video coder the need to adjust the encoding parameters. The buffer-based approach for bit rate regulation is depicted in Figure 3.2. In the next section, we describe the response of the video coder to feedback or picture activity feed-forward messages.

3.4 Adjusting Encoding Parameters for Rate Control

Any attempt to control the output bit rate of a video coder involves trading-off quality and compression efficiency. Reducing the bit rate could be done at the expense of degraded quality. In block-transform video coders, there are four different encoding parameters which could be adjusted to control the output bit

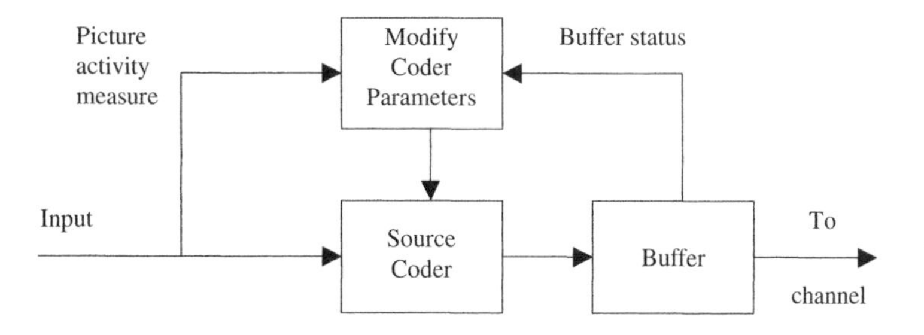

Figure 3.2 Feedback and feed-forward approaches in buffer-based video bit rate control systems

rate. Firstly, the frame rate, which determines the number of encoded frames per second, is one encoding parameter that could be modified to match the bit rate requirements. Since the frame rate control method targets the temporal and not the spatial redundancies of video signals, it is generally used when the quality of individual pictures cannot be compromised. Another possible way to modify the output bit rate is to encode only a spatial portion of each 8×8 block of pixels such as the diagonal coefficients (1×1), (2×2), etc., or only the low-frequency coefficients of a block. Fewer bits are then produced per block at the expense of reduced quality due to the removal of more video data. To optimise quality and preserve the block perceptual fidelity, the DC coefficient, which contains the largest portion of the block energy, has to be coded and AC coefficients could be dispensed for lower output rates.

If the motion aspect of a video scene has an important contribution to the overall quality of video then the spatial video quality could be compromised for a better temporal video quality. In this case, the frame rate is preserved for a coarser quantisation of spatial video details. The third parameter, which can be adjusted for controlling the video bit rate, is the quantisation parameter Qp. This parameter controls the number of bits required to quantise output video codewords, such as transform coefficients. Increasing Qp results in encoding the DCT coefficients with fewer bits, since more zero coefficients would then be obtained (due to quantisation) prior to run-length coding. However, lower Qp values lead to a wider encoding range and hence higher bit rates. Adjusting the quantisation step-size could be done on a frame, GOB or MB basis. Figure 3.3 shows that the number of bits per frame of an H.261 coded sequence at a resolution of 352×240 varies inversely with Qp values.

The fourth encoding parameter that can be manipulated to control the output bit rate of a video encoder is the motion detection threshold. This threshold is set to control the decision of whether an MB in a predicted frame (P-frame) is coded (COD = 0) or skipped (COD = 1). If the threshold increases, the encoder becomes less sensitive to motion and thus the number of coded MBs decreases. Therefore, the number of bits required for encoding a P-frame decreases at the expense of

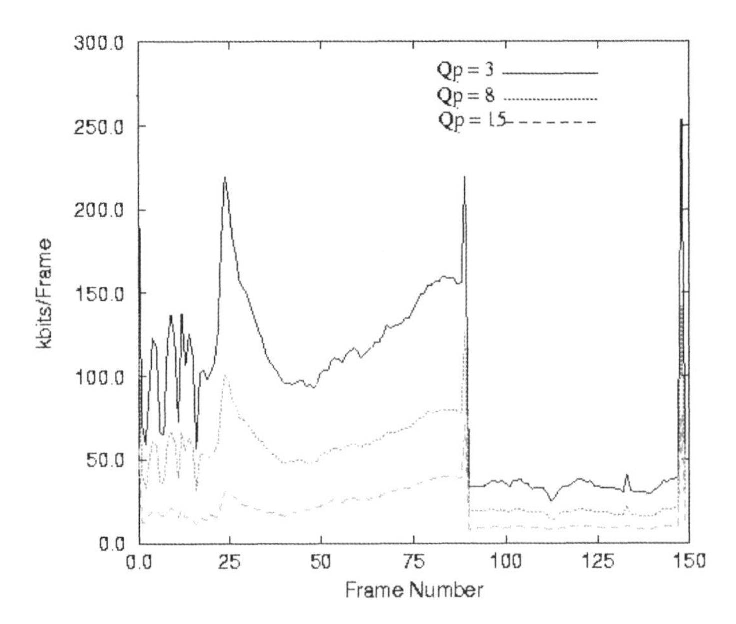

Figure 3.3 Number of bits per frame for a video sequence of 150 frames, with a resolution of 352 × 240, coded with H.261 at different Qp values and a fixed motion threshold of 2.2

lower sensitivity to motion. Conversely, for a lower motion threshold a larger number of MBs will be coded, leading to an improved motion sensitivity but higher bit rates. Similarly, the INTRA/INTER mode decision threshold could also be used to control the output bit rate of each coded MB in a predicted frame. More INTRA coded MBs lead to increased bit rates but improved decoded quality. The improved quality of INTRA coded MBs is mainly due to the absence of prediction in this coding mode. Figure 3.4 shows the output number of bits per frame for the same video sequence as in Figure 3.3, encoded with H.261 at different motion threshold values.

The aforementioned four encoding parameters could be adjusted during the encoding process to control the output bit rate of a video encoder. The adjustment of the parameters is usually done in line with the channel status that is periodically reported to the video source. The regulation of the encoding parameters leads to a variable level of perceptual quality, but this could only have a graceful effect as compared to quality degradation resulting from congestion. Most video communication systems that rely on adjusting the video encoding parameters as part of controlling the output rate adopt preventive flow control techniques (Dagiuklas and Ghanbari, 1992). In these techniques, the rate control system remains active to prevent the network from reaching a state of congestion, hence the name preventive.

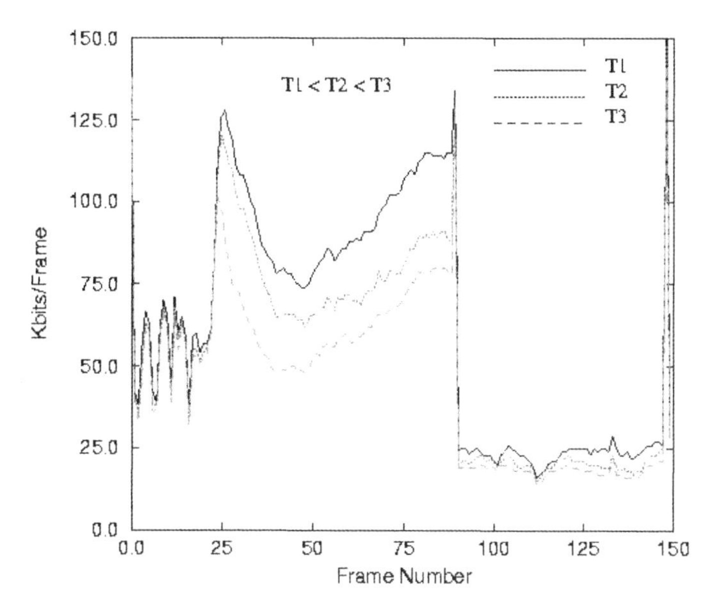

Figure 3.4 Number of bits per frame for a video sequence of 150 frames, coded with H.261 with Qp = 10, for different motion threshold values

3.5 Variable Quantisation Step Size Rate Control

The traditional approach to regulate the output bit rate of a video source is to adjust the quantisation step size of the next frame, GOB or MB, based on the local buffer occupancy that is essentially dictated by the status of the network. However, although varying the quantisation step size affects the output rates, the average number of bits generated for each frame (GOB or MB) is not linearly dependent on the quantisation step size, as shown in Figure 3.5. For instance, when Qp is less than 5, a unity variation can produce two to five times more output video data. Conversely, the same unity change in Qp may generate only a few dozen more bits when the quantisation parameter exceeds 20.

In addition to that, the video content affects the number of bits required to code a video frame. Therefore, classical quantisation rate control techniques provide unpredictable and sometimes highly fluctuating bit rates, thereby increasing the likelihood of local buffer overflow that results in severe data losses in the case of network congestion. In order to produce a stable video output, more sophisticated rate control algorithms have to be employed. In these algorithms, both the buffer fullness and the picture activity have to be used to choose an appropriate quantiser parameter Qp so that the resulting bit rate is close to the target bit rate.

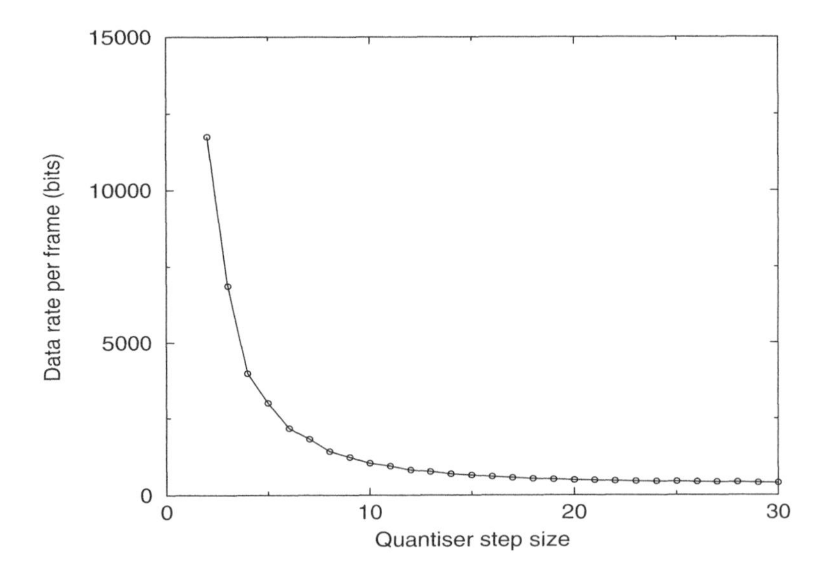

Figure 3.5 Average data rate per frame as a function of quantiser

3.5.1 Buffer-based rate control

One widely accepted buffer-based rate control technique is called the scaleable rate control (SRC) algorithm (ISO/IEC 14496, Annex L) for real-time MPEG-4 video transmissions. SRC is designed to achieve scaleability at various bit rates from 10 kbit/s up to 1 Mbit/s, and various spatial and temporal resolutions. This technique can handle I, P and B frames and can only be applied for single visual object (VO) rate control purposes. The SRC scheme assumes that the encoder rate distortion function can be modelled as:

$$R = X_1 \times S \times Q^{-1} + X_2 \times S \times Q^{-2}$$

where R is the encoding bit count, S is the encoding complexity (mean absolute difference), Q is the quantisation parameter, and X_1 and X_2 are the modelling parameters.

The SRC scheme procedure divides its main processes into four stages: initialisation, computation of the target bit rate before encoding, computation of the quantisation parameter Qp before encoding, and updating the model parameters based on the results obtained from coding the current frame. Firstly, the SRC algorithm checks whether the current frame is an INTRA or INTER frame. For INTRA coded frames, the initialisation part extracts the first and second order coefficients and Qp is set to the value initially specified (by the user or the application). SRC then skips steps 2 and 3 and the rate distortion model parameters are updated based on the encoding results of the current frame. The bits used for the header and the motion vectors are deducted since they are not related to Qp. As a last step, the SRC checks the current buffer occupancy. If it is below 80 per cent the algorithm proceeds to the next frame, otherwise it skips the next frame and updates the buffer occupancy. However, if the current frame is an INTER coded picture then initialisation will be discarded and the algorithm goes to the bit rate computation stage. At this stage, the target bit rate is calculated based on the bits available and the last encoded frame bits. A lower bound of target rate (R/30) is used so that minimal quality is guaranteed. The target rate is adjusted according to the buffer status in order to avoid overflow or underflow. After the target bit rate has been computed, the quantisation parameter computation stage becomes active. Qp is calculated based on the model parameters X_1 and X_2. Qp is limited within the interval [1,31] and can vary by only 25 per cent of the previous Qp value to keep the quality variation under control. After the calculation of Qp for the current frame, the SCR algorithm passes the results to the model updating stage in order to compute the new model parameters, and the procedure continues.

In addition to SRC, another quantisation control scheme adopted in the MPEG-2 video coder is the Test Model 5 (TM5) algorithm [TMOD] for rate control purposes. TM5 describes a procedure for controlling the bit rate by adapting the quantisation parameter of an MB, and it consists of three steps. Firstly, the target bit allocation stage estimates the number of bits available to encode the next picture. This stage is performed before encoding the picture. Then, the rate control stage sets the reference value of the quantisation parameter for each MB by means of a virtual buffer. Finally, the adaptive quantisation stage modulates the reference value according to the spatial activity in the MB to derive the value of the quantisation parameter used to quantise the MB. In the target bit rate allocation stage, TM5 calculates the bit allocation of the next frame using the global complexity measure $X_{i,p,b}$, as indicated in the following formulae:

$$X_{i,p,b} = S_{i,p,b} \times Q_{i,p,b}$$

where S_i, S_p and S_b are the numbers of bits generated by encoding a current I, P or B frame, respectively; and Q_i, Q_p and Q_b are the average quantisation parameters computed by averaging the actual quantisation values used during the encoding process of all the MBs including the skipped ones. After the calculation of $X_{i,p,b}$, the target number of bits for the next picture, namely $T_{i,p,b}$, is calculated in accordance with the overall number of bits (R) assigned to the group of pictures (GOP). If the current picture is the first one in a GOP (INTRA frame) then R is updated as follows:

$$R = G + R$$
$$G = \text{bit rate} \times N_{p,b}/\text{picture rate}$$

where N is the number of pictures in the GOP, N_p and N_b are the numbers of P and B pictures, respectively, remaining in the current GOP, and R is initially set to zero. However, if the current frame is not the first picture in a GOP, i.e. the INTER frame, then R is updated as follows:

$$R = R - S_{i,p,b}$$

where $S_{i,p,b}$ is the number of bits generated in the I, P or B picture, respectively, which was just encoded. After the target number of bits allocated to the next frame has been calculated, stage 1 passes the results to stage 2 on rate control. The rate control stage is based upon the idea of a virtual buffer. Before encoding MB j ($j \geq 1$), the fullness of the appropriate virtual buffer is computed based on the picture type:

$$d_j^{i,p,b} = d_0^{i,p,b} + B_{j-1} - \frac{T_{i,p,b} \times (j-1)}{MB\text{-}cnt}$$

where d_0^i, d_0^p and d_0^b are the initial fullness values of virtual buffers for I, P and B frame types, respectively, B_j is the number of bits generated by encoding all MBs in the picture up to and including MB j, MB-cnt is the number of MBs in the picture, and d_j^i, d_j^p and d_j^b are the fullness values of virtual buffers at MB j for each picture type I, P and B, respectively. The final fullness of the virtual buffer (d_j^i, d_j^p and d_j^b for $j =$ MB-cnt) is used as d_0^i, d_0^p and d_0^b for encoding the next picture of the same type. Then, the reference quantisation parameter Q_j for MB j is computed as follows:

$$Q_j = \frac{d_j \times 31}{2 \times \dfrac{bit\ \ rate}{picture\ \ rate}}$$

where d_j is the fullness of the appropriate virtual buffer. After the buffer has been successfully monitored and its fullness estimated, TM5 proceeds to the third stage to determine the quantisation parameter *mquant* for encoding the MB. To find *mquant*, the spatial activity of the current MB j is measured using the original pixel values of the four luminance frame-organised blocks ($n \in [1,4]$) and the four luminance field-organised blocks ($n \in [5,8]$) as follows:

$$act_j = 1 + \min(vblk_1, vblk_2, \ldots, vblk_8)$$

where

$$vblk_n = \frac{1}{64} \times \sum_{k=1}^{64} (P_k^n - P\_mean_n)^2$$

$$P\_mean_n = \frac{1}{64} \times \sum_{k=1}^{64} P_k^n$$

P_k^n are the sample values in the nth original 8×8 block. Once the MB spatial activity has been determined, the value of act_j is normalised as follows:

$$N\_act_j = \frac{(2 \times act_j) + avg\_act}{act_j + (2 \times avg\_act)}$$

where avg_act is the average value of act_j in the last encoded frame. For the first frame, avg_act is set to 400. Finally, TM5 finds the value of the quantisation parameter $mquant_j$ of MB j as follows:

$$mquant_j = Q_j \times N\_act_j$$

where Q_j is the reference quantisation parameter obtained from the rate control step. The value of $mquant_j$ is clipped to the range [1,31] and is used to code in either the MB or the slice layer.

As is obvious from the two rate control algorithms described above, the quantisation parameter of the current frame (MB or slice) is decided based on the number of bits taken by coding the previous frame. This might not prove enough to ensure a successful rate control scheme for video communications over networks with stringent bandwidth constraints and extremely varying conditions. To achieve smoother output rates of coded video, feed-forward rate control algorithms have to be used, as discussed in the next subsection.

3.5.2 *Feed-forward rate control*

In traditional variable-quantisation rate control algorithms, the quantisation parameter of the next frame is determined based on the number of bits generated by the previous frame. In feed-forward rate control schemes, the quantisation parameter is determined based on the number of bits required to code the prediction error of the current frame, GOB or MB. As described earlier, most of the bits generated by typical block-transform video coding algorithms are spent on transform coefficients and motion vectors, with the number of bits spent on transform coefficients being the most unpredictable. The number of bits required to code the transform coefficients depends on the resulting prediction error (residual matrix) and the quantisation step size. The prediction error per block for the current frame, which is essential for estimating the number of bits required to code the corresponding video data, can be obtained during the motion estimation stage. The quantisation step size can be exploited to estimate the number of bits

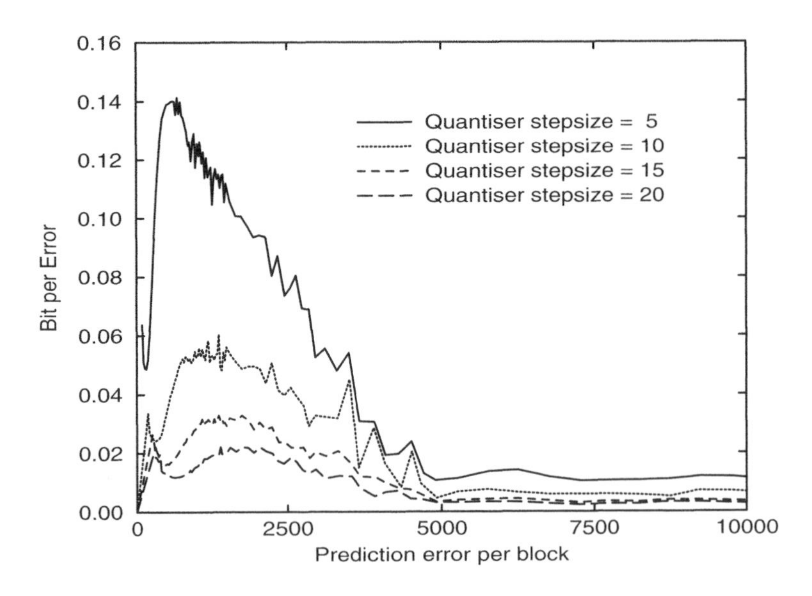

Figure 3.6 Bit per error as a function of prediction error per block for different step sizes

required to code a given prediction error value. Figure 3.6 shows the relationship between the prediction error per block and the average number of bits per error for different quantisation values. These graphs are obtained by using large training sets (Kweh, 1998) taken from five different conventional ITU head-and-shoulder video sequences, including prediction error values, and the resulting number of bits for different quantisation step size Qp values.

In feed-forward algorithms, an initial Qp value of a frame is selected based on Qp and the bit rate of the last coded frame. The objective of choosing Qp value close to the last selected value is to reduce the number of iterations required in the feed-forward rate control technique. With the selected Qp, the number of bits required to code the transform coefficients of the current difference frame is estimated using the prediction error per block and the bit per error curves of Figure 3.6. These estimated bits are then used to estimate the number of bits required to code the motion vectors as well as the coded block pattern for luminance (CBPY). As for the other administrative parameters such as COD, MCBPC, DQUANT for an MB and the pictures headers for a frame, the number of bits spent on them is either constant or relatively negligible. Consequently, the predicted bit rate required to code the current frame is the sum of all these values. This bit rate is then compared to the target bit rate per frame. Qp is increased when the predicted bit rate is higher than the target bit rate of the frame, or *vice versa*. This process is repeated iteratively until the predicted bit rate is equal to the target bit rate or when Qp reaches its maximum allowable value (e.g. 31 in most standard coders). The quantisation value, which yields the closest bit rate, is chosen to code the current frame. This bit rate control algorithm gives an accuracy of ± 15 per cent for the bits used to code the transform coefficients and a less fluctuating bit rate than conventional variable quantisation bit rate control techniques. In order to further smooth out the bit rate fluctuations of feed-forward frame-based rate control algorithm, the quantiser step size is adjusted on an MB level instead of a frame level. In order to maintain a rather uniform video quality, the maximum change in quantisation step size is limited to ± 2 per cent around the chosen Qp value. The following rule defines the MB-based feed-forward rate control algorithm. Let Qp_{frame} be the selected quantiser parameter for the current frame, B_{target} be the target bit rate per frame and B_{total} be the total bits spent until the current MB+ total predicted bits required to code the remaining MB.

If $(B_{total}/B_{target}) > T_1$ and $QP \leqq QP_{frame} + 2$, increase QP, where $T_1 > 1$
If $(B_{total}/B_{target}) < T_2$ and $QP \geqq QP_{frame} - 2$, decrease QP, where $T_2 < 1$ else $QP = QP_{frame}$

This algorithm shows an improved rate control scheme compared to the traditional variable Qp techniques. In order to assess the performance of this control algorithm, a comparative study is presented here with a traditional rate control technique implemented in the H.263 test model (Telenor R&D, 1995). To establish a fair comparison between the traditional rate control scheme and the feed-forward technique, the frame-based technique implemented by Telenor is modified so that the regulation of Qp is achieved on a MB level. Table 3.3 shows the regulated bit rates and PSNR values of four ITU test sequences encoded with the default-mode H.263 coder at a target bit rate of 20 kbit/s, a frame rate of 7.5 frames/s and using both TMN5 and feed-forward rate control algorithms. The achieved bit rates of the feed-forward rate control scheme are very close to the pre-defined target rates and the quality degradation is minimal. Figure 3.7 shows the variations in the output bit rate for the Foreman sequence using both rate control algorithms. The efficiency of the feed-forward rate control algorithm can be seen in the smooth bit rate variations achieved in comparison with the fluctuat-

Table 3.3 Performance of conventional TMN5 and feed-forward MB-based rate control schemes for different sequences coded at 20 kbit/s and 7.5 f/s

	Original TMN5		Feedfor. Controller	
Sequence name (No. of frames)	Actual bit rate (kbits/s)	PSNR (dB)	Actual bit rate (kbit/s)	PSNR (dB)
Foreman (240)	20.33	28.27	20.01	28.14
Carphone (200)	23.06	31.29	20.16	30.86
Suzie (149)	19.91	32.65	20.13	32.81
Salesman (200)	20.86	31.64	19.99	31.57

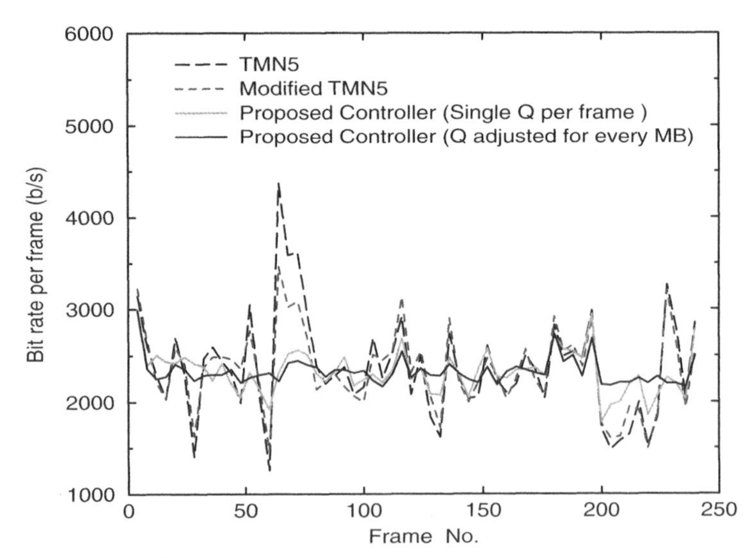

Figure 3.7 Bit rate per frame for the Foreman sequence coded with H.263 at 20 kbit/s and 7.5 f/s using different rate control algorithms

ing rate of its traditional variable-Qp counterpart. Obviously, changing Qp on an MB level achieves the best output bit rate regulation. A penalty to this rate control scheme is, as expected, a less stable perceptual video quality, as shown in the luminance PSNR values of the Foreman sequence in Figure 3.8. These quality fluctuations appear most drastically in periods of high scene activity, where a large number of bits are required for error and motion prediction. The drop in quality due to the bit rate regulation process could be averted by using the region of interest (ROI) coding for rate control purposes, as described in the next section.

3.6 Improved Quality Rate Control Using ROI Coding

In some kinds of video sequences, *a priori* knowledge about the content of the video scene could be exploited for improved coding efficiency by coding the regions of interest (ROI) more accurately than the rest of the video content. For instance, in head-and-shoulder types of video sequence, one tends to concentrate on the face, giving more emphasis to important facial features such as mouth and eyes that are usually most intensively observed. Therefore, it is reasonable to allocate more bits for coding these regions of the scene more accurately at the expense of coarser coding of less important regions (the remaining parts). However, in order to identify the regions of interest in the video scene, *a priori* knowledge about the image must be available (Saghri and Tescher, 1987; Plomplen *et al.*, 1987). In order to be able to employ ROI coding, image segmentation must be employed in the video frames to identify the locations and shapes of these

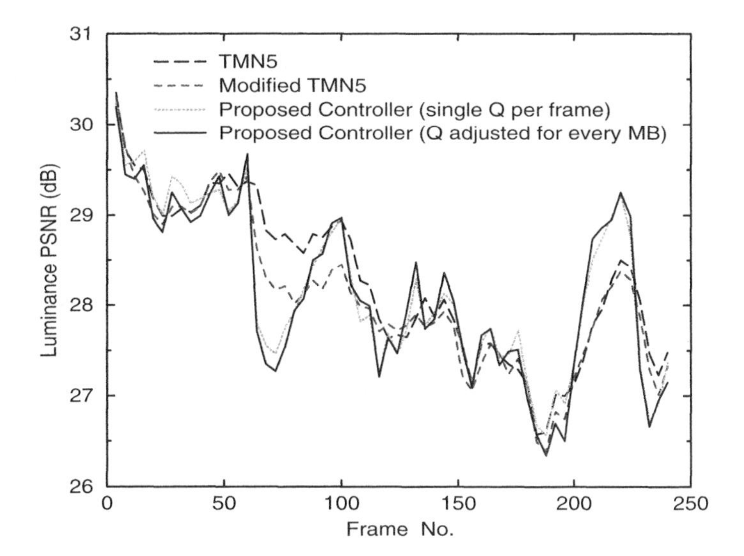

Figure 3.8 *Y*-PSNR values of the Foreman sequence encoded at 20 kbit/s and 7.5 f/s using different rate control algorithms

regions within the video scene, as is the case for object-oriented video compression algorithms such as ISO MPEG-4.

For rate control purposes, the bits required to code both the face region and the background of a frame have to meet the target bit rate requirements. Initially, a smaller quantisation step size is used to code the face region and a coarser quantisation parameter is used for coding the background. The values of these two quantisation step sizes depends on the quantisation parameter Qp set by the rate control algorithm for the next frame in order to meet the target bit rate requirements. Initially, the size of the gap between the two step sizes is set to a minimum by setting Qp_f for the face region at Qp − 2 and that of the background Qp_{nf} at Qp + 6. During the encoding process, if the generated bits are less than the target number of bits then Qp will decrease so that more bits are produced to meet the target bit rate. When Qp is reduced to a threshold value Qp_{lower}, then it will stop to decrease. However, Qp_f will continue to decrease until the target bit rate is met. Consequently, the gap between Qp_{nf} and Qp_f increases. The idea is to hold Qp_{nf} at the value of $Qp_{lower} + 6$. On the other hand, if the generated bits are more than the target bit rate can handle then Qp will increase. When Qp is increased to a threshold value Qp_{upper}, then the gap between Qp_{nf} and Qp_f starts to reduce by holding Qp_{nf} at Qp_{upper}. This simple rate control technique with ROI coding produces satisfactory results, as shown in Table 3.4 where 150 frames of the Miss America sequences are coded at various target bit rates. Both the TMN5 conventional algorithm and ROI coding for face enhancement are employed for rate control purposes. The tabulated results show an improvement in the luminance

Table 3.4 150 frames of the Miss America sequence encoded at different bit rates with two different rate control algorithms

	Face enhanced			TMN5		
Target bitrate (kbit/s)/fr. rate (f/s)	Face PSNR (dB)	Overall PSNR (dB)	Actual bit rate (kbit/s)	Face PSNR (dB)	Overall PSNR (dB)	Actual bit rate (kbit/s)
20/10	34.69	36.82	20.29	32.36	37.89	20.23
17/10	33.37	36.53	17.29	31.51	37.22	17.18
14.4/10	31.83	36.02	14.57	30.52	36.43	14.53
9.6/06	30.96	35.71	9.73	29.77	35.91	9.73

Figure 3.9 Frame of the Miss America sequence encoded at 14.4 kbit/s: (a) conventional variable-Qp TMN5 rate control, (b) ROI coding for enhanced-face rate control

PSNR levels around the face without disturbance to the rate control efficiency. This technique helps regulate the bit rate fluctuations while giving a smoother and sharper perceptual quality around the face area due to ROI coding that favours the facial area. The subjective improvement achieved by this rate control algorithm is depicted in Figure 3.9 which shows a frame of the Miss America sequence encoded at 14.4 kbit/s using the traditional TMN5 and enhanced-face ROI rate control algorithms. On the objective scales, Figures 3.10 and 3.11 show the number of bits per frame and luminance PSNR values, respectively, for 150 frames of the sequence encoded at 20 kbit/s.

Although the above ROI rate control technique achieves its objective in enhancing the perceptual quality of the region of interest in the sequence while maintaining the resultant bit rate close to the target value, it is still in need of improvement, since the bit rate per frame is still highly fluctuating, as shown in Figure 3.10. In order to improve this rate control technique and regulate the output bit rate, the feed-forward algorithm presented in the previous section can be employed to select two quantisation step sizes (as opposed to only one in the previous section) so that the resultant bit rate per frame meets the target value. Initially, the minimum size of the gap between the two step sizes is set to g, i.e.

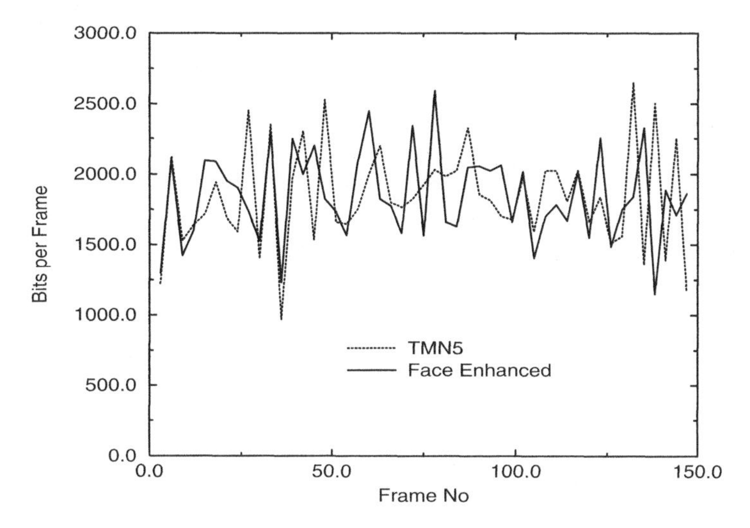

Figure 3.10 Number of bits per frame for the Miss America coded at 20 kbit/s using conventional and ROI rate control algorithms

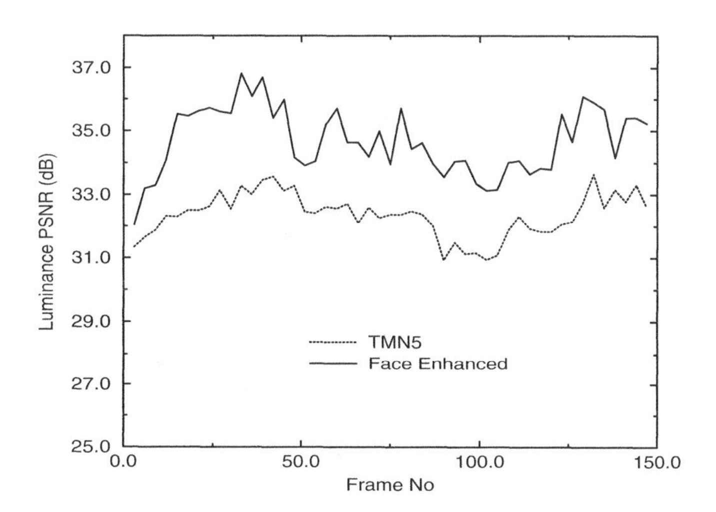

Figure 3.11 *Y*-PSNR for the facial region of 150 frames of the Miss America sequence coded at 20 kbit/s using conventional and ROI rate control algorithms

$$Qp_{nf} = Qp_f + g, \text{ where } g = 0, 1, 2, \ldots, 12$$

The number of bits required to code the current image is first estimated. If the predicted bit rate is greater than the target bit rate, then Qp_f is increased or *vice versa*, until a pair of quantisers capable of producing a resultant bit rate that is close to the target one is found. If Qp_f is greater than a threshold (15) then g is increased for every increment of Qp_f until it reaches 12 which is the maximum value allowed. Alternatively, if Qp_f decreases below a threshold (5), then Qp_f is held at 5 and g starts to decrease. When g is 0, i.e. $Qp_f = Qp_{nf}$ and the predicted bit rate is still less than the target bit rate, then Qp_f will be decreased below 5. Table 3.5 shows the bit rates and luminance PSNR values taken for four video sequences encoded at target bit and frame rates using the conventional and face-enhanced ROI feed-forward rate control algorithms. It can be seen that, due to ROI coding, a gain of about 2–3 dB in luminance is achieved for the face region of each coded sequence in the table. In Figure 3.12, the bit rate per frame is plotted for the sequence Foreman. Using the enhanced-face rate control algorithm with feed-forward bit rate prediction, the observed smooth output rate is comparable to that obtained with a stabilised bit rate algorithm. Figure 3.13 shows the luminance PSNR values of the Foreman sequence using three different rate control algorithms. It can be seen that the overall PSNR of the stabilised rate control algorithm is generally always higher than that of ROI coding, as shown in Figure 3.13(a). However, the PSNR values around the face are always better than those achieved by a stabilised bit rate algorithm due to the enhanced face coding of the algorithm, as shown in Figure 3.13(b).

3.7 Rate Control Using Prioritised Information Drop

The output bit stream of a standard video coding algorithm consists of sets of fixed and variable-length codewords (VLCs). Each VLC represents a particular piece of information related to the temporal and spatial details of the video sequence.

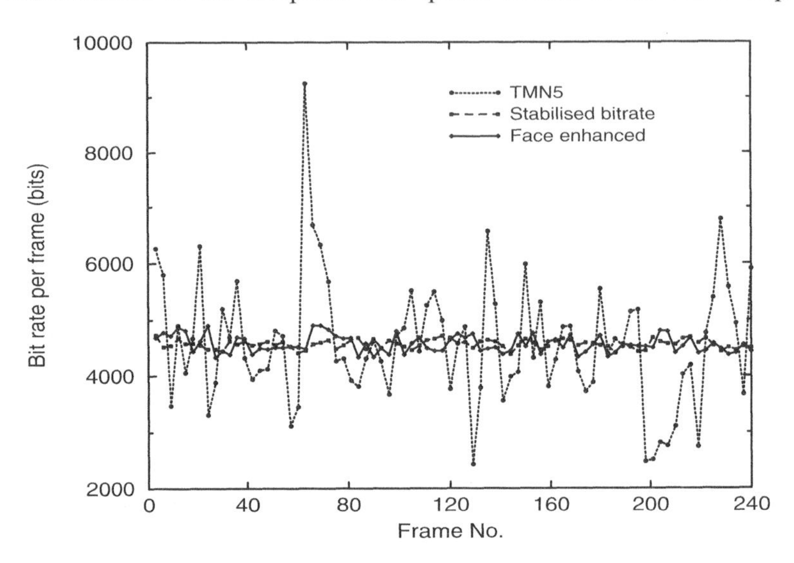

Figure 3.12 Bit rate per frame for Foreman coded at 48 kbit/s using three different rate control algorithms

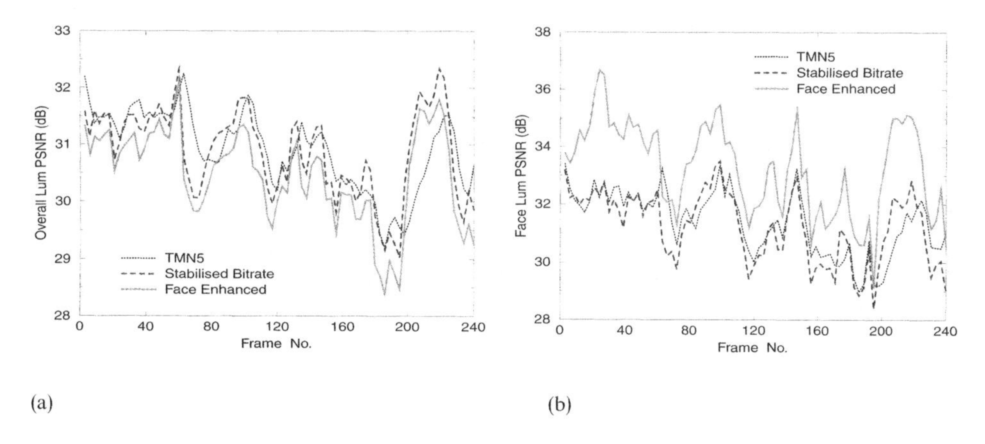

Figure 3.13 Y-PSNR values of the Foreman sequence coded at 48 kbit/s using three different rate control algorithms: (a) overall PSNR (b) facial region PSNR

Table 3.5 Bit rates and Y-PSNR values for four ITU sequences coded at target bit and frame rates using conventional TMN5 and ROI (with feed-forward prediction) rate control algorithms

	Enhanced face rate controller			TMN5		
Target bit rate (kbits/s)/ frame rate (f/s)	Face PSNR (dB)	Overall PSNR (dB)	Actual bit rate (kbit/s)	Face PSNR (dB)	Overall PSNR (dB)	Actual bit rate (kbit/s)
Foreman (0–240) 48 kbit/s, 10 f/s	33.81	30.35	48.00	31.47	30.91	48.14
Carphone (0–200) 32 kbit/s, 6.25 f/s	35.03	32.35	32.04	32.52	33.73	35.05
Suzie (0–149) 28 kbits, 6.25 f/s	36.15	34.81	28.15	33.12	34.56	28.01
Miss America (0–149) 14.4 kbit/s, 6.25 f/s	40.03	37.41	14.50	38.16	37.89	14.35

These video codewords are ordered according to the syntax of the video coding algorithm and then sent to a local buffer before transmission. In the case of network congestion, these buffered codewords could be subject to excessive time delays. Since real-time video is highly delay-sensitive, the delayed video codewords are arbitrarily dropped off the local buffer if their buffering time exceeds a certain pre-defined threshold. Since the delayed video data is dropped at the encoder side, the synchronisation between the video encoder and decoder is maintained by sending a non-coded flag (COD = 1) for each dropped MB. This means that the delayed MB data is then replaced by a single bit which requests the decoder to skip the corresponding MB during the video reconstruction process. However, the synchronisation could be lost when the decoder falls on an unacceptable value of an MV component for instance. This prompts the decoder to skip the information of the current frame until the next error-free synch word in the bit stream. If the compressed video data of a predicted frame is arbitrarily, i.e. without any preference, dropped off the local buffer at the encoder side (for a reported network congestion for instance), both MV and transform coefficients will be lost. Since the INTER mode employs differential coding for predicting motion data, dropping an MV has an accumulative negative impact on the decoded video quality. This could be perceived in both the spatial domain (due to the spatial correlations of MVs) and the time domain (due to predictive coding). Figures 3.14 and 3.15 show the subjective and objective effects, respectively, of dropping INTER MBs off the local buffer of the video encoder to simulate the effect of a network congestion. MV and run-length codewords representing the transform coefficients of video data blocks are randomly and unselectively dropped at a rate of 5 per cent. This loss rate represents the bit error ratio at which MVs and run-length codewords (AC coefficients) are dropped. It is calculated as the ratio of the number of dropped bits over the total number of bits in the coded video stream. It can be observed from parts (b) of Figures 3.14 and 3.15 that the network state of congestion results in a disastrous degradation of decoded video quality within a few seconds of video transmission. Eventually, this effect is further aggrevated with more active sequences and longer prediction intervals.

Because of the differential coding employed for MV prediction, the values of the MV predictors are very important to predict the motion of subsequently neighbouring MBs. This means that MVs are more sensitive to errors than AC coefficients which have a smaller contribution to temporal video quality. On the other hand, transform coefficients occupy a larger portion of the coded bit stream, as indicated in Section 3.2; therefore, dropping these coefficients in the case of congestion helps reduce the flow rate of the video encoder with only a graceful degradation of the reconstructed video quality. Consequently, the output video parameters (e.g. MVs and DCT coefficients) of a standard video coding algorithm

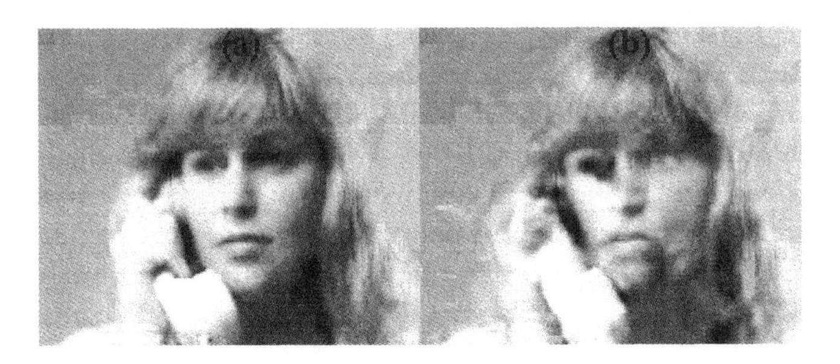

Figure 3.14 Frame 150 of the Suzie sequence encoded with H.263 at 55 kbit/s: (a) error-free, (b) subject to 5 per cent drop rate of MV and AC coefficients of P-frames due to network congestion

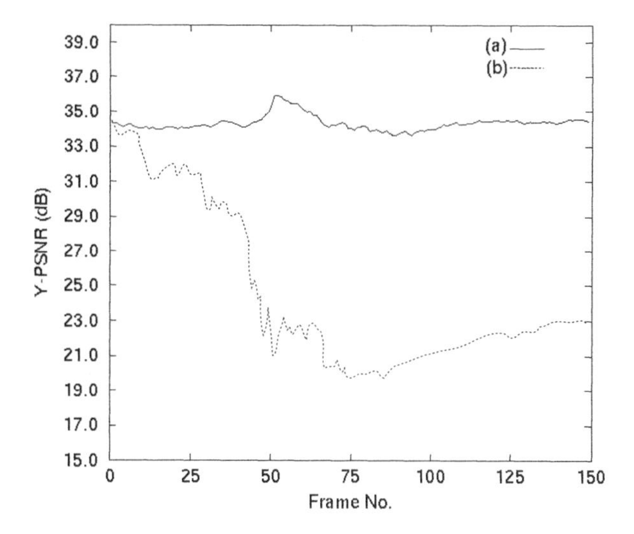

Figure 3.15 *Y*-PSNR graphs of the Suzie sequence encoded with H.263 at 55 kbit/s: (a) error-free, (b) subject to 5 per cent loss on MV and AC coefficients of P-frames due to network congestion

have different sensitivities to information loss. This difference could be better observed by evaluating the effect of information drop on one type of video parameters at a time. This is similar to partitioning the coded video data into two parallel sub-streams, each containing one type of video parameters, and applying the information drop on one sub-stream at a time. Figures 3.16 and 3.17 depict the difference in the sensitivities to information loss of MVs and DCT coefficients of a P-frame on the subjective and objective scales, respectively. These sensitivity measurements are obtained by dropping variable-length codewords of a single sub-stream at one time (either MVs or DCT run-length codewords) and the same bit error ratio is used in both cases. In other words, each case consists of dropping 5 per cent of the data bits off one sub-stream while keeping the other sub-stream intact. Only the data bits of MBs are dropped, and the affected MBs are coded by setting the COD bits. Using the same information loss rates, it is shown both objectively and subjectively that the MV sub-stream is much more sensitive to information loss than the AC coefficient sub-stream.

When the video encoder is notified of bad network conditions (e.g. congested links), it can reduce its output to the channel by dropping video parameters to help relieve the network from excessive traffic loads. Since the INTRADC coefficient contains a high portion of the video block energy, it is not advisable to drop it for flow control purposes. The study of the information drop sensitivities of output video parameters reveals that dropping AC coefficients would have a minor effect on the reconstructed quality as compared to motion data. For that reason, some levels of priority could be assigned to various types of video parameters based on

Figure 3.16 Frame 150 of the Suzie sequence encoded with H.263 at 55 kbit/s and subject to 5 per cent information drop: (a) on the sub-stream of DCT coefficients of a P-frame, (b) on the MV sub-stream

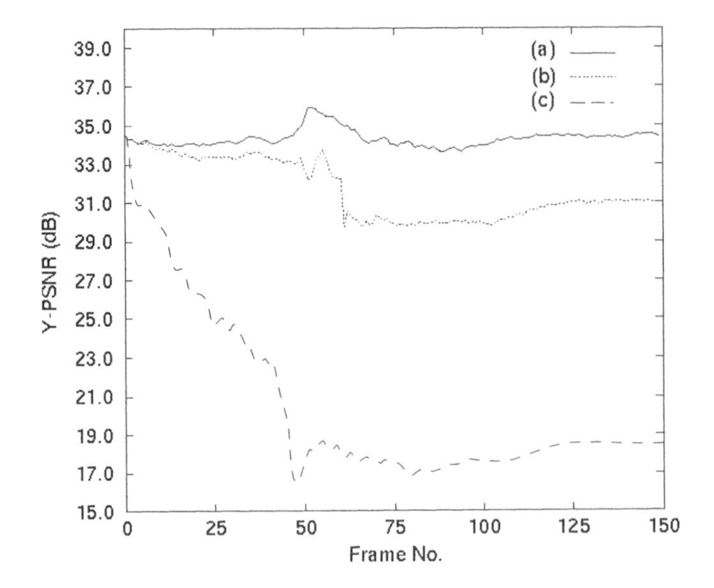

Figure 3.17 *Y*-PSNR graphs for the Suzie sequence encoded with H.263 at 55 kbit/s; (a) error-free, (b) 5 per cent drop applied on the sub-stream of DCT coefficients of a P-frame, (c) 5 per cent drop applied on the MV sub-stream

their sensitivity to loss and their contribution to overall video quality. Assigning these priorities must be done based on a complete study of the sensitivity of individual video parameters to information drop (Leicher, 1994). This technique has to account for all the parameters of a video frame including the administrative data used to enable the synchronisation of the decoder. The prioritisation of video parameters could then become a dynamic process that takes into account both the sensitivity of video parameters and the reported channel conditions at one instant of time. When the network is reported to be in good condition, the video coder guarantees the transmission of all the coded information by setting all priority

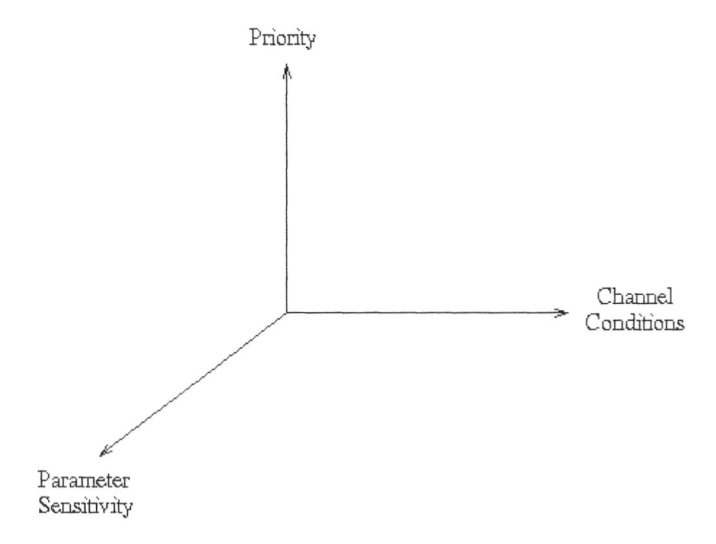

Figure 3.18 Time-varying factors that affect the dynamic prioritisation algorithm of video parameters

levels over a certain threshold value. However, in deteriorated network conditions, the encoder assigns the priority levels to ensure the transmission of critical video data that contributes the most to the reconstructed video quality (Albanese *et al.*, 1996). Figure 3.18 depicts a 3-D reference model that illustrates the dependency of the dynamic prioritisation scheme on the time-varying channel conditions and the sensitivity of video parameters to information loss.

To regulate its output bit rate, the encoder makes an estimate of the number of bits it has to drop by analysing the statistics data of the channel feedback reports. Then, the encoder drops the predicted frames parameters off the local buffer based on the assigned priorities.

Let Nb be the number of bits to discard as estimated based on the channel feedback reports, and let P be the set of various parameters in a video MB. For every $i \in P$ starting with the lowest priority parameter, let $Nb(i)$ be the number of bits belonging to parameter i, and let $Limit(i)$ be the number of lost bits belonging to parameter i.

Then the prioritised information loss can be illustrated in the following simple pseudocode.

$$If\,((Limit(i-1) = Nb(i-1))\ or\ (i = 1))\ then$$

$$If\,((Nb - \sum_{k=1}^{i-1} Nb(k)) \geq Nb(i)\ then$$

$$Limit(i) = Nb(i)$$

Else

$$Limit(i) = Nb - \sum_{k=1}^{i-1} Nb(k)$$

Else

$Limit(i) = 0;$

This algorithm runs as a preventive congestion control mechanism. It controls the bit rate of a video coder while minimising the effect of information loss on the video quality. Figures 3.16(a) and 3.17(b) show that a 5 per cent information drop imposed on the AC coefficients of every P-frame of a 150-frame Suzie sequence does not cause serious damage to the decoded video quality on either the subjective or the objective scales. Although 5 per cent of the information is dropped off the local buffer, the luminance PSNR values show that the quality is maintained within a range of about 2 dB after 150 frames into the sequence time. Therefore, the bit rate is controlled without adjusting any video encoding parameter, thereby achieving a consistent quantisation process regardless of the network state.

3.8 Rate Control Using the Internal Feedback Loop

In the previous section, dropping video parameters is performed in an open loop structure. In other words, the locally reconstructed picture memory is not updated in a way that maintains the match between the reconstructed frame at the encoder and the corresponding frame at the decoder. A mismatch between these two frames leads to the accumulative damage of the decoded video frames.

Because of the prediction used in INTER coding, a dropped MB propagates errors in other MBs located within the same frame and also the next video frames. The picture damage becomes more disastrous in degraded channel conditions, whereby a large number of MBs needs to be dropped for rate control purposes. To stop the accumulation of errors and reduce their effects on the perceptual quality, a feedback loop is introduced in the encoder (Sadka, Eryurtlu and Kondoz, 1996) in order to update the memory of the locally reconstructed frames. The block diagram of the feedback-controlled H.263 video encoder is depicted in Figure 3.19.

ILB senses the network conditions and decides whether buffered parameters are to be discarded or fed to the network (Maglaris *et al.*, 1988). When the network is reported overloaded, ILB drops a certain number of buffered video MBs and updates the picture memory in order to create a match between the locally reconstructed picture and the corresponding remotely decoded one (i.e. after the drop). By the time the encoder is notified of degraded channel conditions through feedback reports, time will elapse before the necessary information drop is performed, and modifications to local picture memory are accordingly delayed. Due

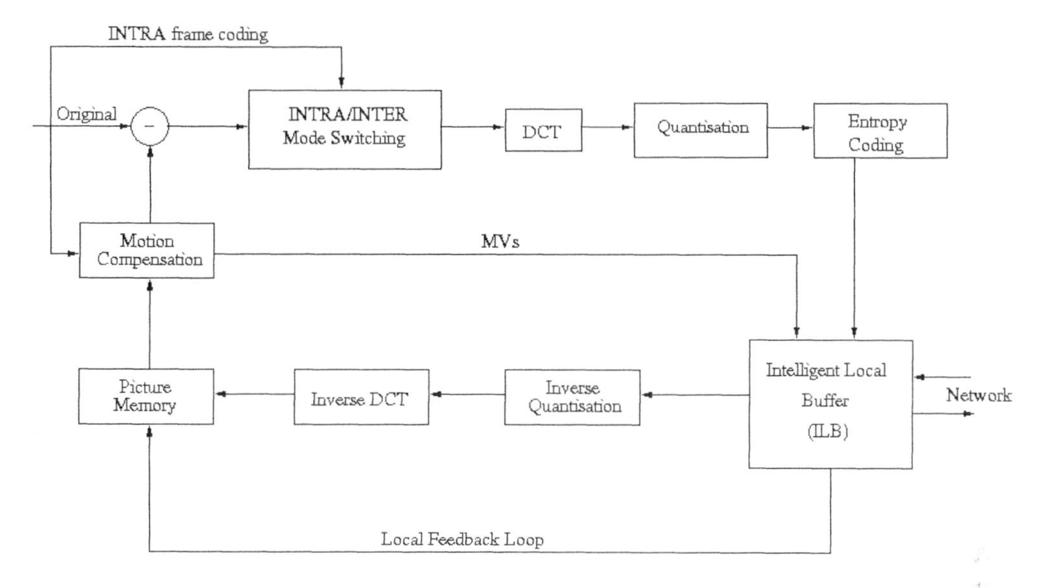

Figure 3.19 Feedback controlled video coder for output rate control

to this delay, the errors will persist in the decoded video sequence for a certain while until the video sequence is refreshed with a better quality frame. Assuming a time delay of one frame, the picture damage will appear for a period of 40 ms (assuming a frame rate of 25 f/s) until the frames are refreshed by means of the feedback-controlled algorithm. When no motion compensation is employed, the MBs do not contain any motion data and therefore the quality improvement brought by this technique becomes marginal.

If the video encoder has a preventive approach to network congestion, ILB has the role of estimating the required number of bits to be discarded as a trade-off between video quality and rate/congestion control. To illustrate the efficiency of the local feedback loop technique in controlling the video rate without compromising quality, we will consider a MV-only information drop. When ILB drops an MV, the consequence would be as if a (0,0) MV was sent to the decoder. This results in replacing the dropped MB by the coinciding one in the previous frame (and adding the residual matrix, if any). Table 3.6 shows that the colour data is less sensitive to information loss than luminance. The chrominance components U (Cb) and V (Cr) may accidentally give better PSNR values for both P and B-frames when feedback is applied at high loss rates. The loss rates estimate the percentage of lost MV bits to the total number of bits of MV data.

When PB-frame mode is used, it is expected that the compressed video data will become more sensitive to loss due to the high compression ratios of B-frames. This sensitivity results in degraded picture quality, as illustrated in Figure 3.20(a). However, with the feedback control mechanism, errors do not last for longer than 40 ms and error spatial/temporal accumulation is precluded, as shown in Figure 3.20(b). Although PB-frame mode makes the coder less robust to information loss,

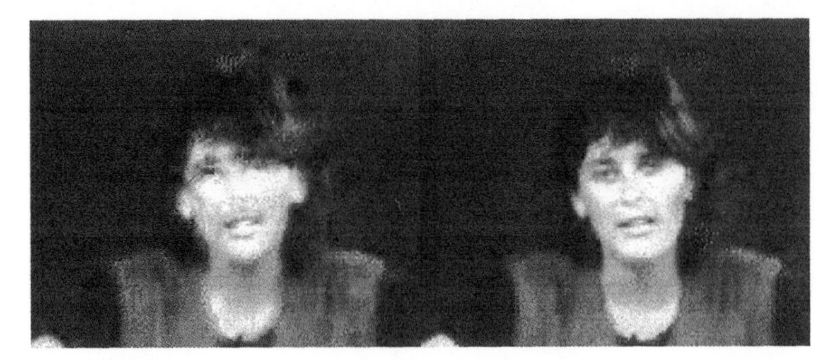

Figure 3.20 Miss America frame no. 148: (a) 20 per cent MV drop, (b) 20 per cent MV drop with feedback

it can be used with the feedback control scheme to further reduce the bit rate without causing significant damage to the video quality. When used with its negotiable options, H.263 becomes more sensitive to errors. However, the negotiable options contribute to the reduction of the output bit rate, thereby ensuring a better coder performance in congestion control terms. The results shown in Table 3.6 and Figure 3.21 reveal that this technique adapts perfectly to rate control scenarios. The output bit rate of H.263 is reduced by more than 2 kbit/s (80 bits/frame on average) for a total output bit rate of 15 kbit/s. For high activity scenes, experiments showed that the bit rate reduction falls in the range of tens of kbit/s while a good drift-free video quality is still maintained. The higher bit rate of the feedback-loop encoder compared to the ordinary H.263 one is mainly due to the updated content of the locally reconstructed picture memory. Encoding some MBs, which have their best-match MBs dropped, becomes more bit-costly. Due to reduced correlations, these MBs are either INTRA coded or INTER coded, with a larger number of bits required to code longer MVs or larger TCOEFF amplitudes.

3.9 Reduced Resolution Rate Control

Traditional variable-quantisation rate control algorithms proved efficient in reducing the fluctuations of the output bit rate of a video encoder. However, in applications where the bit rate budget is extremely low, the traditional rate control algorithms would fail to produce acceptable results since the quantisation step size cannot be increased beyond an upper bound (i.e. 31). In this case, the variable Qp rate control techniques could be used in conjunction with the reduced resolution scheme to achieve a more efficient bit rate regulation process. The reduced resolution technique consists of down-sampling each MB in the prediction error before it is encoded, and up-sampling the reduced-resolution reconstructed block at the decoder in order to produce the motion compensated picture. Firstly, motion estimation is performed on an MB basis. Then the resulting prediction

Table 3.6 SNR and bit rates for the Miss America sequence using the feedback-loop rate control mechanism

MV loss %	Y-PSNR P(B) (dB)	Cb PSNR P(B) (dB)	Cr PSNR P(B) (dB)	Total bit rate (kbit/s)
0	37.18 (37.21)	37.91 (37.85)	36.69 (36.46)	15.38
20%	31.0 (31.12)	37.14 (37.16)	33.86 (33.94)	12.89
20% + Feedback	36.29 (37.14)	36.29 (37.14)	36.68 (36.74)	13.18

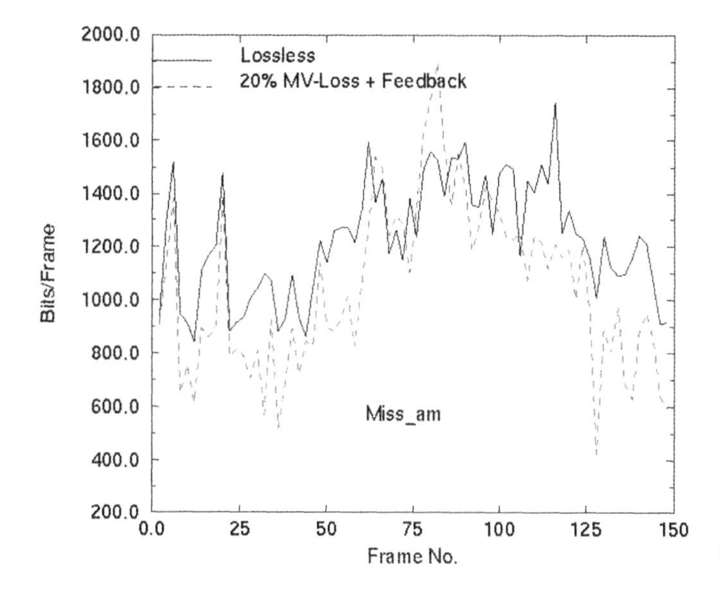

Figure 3.21 Number of bits per frame for 150 frames of the Miss America sequence using the feedback-loop rate control mechanism

error image is reduced by half in each dimension using the bilinear pixel interpolation which is also used for half-pixel motion vector search in ITU-T H.263 video coding standard, as illustrated in Figure 3.22.

Each down-sampled 8×8 luminance block is transformed using an 8×8 DCT, while the 4×4 chrominance blocks are transformed using a 4×4 DCT. The transformed coefficients are quantised, raster-scanned in a zigzag order and then run-length encoded. At the decoder, the transformed coefficients are recovered and inverse-transformed to reconstruct the 8×8 luminance blocks. Each reconstructed 8×8 block is then up-sampled to produce the 16×16 MB using the bilinear filter depicted in Figure 3.23. Finally, the up-sampled MB is added to the motion compensated MB to obtain the final reconstructed MB, as shown in Figure 3.24 for a modified H.263 decoder with reduced-resolution capabilities. Similar processing is carried out for the chrominance blocks while taking into consideration the resolution differences.

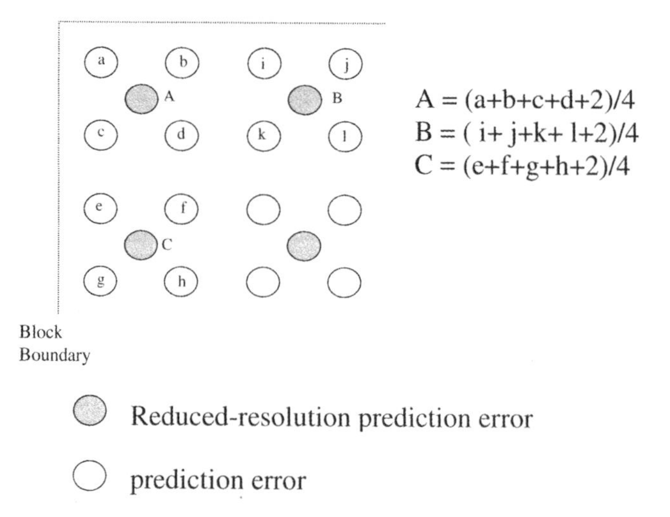

Figure 3.22 Down-sampling of prediction error using pixel interpolation

For rate control purposes, the reduced resolution coding is designated for use under very tight bit rate budget considerations. Since the reduced resolution frame results in a lower number of coefficients, the output bit rate per frame is reduced, while keeping a constant quantisation parameter and a constant frame rate. If this coding mode is to operate on an MB basis, then an additional bit per MB is required to indicate its use. This incurs an extra bit in the bit stream for every coded MB, giving a maximum of 99 bits per frame for a QCIF resolution sequence. For this reason, the down-sampling mode is usually decided on a frame-by-frame basis, and therefore a single extra bit per frame is then required in the frame header to indicate its use.

For extremely low bit rate budgets, both variable-Qp and reduced resolution rate control algorithms operate simultaneously to guarantee a smooth output bit rate without compromising the perceptual quality of decoded video. The quantiser is selected with a bit rate prediction algorithm. For instance, the feed-forward rate controller described in Section 3.5.2 could be used to estimate the quantiser step size of a video frame. If the estimated bit rate still exceeds the target bit rate of a frame after applying the feed-forward rate controller, then the reduced resolution controller is switched on and a new quantisation step size is estimated, as shown below:

If $(Qp_{estimated} > Qp_{threshold})$
{
Switch to reduced resolution mode;
Select new Qp;
}

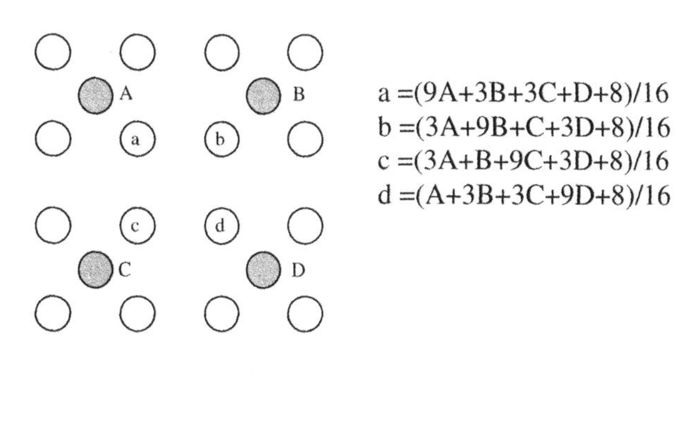

Reduced-Resolution reconstructed prediction error

Reconstructed prediction error

(a)

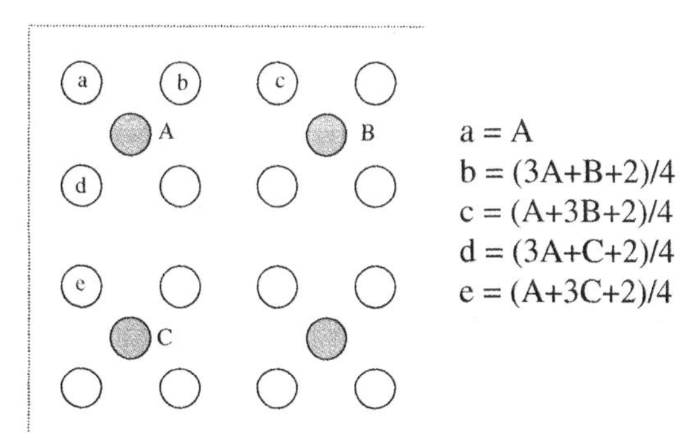

Block Boundary

Reduced-Resolution reconstructed prediction error

Reconstructed prediction error

(b)

Figure 3.23 Up-sampling of the reconstructed block prediction error: (a) inside a block, (b) at a block boundary

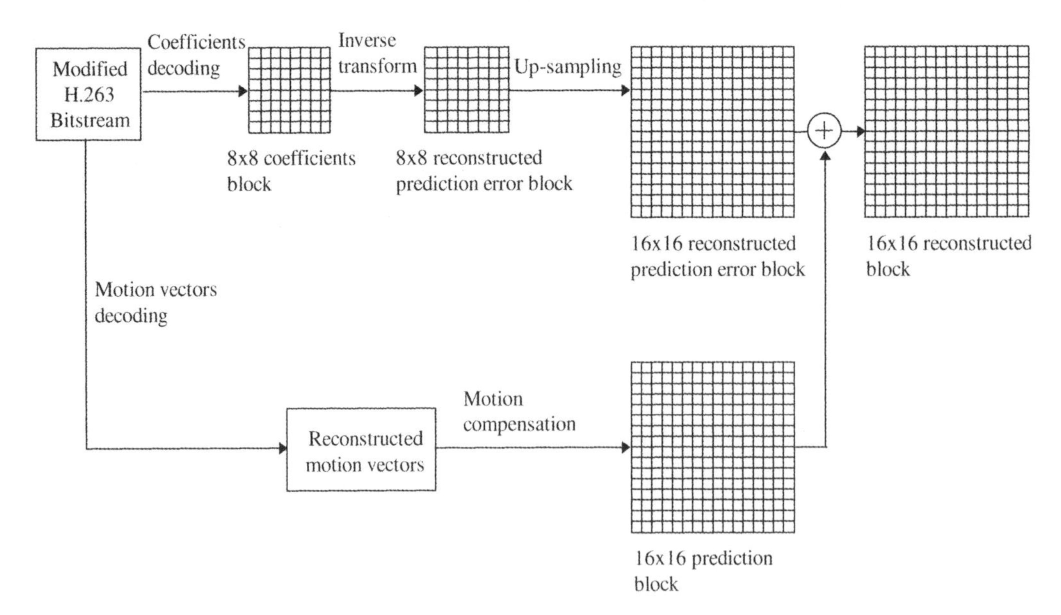

Figure 3.24 H.263 decoder with reduced resolution processing

This switching strategy performed on a frame basis reduces the complexity that is due to unnecessarily using the reduced resolution mode when the estimated bit rate falls below the target one. As already described, the feed-forward algorithm tries to assign a fixed number of bits to each frame by selecting a quantisation step size Qp that can achieve the target bit rate. Consequently, a fixed frame rate is expected if each frame can be coded with the assigned bit rate. Figure 3.25 shows the bit rate per frame for 150 frames of the sequence Silent Voice. It is a conven-

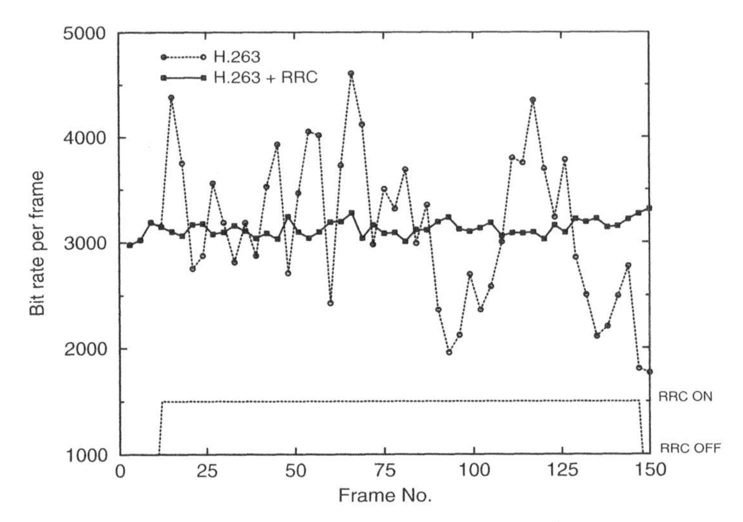

Figure 3.25 Bit rate per frame for 150 frames of the Silent Voice sequence coded at 40 kbit/s and 10 f/s

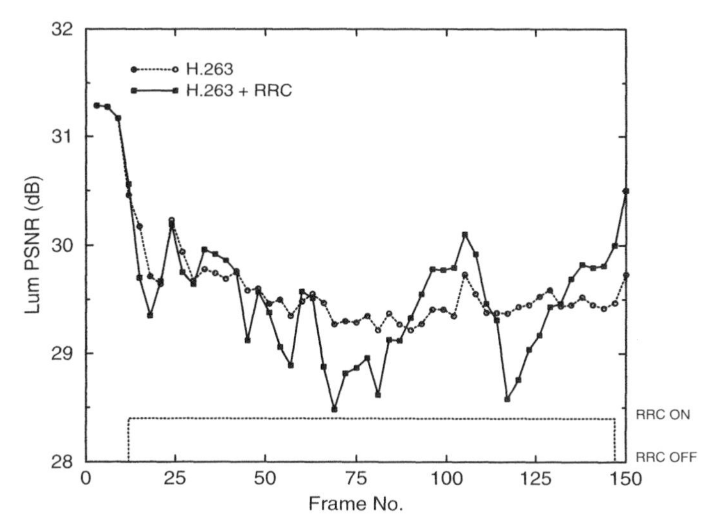

Figure 3.26 Y-PSNR values for 150 frames of the Silent Voice sequence coded at 40 kbit/s and 10 f/s

tional MPEG-4 highly active test sequence containing a woman using sign language to communicate and rapidly moving her arm throughout the whole sequence. The background of the sequence is also very detailed. For fair comparison, the feed-forward frame-based rate control algorithm is implemented in both the original H.263 standard encoder and the reduced resolution modified one. As can be seen, the average bit rate per frame is almost stable when the reduced resolution rate controller is used. Figure 3.26 shows the corresponding luminance PSNR values for both coders. The objective quality obtained using the original H.263 coder is almost always higher than that of the modified H.263 with reduced resolution capabilities. However, the objective quality of the modified coder outperforms that of the original coder during the second half of the sequence. This happens when the original coder starts to run out of bits due to overspending of the available bit rate at the beginning of the sequence. On the subjective scale, the reconstructed sequence of the reduced resolution decoder shows some blurriness due to the up-sampling process. However, the picture quality is smoother with fewer blocking artefacts, resulting from the smaller area of the quantised blocks. It can also be noticed that the reduced resolution controller starts being active from the 10th frame of the sequence and remains switched on throughout the whole encoding process.

3.9.1 Reduced resolution scheme with adaptive frame rate control

For more efficient rate control capabilities, the reduced resolution technique could also be combined with a variable frame rate mechanism that can drop some original frames in case of stringent bit rate budget restrictions. Initially, the number of bits (W) in the buffer is set to zero. The first frame of the sequence is INTRA coded with a quantisation parameter INTRAQ set to 15.

Let B' be the number of bits spent on the previous encoded frame, R the target bit rate per second, G the frame rate of the original video sequence (f/s), F the target frame rate (f/s), and M the threshold for frame skipping ($M = R/G$).

After coding each frame, the number of bits in the local encoder buffer is

$$W = W + B'$$

The number of frames that need to be skipped can then be determined using the following scenario:

$$
\begin{aligned}
&Frame\_skip = 0;\\
&While\,(W > M)\\
&\{\\
&\quad W = W - R/G;\\
&\quad Frame\_skip{+}{+};\\
&\}
\end{aligned}
$$

The next "frame_skip" frames of the original video sequence are skipped and the target bit rate for the next frame is $B = R/F + \Delta$, where $\Delta = W/F$.

Using the bit rate B, the feed-forward rate controller is used to find the appropriate Qp for the next frame. This variable frame rate algorithm is applied on both the original H.263 and modified H.263 with reduced resolution coding capability. Figure 3.27 shows the bit rate per frame for both the original and

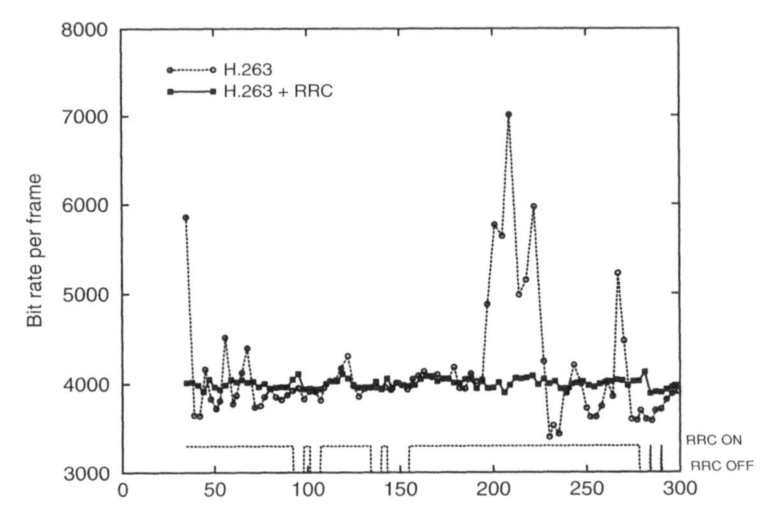

Figure 3.27 Bit rate per frame of Silent Voice coded at 40 kbit/s using variable frame rate control for both the original H.263 coder and the modified coder with reduced resolution mode

modified (with reduced resolution) H.263 using the variable frame rate algorithm described above. Since the first frame is INTRA coded, both coders skip the first 40 frames before they start to encode the first INTER frame. Consequently, a large number of MBs in the 41st frame cannot be predicted from the first reconstructed frame and need to be INTRA coded. The use of the reduced resolution controller helps reduce the output bit rate of the modified coder, whilst the original TMN5 fails to reduce its bit rate since the estimated Qp has already attained its maximum (i.e. 31). Whenever an overspending of bit rate is detected, more frames are skipped to empty the output buffer before the next frame is coded. As a result, the decoded sequence becomes jerkier, whereas the same frame rate could be maintained using the reduced resolution mode of the modified H.263. The corresponding luminance PSNR plots depicted in Figure 3.28 show that whenever the reduced resolution mode is switched off, the PSNR returns immediately to the same level, or even higher for some frames, as the original H.263. On the subjective scale, the coder with the reduced resolution mode shows a better perceptual quality with a smoother movement, while the original coder shows high jerkiness as certain hand signs are skipped.

3.10 Rate Control Using Multi-layer Coding

Multi-layer video coding consists of forming the output stream from a number of layers of different bit rates, frame rates and possibly resolutions (as seen in the previous section for full-size video frames) to achieve a scaleable output. The

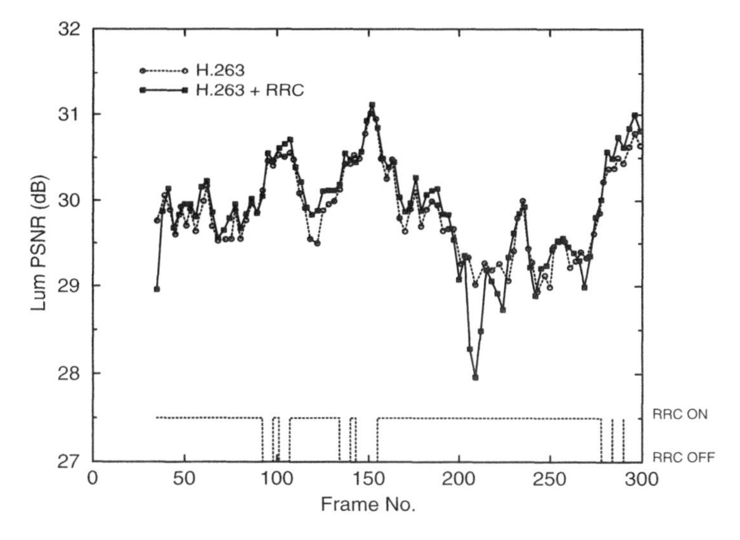

Figure 3.28 *Y*-PSNR values of the Silent Voice sequence coded at 40 kbit/s using variable frame rate control for both the original H.263 coder and the modified coder with reduced resolution mode

resulting video layers consist of one base layer (BL) and a number of enhancement layers (EL), all of which have different contributions to decoded video quality. The base layer is essential for the reconstruction of the video sequence, while enhancement layers help improve the perceptual quality but their absence causes a graceful deterioration of the received video quality. The enhancement layers could then be used as a trade-off between quality and compression efficiency in order to control the output bit rate of the video coder.

An adaptive two-layer video coding technique has been proposed and applied to H.261 for transmission over ATM networks (Ghanbari, 1992). The base layer consists of a low quality image generated at a very low bit rate while the second layer is the difference between the input original picture and the output of the base layer coder. In order to ensure graceful degradation, data from the base layer is transmitted with high priority using the guaranteed bandwidth of an ATM network. However, packets from the second layer are transmitted through the non-guaranteed channel and are vulnerable to loss if congestion arises. If the second-layer packets are received, the decoded picture quality will be enhanced.

Video scaleability techniques have also been applied on the H.263 video coding standard, where the base layer and a single enhancement layer presented different temporal and spatial scaleabilities (Stuhlmuller, Link and Girod, 1999). In this scaleability scheme, the base layer consists of QCIF frames coded at a frame rate f_{BL} whereas the enhancement layer consists of CIF frames coded at twice f_{BL}. Moreover, the scaleability of the encoder is further improved by protecting the BL information with a more powerful error protection scheme than that of the EL, thereby ensuring a guaranteed minimal perceptual quality in the case of video packet losses.

ISO MPEG-4 video coding algorithm also supports the multi-layer stream for temporal scaleability purposes. The coded output consists of a base layer and one or more enhancement layers for each visual object plane in the scene. The composition of each VOP at the decoder consists of essentially decoding its base layer and progressively improving the picture quality by decoding the stream of the enhancement layer(s). The base and enhancement layers could be encoded at different VOP rates to achieve the required temporal scaleability of the coded output.

In addition to the temporal scaleability created by multi-layer coding, MPEG-4 supports spatial scaleability that is due to the object-oriented coding of video frames. Each video object could then be encoded at different quantisation levels and spatial resolutions, the values of which are decided in accordance with the time-varying bandwidth requirements. The frame rate, the quantisation parameter and the resolution of a layer (base or enhancement) for each of the arbitrary-shape objects are all parameters that could be exploited to control the aggregate flow of the video source while giving priority and emphasis to regions of interest within the video scene. To achieve a higher compression ratio and a smoother motion prediction, the advanced prediction mode (APM) could be used. Table 3.7 shows

Table 3.7 Temporal scaleability for the Foreman sequence in the multi-layer MPEG-4 video coder where Qp = 12 for both layers

BL at 5 f/s/(BL & EL) at 10 f/s					
APM (BL)	APM (EL)	BL *Y* PSNR (dB)	BL bit rate (kbit/s)	BL & EL *Y* PSNR (dB)	BL & EL bit rate (kbit/s)
—	—	31.33	29.17	31.35	49.03
✓	—	31.49	28.7	31.53	48.09
—	✓	31.33	29.17	31.41	48.69
✓	✓	31.49	28.7	31.53	47.83

the luminance PSNR values and bit rates for both the BL coded at 5 f/s and the combined BL and EL at 10 f/s for 150 frames of the Foreman sequence with a quantisation parameter of 12 for INTER and INTRA frames of both layers. The results show that coding only the base layer at 5 f/s achieves a considerable reduction in bit rate as opposed to coding the sequence at full temporal resolution. Although the spatial picture quality is not degraded due to coding only the base layer, the temporal picture quality is impaired due to the reduced frame rate leading to the dropping of some video frames and hence resulting in a coarser motion prediction process. Moreover, the results show that when the advanced prediction mode is employed in both layers, the multi-rate coder produces the lowest bit rate at the best average *Y*-PSNR value. In video sequences incorporating more than one object, a base layer and at least one enhancement layer could be generated for each VOP. The layers of each VOP could then be coded with different temporal resolutions and quantisation parameters, depending on the importance of the corresponding VOP and its contribution to overall video quality.

It should be noted that single-layer (SL) coding produces slower bit rates than multi-layer (ML) coding at the same aggregate frame rate and the same quantisation step size as shown in Table 3.8. This is mainly due to the additional bit overhead required to code the administrative data of the enhancement layers. Again, with lower temporal resolutions used for the base layer, the perceptual quality of individual pictures is not dramatically deteriorated, especially for low to medium activity scenes and low frame rates. If two or more enhancement layers are used then the multi-layer coder performance is improved and the output stream becomes more scaleable at the expense of increased coding complexity and aggregate flow rate (due to additional layers overhead). In the case of network congestion, the multi-layer coder could dispense of the low-priority enhancement layers to guarantee a minimum level of perceptual quality achieved by the delivery of the high-priority base layer. In this case, the multi-layer coder has to adjust the quantisation parameter, spatial and temporal resolutions of the base layer and the number of coded VOPs in the video scene in accordance with the output bit rate

Table 3.8 PSNR and bit rate values for single layer, base layer and multi-layer streams for the Foreman sequence coded by MPEG-4 with Qp = 13

BL/(BL & EL) frame rates (f/s)	PSNR (dB)			Bit rate (kbit/s)		
	ML	BL	SL	ML	BL	SL
7.5/15	31.04	30.97	31.01	57	32	47
5/10	31.00	30.94	31.00	44	25	37.5
3/6	30.92	30.89	30.96	33	28	18

requirements dictated by the reported congestion conditions. Furthermore, the scaleability of the multi-rate coder could be enhanced by means of the advanced prediction mode as shown in Table 3.7 and the use of PB-frame mode. The latter contributes to achieving higher compression ratios for the base layer, thereby increasing the likelihood of delivering its bit stream in bad network conditions. On the other hand, the rate control algorithms described in previous sections could also be employed to reduce the bit rate fluctuations of base and enhancement layers streams.

3.11 Fine Granular Scaleability

To compensate for the unpredictability and variability in bandwidth between the sender and receiver(s) of a network such as the Internet, streaming solutions usually resort to achieving scaleability using one of a variety of layered video coding methods (Tan and Zakhor, 1999; McCanne, Vetterli and Jacobson, 1997). In addition to proprietary and previously standardised multi-layer video coding methods, the video activities of the ISO MPEG-4 standard have taken the Internet applications into immense consideration and produced a framework for a new compression tool for scaleability purposes. MPEG-4 has adopted a new scaleable compression tool, namely fine granular scaleability (FGS), for video streaming applications (Chen *et al.*, 1998; ISO/IEC 14496-2). FGS exploits the multi-layer structure (Section 2.5.9) of MPEG-4 to achieve quality (SNR), temporal or hybrid SNR/temporal scaleability. This scaleability technique has been proposed for MPEG-4 video streaming applications over the Internet with both unicast and multicast transmissions (Radha *et al.*, 1999; Radha, 2000).

In all the multi-layer scaleability techniques described in the previous section, scaleable video consists of a base layer (BL) and one or multiple enhancement layers (EL). The BL represents the minimal amount of data required for the reconstruction of the video sequence at the decoder side. The ELs represent additional information that contributes, if correctly received and decoded, to the enhancement of the decoded video quality. On the other hand, FGS requires only a single enhancement layer decoder at the receiver and confines the effect of packet losses to only the enhancement-layer picture that is suffering from the loss, hence the low complexity of this scaleability method and its resilience to packet losses. The BL is coded using a DCT-based compression scheme while the EL could be coded using either DCT, or embedded zero-tree wavelet (EZW) compression or matching-pursuit based methods (Chen *et al.*, 1998). Since bit-plane DCT and EZW methods presented similar results (Van der Schaar, Chen and Radha, 2000), the EL is encoded using the embedded DCT for consistency purposes in reference to the MPEG-4 compliant DCT-based compression scheme of the BL. Using the DCT transform at both the BL and EL enables the encoder to perform a simple residual computation of the FGS EL. In the BL, DCT coefficients are coded with run-length and non-zero level types of code, while the bit-planes of the FGS EL are coded with run-length codes only since the non-zero levels (amplitudes) can only be 1.

Fine granular scaleability is a technique that was primarily used for still image coding, whereby images could be decoded progressively as more data is received. As soon as the data corresponding to the BL is received, the video images can be reconstructed. With more data arriving to the decoder, the quality of the decoded video is progressively enhanced until the full amount of data is received, decoded and displayed. The granularity is achieved by decreasing the amount of prediction in the pictures of the enhancement layer thereby reducing the coding efficiency of the video coding scheme. Therefore, in order to maintain a good balance between the coding efficiency and the scaleability structure, FGS uses only two layers: a BL coded at a bit rate R_{BL}, which is the minimum bandwidth available *Rmin*, and a single enhancement layer coded with fine granularity at a maximum bit rate *Rmax*. The encoder needs only to know the range of bandwidth [*Rmin*, *Rmax*] required to code the video content rather than knowing the exact bit rate at which the video is to be coded. This would then enable the scaleability controller (such as a streaming server in Internet applications) to flexibly send any desired portion of the EL, along with the BL data, as shown in Figure 3.29, without having to perform any of the MB or frame-based real-time complex rate control algorithms described earlier in this chapter.

Using the information sent to the encoder by a specific system module that can determine (either offline or in real-time) the bandwidth interval [*Rmin*, *Rmax*], the base layer is coded at the rate $R_{BL} \leq Rmin$. Subsequently, the residual image is computed and the EL is obtained by over-coding the residual image at a bit rate $(Rmax - R_{BL})$ using fine granularity. The streaming server monitors the varying network conditions, and therefore the available bandwidth R can be estimated prior to EL transmission. Based on this estimate, the EL is transmitted using a bit rate R_{EL}, (equal to $\min(R - R_{BL}, Rmax - R_{BL})$), as shown in Figure 3.30 for a video streaming system.

In addition to the SNR (quality) scaleability described above, FGS supports temporal scaleability. In the former case, the overall frame rate of the transmitted video is equal to that of the base layer regardless of the available bandwidth. However, with temporal scaleability, FGS allows for a smoother motion represen-

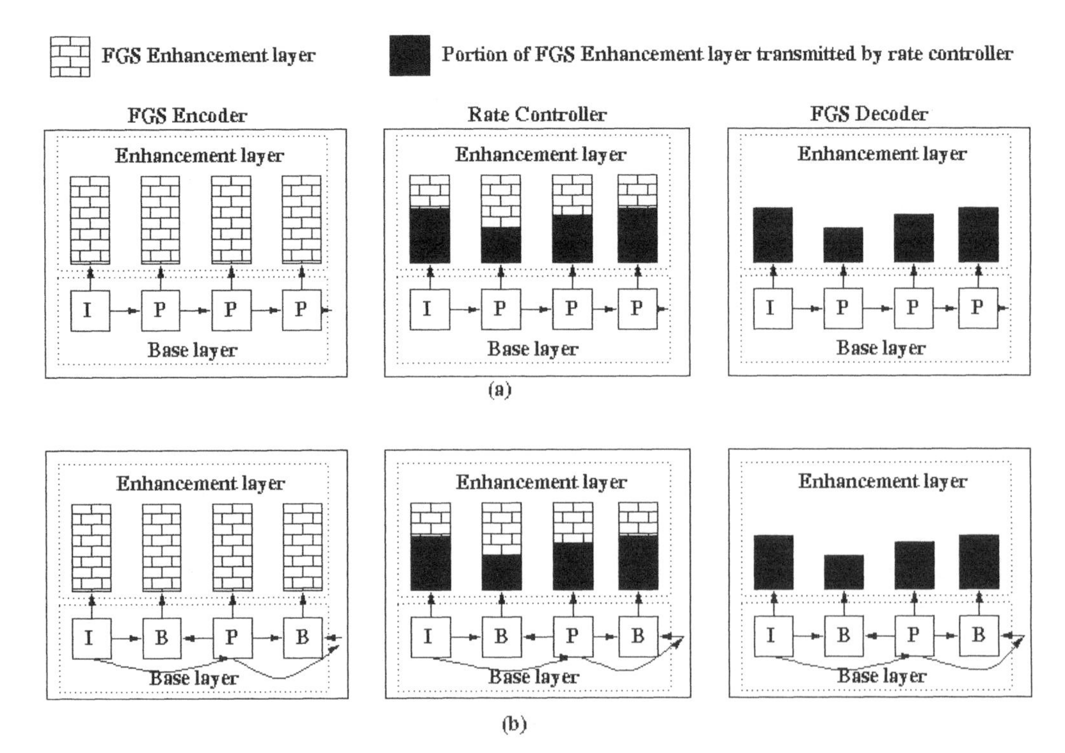

Figure 3.29 FGS-enabled MPEG-4 encoder used for rate control purposes: (a) without B-frames in the base layer, (b) with B-frames in the base layer

tation. In this case, the enhancement layer includes motion-compensated residual frames referred to as the FGS temporal (FGST) frames. Each FGST picture is predicted from BL frames that do not temporally coincide with this corresponding FGST frame. Consequently, both SNR-scaleability (FGS) and FGST frames could be encoded and decoded using the same enhancement layer encoder/decoder since these frames are never encoded/decoded simultaneously. Moreover, FGS could also support hybrid SNR-temporal scaleability using the same enhancement layer. This enables the user to trade-off between the motion smoothness provided by temporal scaleability and the quality improvements achieved by SNR scaleability. Complexity-wise, there is no complexity overhead associated with the single-layer FGS scaleability solution, as compared to the multi-layer SNR-temporal scaleable video coding. The three types of scaleability supported by FGS are illustrated in Figure 3.31.

3.12 Conclusions

In this chapter, a variety of rate control algorithms employed in video communications have been presented and analysed. Video coding algorithms support a

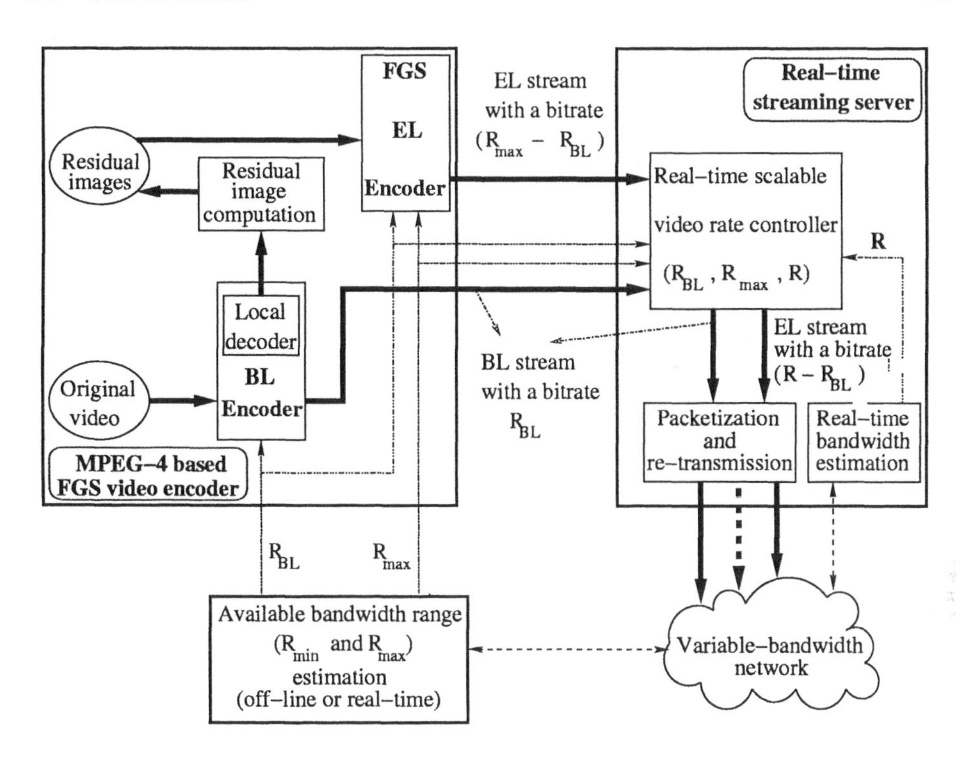

Figure 3.30 Unicast video streaming system using fine-granular scaleability

constant decoded quality only with variable bit rates, and constant bit rates could only be achieved with a variable decoded video quality. The bit rate variability of a video stream is attributed to a number of factors. Firstly, a digital video sequence incorporates a large number of temporal, spatial and statistical redundancies, the removal of which affects the rate of the coded bit stream. Since the compression process is a trade-off between the coding bit rate and quality, the suppression of the inherent redundancies is a factor that dictates the variability in bit rate to achieve this trade-off. Moreover, the video encoding parameters, such as the quantisation step size Qp, the motion threshold and the use of variable-length codes, are all factors that contribute to the variability in the output bit rate of the video encoder. When the video source is allocated a fixed amount of bandwidth throughout a communication session, the bit rate variability could be smoothed out by means of an internal buffer that can regulate the rate of transmission of video data at the expense of a certain incurred delay. However, when the bandwidth requirements of the network are time-varying, the output bit rate of a video coder must be adapted to the changing bandwidth conditions at any instant of time. For this purpose, rate control algorithms have to be used for throughput and quality optimisation.

Some rate control algorithms rely on changing the video encoding parameters

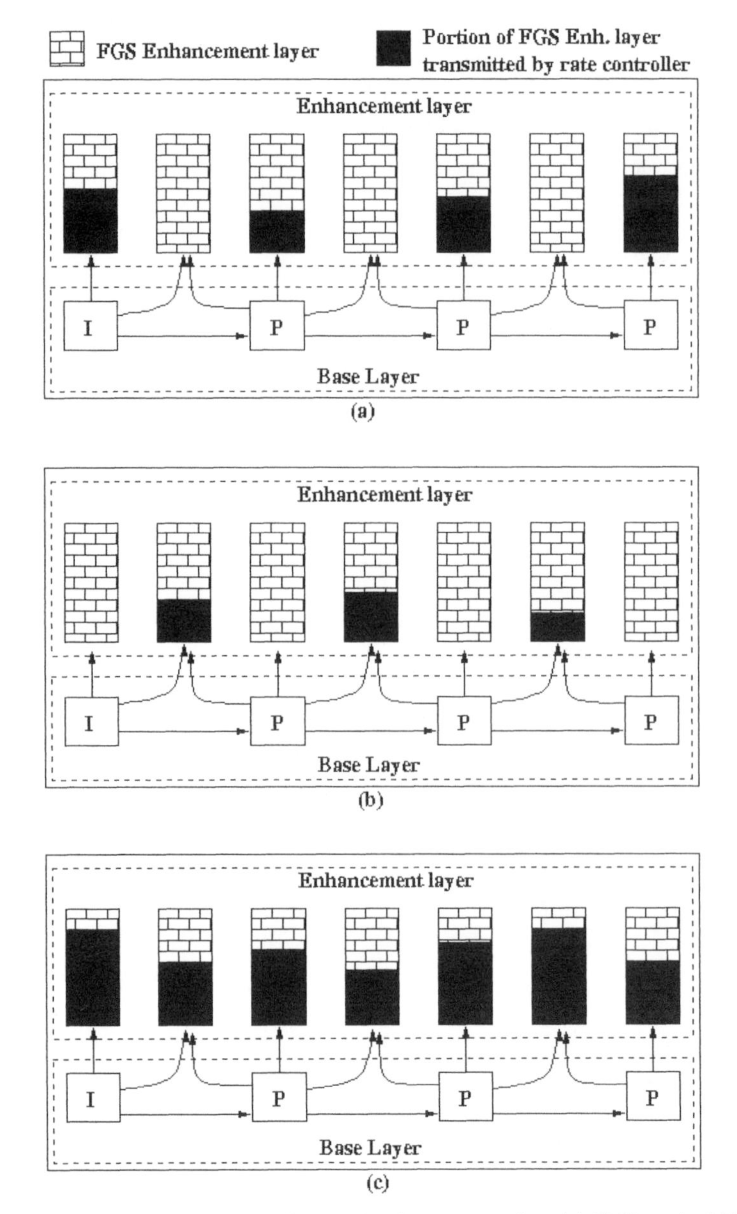

Figure 3.31 FGS-based rate controller mechanism supporting: (a) SNR scaleability, (b) temporal scaleability, (c) hybrid SNR-temporal scaleability

in accordance with the reported channel status in order to obtain the required rate that the channel can accommodate at a particular instant of time. The encoding parameter that is most widely used for rate control purposes is the quantisation parameter Qp which could be changed on an MB, GOB or frame basis. In most rate control techniques employed today, the quantisation step size is determined based on the estimated number of bits available and the previous Qp value. The bit rate requirement of an MB or a frame is not considered in the calculation of its corresponding Qp value; Qp is determined based on its value in the previous MB or frame and the estimated bandwidth available. To make the calculation of Qp more content-based, a frame or an MB is dummy-encoded to check its motion activity and spatial details before a quantisation value is decided in line with the remaining estimated bandwidth. The techniques that support this Qp evaluation paradigm are referred to as feed-forward rate control algorithms. It has been demonstrated in this chapter that feed-forward rate control algorithms achieved an improvement over traditional rate control algorithms in smoothing out the bit rate fluctuations of a video coder. In addition to the rate control capabilities of feed-forward techniques, quality improvement is made possible through the use of region-of interest (ROI) coding, mainly in head-and-shoulder types of sequence.

In addition to changing encoding parameters, other rate control algorithms operate by exploiting the difference in error sensitivity between various video parameters such as administrative data, motion vectors, and transform coefficients. The presented study demonstrated that transform coefficients, which occupy the largest portion of the video stream bandwidth, are the least sensitive to errors. Therefore, levels of priority could be assigned to various video parameters in order to mitigate the effects of network congestion by first dropping the least sensitive segment of the stream. This results in curing the congestion without a considerable damage to the overall video quality. Moreover, an internal feedback rate control mechanism was presented whereby a feedback loop was introduced inside the video encoder. This feedback loop is used to map the changes incurred on the coded bit stream due to information drop (resulting from congestion) onto the local picture memory. As the internal picture buffer is updated, the dropped information is no longer used in any further prediction process, thereby limiting the propagation of errors that stems from the loss of video information in congested links. The feedback loop rate control mechanism proved very efficient in securing an acceptable video quality and maintaining a controlled rate scenario even under harsh network congestion conditions. Furthermore, rate control could be achieved by down-scaling/upscaling the resolution of the video content in accordance with the bandwidth requirements. Fixed and variable frame reduced-resolution rate controllers have been presented, and both subjective and objective results have been produced for comparison against traditional rate control schemes. On the other hand, multi-layer rate control algorithms have been examined to achieve temporal scaleability in video coders. Results for both the base and enhancement layers coded at different bit and frame rates were tabulated to reflect the temporal scaleability that could be achieved by multi-layer rate control techniques. Finally, a new ISO MPEG-4 scaleability technique, namely fine-granularity scaleability, was explained and examined for its suitability to provide SNR (quality), temporal and hybrid SNR-temporal scaleability structures, especially in video streaming applications. The FGS enhancement layer coder could be a bitplane DCT-based compression algorithm, and therefore it places no additional complexity on the DCT MPEG-4 based compression scheme used to encode the base layer. FGS is very useful in achieving a progressive quality improvement by guaranteeing a minimal quality level presented by sending the base layer data using, at most, the minimum available bandwidth. The portion of the EL data that can then be transmitted in real time is defined by the available network bandwidth less the BL rate.

3.13 References

Albanese, A., Bloemer, J., Edmonds, J., Luby, M., and Sudan, M., Priority encoding transmission, *IEEE Trans. on Information Theory*, **42**, 1737–1747, Nov. 1996.

Bolot, J. C., and Turletti, T., A rate control mechanism for packet video in the Internet, *Proc. IEEE Infocom*, Toronto, Canada, June 1994.

Bolot, J. C., Turletti, T., and Wakeman, I., Scalable feedback control for multicast video distribution in the Internet, *Proc. IEEE Infocom*, Toronto, Canada, June 1994.

Chen, Y., Dutour, C., Radha, H., Cohen, R., and Buteau, M., Request for the fine granular video scalability for media streaming applications, Contribution to 44th MPEG meeting, Dublin, Ireland, July 1998.

Cote, G., Erol, B., Gallant, M., and Kossentini, F., H.263+ : video coding at low bit rates, *IEEE Trans. on Circuits Systems Video Technology*, **8**, No. 7, 849–866, Nov. 1998.

Dagiuklas, A., and Ghanbari, M. Preventive flow control method for packet video, *IEE Proc. on Communications*, **143**, No. 2, 98–196, Apr. 1996.

Ghanbari, M., An adapted H.261 two-layer video codec for ATM networks, *IEEE Trans. on Communications*, **40**, No. 9, 1481–1490, Sept. 1992.

Girod, B., Scalable video for multimedia systems, *Computers & Graphics*, **17**, No. 3, 269–276, 1993.

Horn, U., and Girod, B., Scalable video transmission for the Internet, *Computer Network and ISDN Systems*, **29**, 1833–1842, Nov. 1997.

ISO/IEC 14496 Information technology – Generic coding of audio-visual objects – Part 2: Visual (Annex L – Rate control), 1999.

ISO/IEC 14496-2, MPEG-4 Video FGS v. 4.0, N3317, Proposed Draft Amendment (PDAM), Noordwijkerhout, the Netherlands, Mar. 2000.

Jacobson, V., Congestion avoidance and control, *Proc. ACM Sigcomm*, Stanford, CA, 314–329, Aug. 1988.

Kumar, V., *Mbone: interactive multimedia on the Internet*, New Riders Publishing, Indiana, 1996.

Kurose, J., Open issues and challenges in providing QoS guarantees n high speed networks, *Computers Communications Review*, **23**, 6–15, Jan. 1993.

Kweh, T. H., Improved quality block-based low bit rate video coding, PhD thesis, University of Surrey, 1998.

Leicher, C., Hierarchical encoding of MPEG sequences using priority encoding transmission (PET), Technical report. TR-94-058, International Computer Science Institute, Berkeley, Nov. 1994.

McCanne, S., Vetterli, M., and Jacobson, V., Low-complexity video coding for receiver-driven layered multicast, *IEEE JSAC*, **16**, No. 6, Aug. 1997, 983–1001.

Maglaris, B., Anastassiou, D., *et al.*, Performance models of statistical multiplexing in packet video communications, *IEEE Transactions on Communications*, **36**, No. 7, 834–843, July 1988.

Perra, C., Pinna, M., and Giusto, D. D., H.263+ rate control at fixed objective quality, *Proc. of the Packet Video Workshop 2000*, Sardinia, Italy, May 2000.

Plompen, R. H., Groenveld, J. G., Booman, F., and Boekee, D. E., An image-knowledge based video codec for low bit rates, *Proc. SPIE*, **804**, 379–384, 1987.

Radha, H., Chen, Y., Parthasarathy, K., and Cohen, R., Scalable Internet video using MPEG-4, *Signal Processing: Image Communication*, **15**, 95–126, Sept. 1999.

Radha, H., Fine granularity scalability: a new framework for unicast and multicast video streaming over the Internet, *IP Multicast Summit 2000* (mcast2000), San Francisco, Feb. 2000.

Radha, H., Chen, Y., Parthasarathy, K., and Cohen, R., Scalable Internet video using MPEG-4, *Signal Processing: Image Communication*, **15**, No. 1–2, 95–126, 1999.

Sadka, A. H., Eryurtlu, F., and Kondoz, A. M., Rate control feedback mechanism for packet video networks, *IEE Electronics Letters*, **32**, No. 8, 716–717, Apr. 1996.

Saghri, J. A., and Tescher, A. G., Knowledge-based image bandwidth compression and enhancement, *Proc. SPIE*, **804**, 201–216, 1987.

Stuhlmuller, K., Link, M., and Girod, B., Scalable Internet video streaming with unequal error protection, *Proc. Packet Video Workshop '99*, Apr. 1999.

Tan W., and Zakhor, A., Real-time internet video using error resilient scalable compression and TCP-friendly transport protocol, *IEEE Trans. on Multimedia*, **1**, No. 2, 172–186, June 1999.

Telenor R&D, H.263 video codec test model, Nov. 1995.

Test Model 5, http://www.mpeg.org/MPEG/MSSG/tm5/Overview.html

Van der Schaar, M., Chen, Y., and Radha, H., Embedded DCT and wavelet methods for fine granular scalable video: analysis and comparison, *IVCP 2000, Proc. SPIE*, **2974**, 643–653, Jan. 2000.

Wang, L. Rate control for MPEG video coding, *Signal Processing: Image Communications*, **15**, 493–511, 2000.

4

Error Resilience in Compressed Video Communications

4.1 Introduction

Compressed video streams are intended for transmission over communication networks. With the advance of multimedia systems technology and wireless mobile communications, there has been a growing need for the support of multimedia services such as mobile teleconferencing, telemedicine, mobile TV, distance learning, etc., using mobile multimedia technologies. These services require the real time transmission of video data over fixed and mobile networks of varying bandwidth and error rate characteristics. Since the coded video data is highly sensitive to information loss and channel bit errors, the decoded video quality is bound to suffer dramatically at high channel bit error ratios (BER). This quality degradation is exacerbated when no error control mechanism is employed to protect coded video data against the hostility of error-prone environments. A single bit error that hits a coded video stream could lead to disastrous quality deterioration for extended periods of time. Moreover, the temporal and spatial predictions used in most of the video coding standards today render the coded video stream rather more vulnerable to channel errors. This vulnerability is represented by the rapid propagation of errors in both time and space and the quick degradation of the reconstructed video quality. To mitigate the effects of channel errors on the decoded video quality, error-handling schemes must be efficiently applied at both the video encoder and decoder.

Since real-time video transmissions are sensitive to time delays, the issue of re-transmitting the erroneous video data is totally ruled out. Therefore, other forms of error control strategy must be employed to mitigate the effects of errors inflicted on coded video streams during transmission. Some of these error control schemes employ data recovery techniques that enable decoders to conceal the effects of errors by predicting the lost or corrupted video data from the previously reconstructed error-free information. These techniques are decoder-based and incur no changes on the transport technologies employed. Moreover, they do not place any redundancy on the compressed video streams and are thus referred to as zero-redundancy error concealment techniques (Wang and Zhu, 1998). Other error control schemes operate at the encoder and apply a variety of techniques to enhance the robustness of compressed video data to channel errors. These are known as error resilience techniques, and they are widely used in video communications today (Redmill *et al.*, 1998; Talluri, 1998; Soares and Pereira, 1998; Weng *et al.*, 1998). The last type of error control mechanism operates at the transport level and tries to optimise the packet structure of coded video frames in terms of their error performance as well as channel throughput. These techniques are the most complex as they depend on the networking platforms over which coded streams are intended to travel and the associated network and transport protocols (Guillemot *et al.*, 1999; Parthsarathy, Modestino and Vastola, 1997). In this chapter, we cover a variety of the error concealment and resilience techniques used in video communications today, and the transport-based error control schemes will be examined in the next chapter.

4.2 Effects of Bit Errors on Perceptual Video Quality

The error performance of most video coding standards is degraded mainly due to two major factors, namely the motion prediction and the bit rate variability discussed in Section 3.2. In the motion prediction process of ITU-T H.263, for instance, motion vectors (MV) are sent in differential coordinates in both pixel and half-pixel accuracies. In other words, each MV is sent as the difference between the estimated MV components and those of the median of three candidate MV predictors belonging to MBs situated to the top, left and top-right of the current MB. If an error corrupts a particular MB, the decoder would be unable to correctly reconstruct a forthcoming MB whose MV depends on that of the affected MB as a candidate predictor. Similarly, the failure to reconstruct the current MB because of errors prevents the decoder from correctly recovering forthcoming MBs that depend on the current MB in the motion prediction process. The accumulative damage due to these temporal and spatial dependencies might be caused by a single bit error, regardless of the correctness of subsequent information.

Similarly, the variable bit rate nature of coded video streams is another predicament for error robustness in compressed video communications. If a variable-length video parameter is corrupted by errors, the decoder will fail to figure out the original length of this parameter, thereby losing its synchronisation. The effects of a bit error on the decoded video quality can be categorised into three different classes, as follows.

A single bit error on one video parameter does not have any influence on segments of video data other than the damaged parameter itself. In other words, the error is limited in this case to a single MB that does not take part in any further prediction process. One example of this category is encountered when an error hits a fixed-length INTRADC coefficient of a certain MB which is not used in the coder motion prediction process. Since the affected MB is not used in any subsequent prediction, the damage will be localised and confined only to the affected MB. Moreover, the decoder will not lose synchronisation, since it has skipped the correct number of bits when reading the erroneous parameter before moving to the next parameter in the bit stream. This kind of error is the least destructive of the three to the quality of service.

The second type of error is more problematic because it inflicts an accumulative damage in both time and space due to prediction. When the prediction residual of motion vectors is sent, bit errors in motion code words propagate until the end of the frame. Moreover, the error propagates to subsequent INTER coded frames due to the temporal dependency induced by the motion compensation process. This effect can be mitigated if the actual MVs are encoded instead of the prediction residual. As illustrated in Figure 4.1 for the 30 frames of the Foreman sequence encoded at 30 kbit/s, the quality of the decoded picture can be improved for error rates higher than 10^{-4} when the actual MV values are transmitted. At lower error rates, the quality drops slightly, since the compression efficiency is decreased when no MV prediction is used. The damage to the picture quality depends on the number of successive frames that are INTER coded following the bit error position. Thus, PSNR values tend to decrease with time due to error accumula-

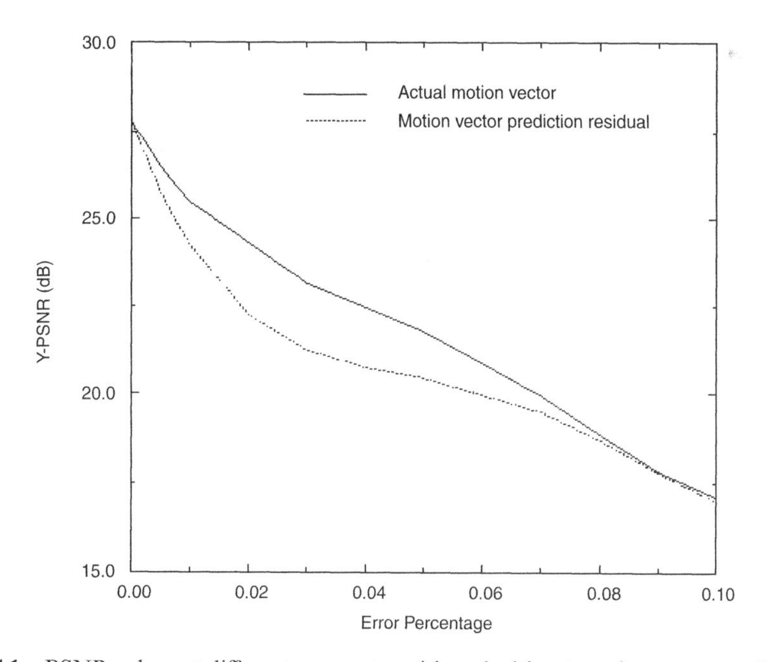

Figure 4.1 PSNR values at different error rates with and without motion vector prediction

tion. This category of error is obviously more detrimental to the quality of decoded video than the first one; however, it does not cause any state of de-synchronisation, since the decoder flushes the correct number of bits when reading the erroneous motion code words.

The worst effect of bit errors occurs when the synchronisation is lost and the decoder is no longer able to figure out to which part of a frame the received information belongs. This category of error is caused by the bit rate variability characteristic. When the decoder detects an error in a variable length code word (VLC), it skips all the forthcoming bits, regardless of their correctness, in the search for the first error-free synch word to recover the state of synchronisation. Therefore, the corruption of a single bit is transformed into a burst of channel errors. The occurrence of a bit error in this case is manifested in two different scenarios. The first scenario arises when the corrupted VLC word results in a new bit pattern that is a valid word in the Huffman table corresponding to that specific parameter. In this case, the error cannot be detected. However, the resulting VLC word might be of a different length, causing the decoder to skip the wrong number of bits before moving forward to the next piece of information in the bit stream, thereby creating a loss of synchronisation. This situation remains until an invalid code word is detected, implying the occurrence of an error and causing the decoder to stop its operation and search for the next error-free synch word. The second scenario appears when the corrupted VLC word (possibly in conjunction with subsequent bits) results in a bit pattern that is not deemed legitimate by the Huffman decoder. In other words, the decoder fails to detect any valid VLC word for a particular video parameter within a segment of the bit stream that corresponds to the maximum length of the corrupted code word. In this case, the decoder signals the occurrence of an error, skips all the forthcoming bits and resumes decoding at the next intact synch word. Figure 4.2 illustrates these two scenarios. Figure 4.3 demonstrates the importance of synchronisation of an H.263 decoder to the reconstructed video quality. The H.263 decoder is modified in a way that ensures resynchronisation just after the position of error. Therefore, the decoder is able to detect an error in a video parameter and look for the next error-free synch word. In other words, only video parameters such as MVs and DCT coefficients are corrupted without the decoder losing its synchronisation (Figure 4.2(b)). Administrative information such as COD, MCBPC, CBPY, synch word, etc., affect the synchronisation of the decoder although they might be fixed-length coded. If one of these control parameters is corrupted by errors, there is no means for the video decoder to detect it until it falls on an invalid Huffman code word later in the bit stream. This loss of synchronisation leads to a dramatic drop of perceptual quality. It is evident that, with maintained synchronisation, the average PSNR values are significantly higher for error rates above 10^{-4}, again for the Foreman sequence encoded with H.263 at 30 kbit/s. Consequently, the synchronisation information is very sensitive to errors and hence very crucial for the correct decoding of a compressed video stream. Therefore, a block-based video decoder must be made

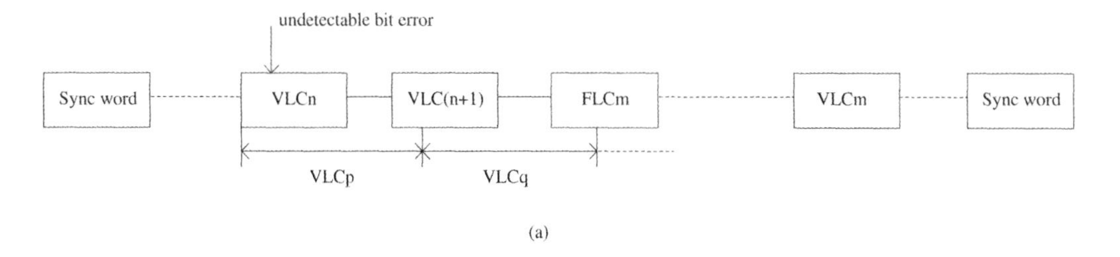

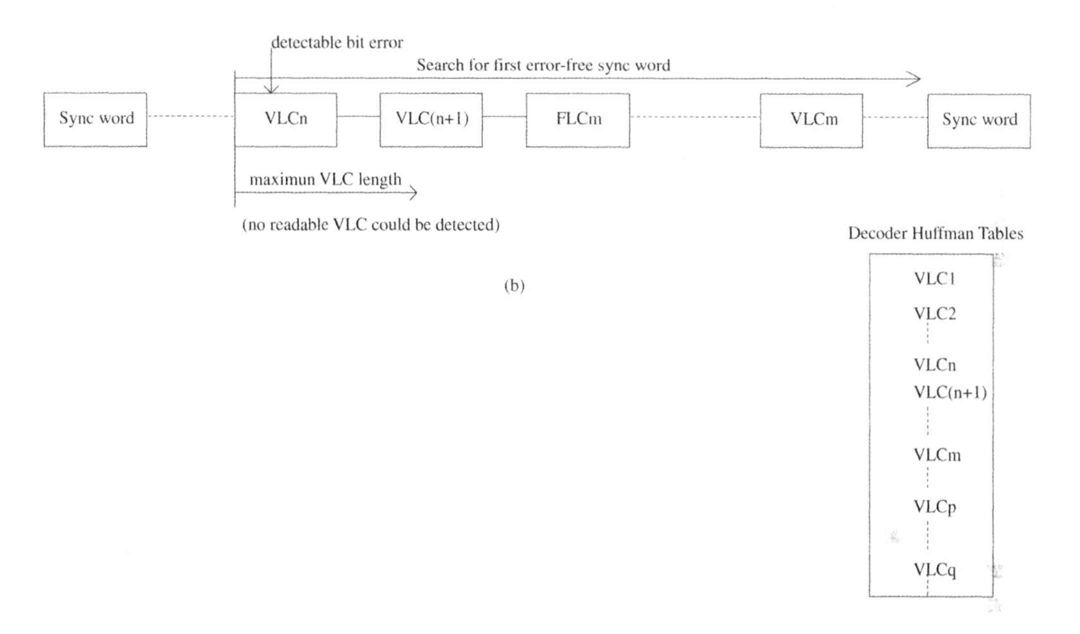

Figure 4.2 Bit errors leading to loss of synchronisation in the video decoder

robust enough to detect the channel errors and resynchronise at the correct bit pattern very quickly and with minimal quality loss.

4.3 Error Concealment Techniques (Zero-redundancy)

Error concealment or post-processing error control consists of a mechanism by which only the decoder fulfills the task of error control (Wang and Zhu, 1998). The encoder does not add any redundant bits onto the application layer coded stream for error protection purposes. On the other hand, no transmission or transport level mechanism is adopted in these techniques to reduce the severity of artefacts resulting from transmission errors. Error concealment techniques are purely

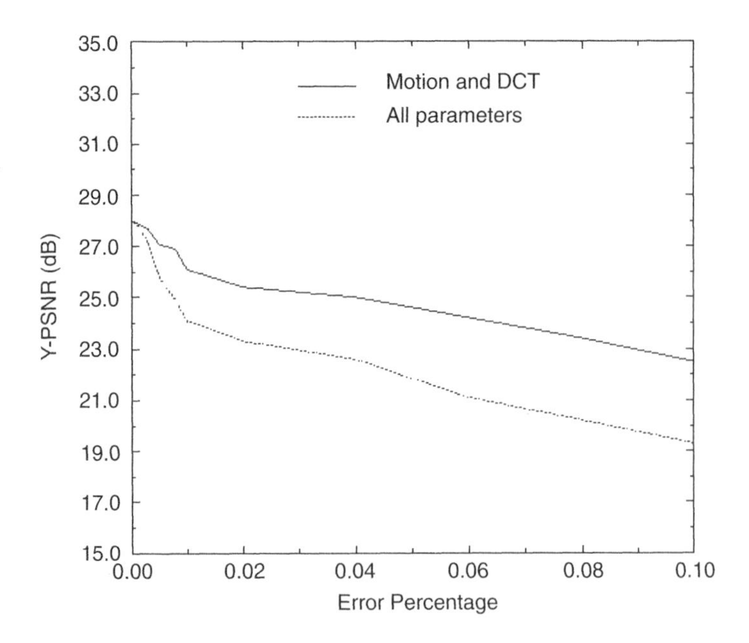

Figure 4.3 PSNR values at different error rates with and without loss of synchronisation

decoder-based, whereby the video decoder attempts to benefit from previously received error-free video information for the approximate recovery of lost or erroneous data without relying on additional information from the encoder. Some error concealment techniques are combined with other error control schemes to provide an interactive error handling mechanism in a video communication system (Wada, 1989). In this technique, the encoder relies on some kind of feedback channel signalling from the decoder that includes information about the corrupted MBs. In addition to post-processing error concealment, the encoder contributes to the error control mechanism by avoiding the use of damaged MBs in any further prediction process. However, in this section, we limit the discussion of error concealment to these techniques that are restrictively decoder-based and hence redundancy-free. In these error concealment algorithms, several techniques such as spatial and temporal interpolation, filtering and smoothing of available video data could be employed to estimate and sometimes predict missing video information such as coded shape data (Shirani, Erol and Kossentini, 2000), motion vectors, transform coefficients and administrative bits (Chhu and Leou, 1998; Lam and Reibman, 1995).

For an error concealment technique to be activated, an error detection mechanism is required to indicate to the decoder the occurrence of errors. In the previous section, it was shown that the error detection is signalled by the loss of synchronisation due to error-corrupted VLC parameters. In addition to loss of synchronisation, the video decoder claims an error when the number of AC coefficients of any 8×8 block of pixels is found to have exceeded 63 or when the decoded MV component or quantisation parameter is outside the acceptable range ([1,31] for the latter). However, transmission errors could also be detected using transport level headers such as checksum, parity bits, CRC (Cyclic Redundancy Checks) codes, for bit errors or sequence numbers, temporal references, etc., for packet erasures. These codes are attached normally to packets, as defined by the transport protocol, and their values are indicators as to whether transmission errors have occurred.

Error concealment techniques take advantage of the human eyes tolerance to distortion in the high-frequency components more than the low-frequency components of a video frame. Some techniques rely on multi-layer video coding to send low-frequency DC coefficients and motion vectors in the base layer and high-frequency AC coefficients in the enhancement layer (Kieu and Ngan, 1994). When the high-frequency components of the more error-prone enhancement layer are corrupted, the concealment technique recovers their values by using the DCT coefficients of the corresponding motion-compensated MBs in the previous frame. All of these techniques, however, make use of the spatial and/or temporal correlations between damaged MBs and their neighbouring MBs in the same and/or previous frame to achieve concealment (Lam and Reibman, 1995). Some of these techniques apply to INTRA coded MBs to recover the INTRADC coefficients of error-affected MBs, whereas other techniques apply only to INTER coded MBs to recover the corresponding motion data. Techniques have been proposed for the error concealment of the damaged shape data of MPEG-4 video coded sequences (Shirani, Erol and Kossentini, 2000). Error concealment methods attempt to reduce the visual artefacts in segments of a video stream that lie between two error-free synch words. If a synch word is inserted once every GOB, then a damaged MB leads to the corruption of a whole slice of video (assuming that a synch word is inserted at the beginning of each GOB). In this case, error concealment must be applied to reduce the effects of errors on the whole slice rather than on the affected MB only. In some transport schemes, the order of transmission of coded MBs is changed by means of interleaving. Despite the processing delay incurred by this technique and controlled by the interleaving depth, the use of interleaving allows the errors to disperse within the spatial area of a video frame, hence causing damage only to spatially disjointed blocks and reducing the likelihood of damaging a whole row of MBs. It is obvious that the choice of interleaving depth is a trade-off between the associated delay and the spreading factor of error-affected MBs or else the efficacy of the concealment technique.

4.3.1 Recovery of lost MVs and MB coding modes

If the coding mode of the damaged MB is known to be INTER, then the simplest concealment method is to replace the erroneous MB by the spatially coinciding

MB in the previous frame. This technique, despite its simplicity, might sometimes prove inefficient, as it leads to some inaccurate concealment results (annoying visual artefacts) especially in the presence of large motion in the video scene. Alternatively, if the motion data has been received free of errors, then the affected MB could be replaced by the motion-compensated MB, i.e. the MB pointed at by the actual motion vector of the lost MB. The latter technique could yet lead to fine concealment results when error-free motion data is available. However, in many circumstances, the motion vector of the error-damaged MB is also corrupted by transmission errors, and therefore the recovery of the erroneous MV is necessary for the reconstruction of the damaged INTER coded MB. This situation gets even worse when the coded/uncoded flag (COD) and/or the modes of coded MBs are also corrupted.

If the motion data of a particular MB is corrupted, the most straightforward and simplest technique to restore its MV is to force a zero vector. Therefore, this is equivalent to assuming that the spatially corresponding MB in the previous frame was the best match MB in the motion estimation process at the encoder. If the transform coefficients of the damaged MB have also been corrupted by errors, then error concealment is similar to replacing the erroneous MB by the spatially coinciding MB in the previous frame as indicated above. This method gives good concealment results in relatively small motion video sequences. Another method is to replace the lost MV by the MV of the spatially corresponding MB in the previous frame. A third method suggests using the average of MVs from the spatially adjacent MBs. However, if an MB is damaged by errors, adjacent MBs to the right (H.261) and below (H.263 and MPEG-4) are also affected due to motion prediction which uses three candidate predictors, as described in Chapter 2. Therefore, the MVs of only the left and top neighbouring MBs are used in the error concealment process. In some cases, instead of using the average, the median of MVs of spatially adjacent MBs is used to predict the lost or error-damaged MV. It has been found through experimentation that the last method yields satisfactory results and produces the best reconstruction results of all the available MV recovery methods (Narula and Lim, 1993). Optimal concealment techniques combine these four methods and choose the method that essentially leads to the smallest boundary matching error (sum of boundary variations between recovered MB and neighbouring ones). A more sophisticated technique for recovering a lost MV consists of predicting its value from MVs of spatially adjacent MBs in the previous frame. The MV that best moves its corresponding MB in the direction of the damaged MB (MB with lost MV) is used as the value of the lost MV. This method is based on the assumption that if a portion of the picture in the previous frame is moving into the direction of the damaged MB then it is likely that it will continue to move in the same direction into the next frame. This method obviously fails when errors occur on the edge blocks or the boundaries of an object. Figure 4.4 shows the subjective quality obtained by three different MV recovery techniques. On the other hand, if the coding mode is damaged, the affected MB is

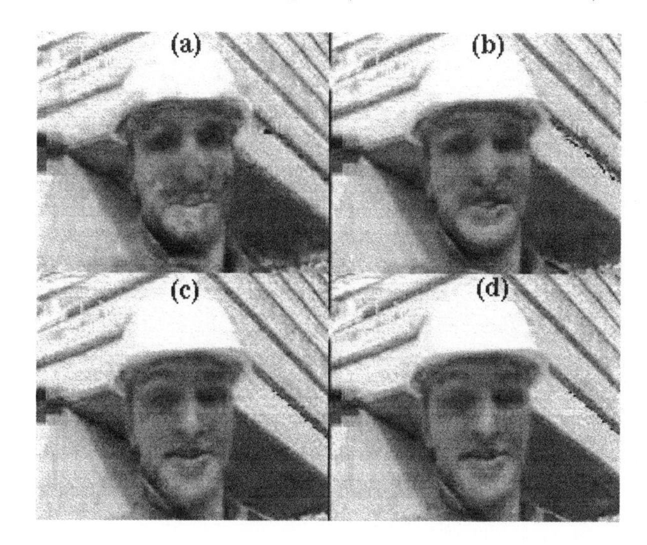

Figure 4.4 One-hundredth frame of Foreman coded with H.263 and subject to random errors with BER = 0.01 per cent: (a) no concealment, (b) zero-MV technique, (c) MV of spatially corresponding MB in previous frame, (d) MV of MB in previous frame that best moves in the direction of the lost MV

treated as an INTRA coded block. The MB is then recovered using information from spatially adjacent undamaged MBs only. The reason for that is to avoid any error in predicting a coding mode in such cases as a scene change, for instance.

4.3.2 Recovery of lost coefficients

Lost coefficients in a damaged block can be interpolated from spatially corresponding coefficients in adjacent blocks. One method is to interpolate each lost coefficient from its corresponding coefficients in its four neighbour blocks. When only some coefficients in a block are damaged, coefficients in the same block could be used for the interpolation of the lost coefficient value. However, if all coefficients of a block are lost then this frequency-domain interpolation is equivalent to interpolating each pixel in the block from the corresponding pixels in four adjacent blocks rather than the nearest available pixels. Since the pixels used for interpolation are eight pixels away from the lost pixel value in four separate directions the correlation between these pixels and the missing pixel is likely to be small, and therefore the interpolation may not be accurate. To improve the prediction accuracy, the missing pixel values could be interpolated from the four one-pixel wide boundaries of the damaged MB. The pixels in all of the four one-pixel wide boundaries could be used, or alternatively only those pixels in the two nearest boundaries, as shown in Figure 4.5. The spatial interpolation of lost coefficients is more suitable for INTRA coded blocks. For INTER coded blocks, the interpola-

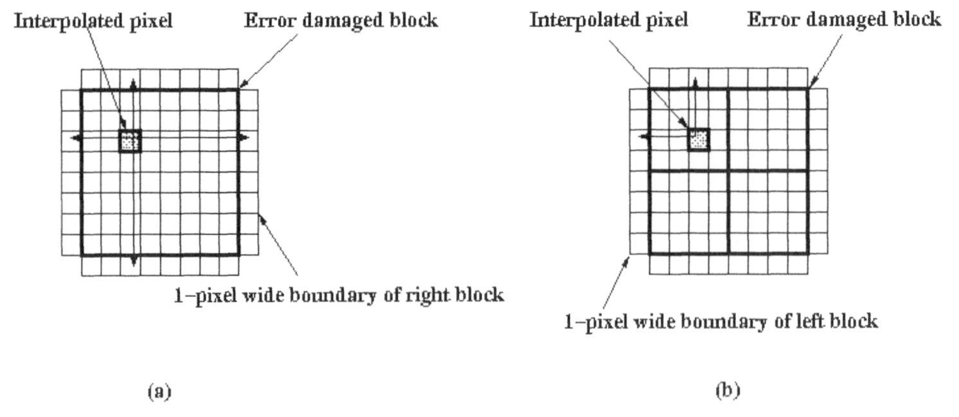

Figure 4.5 Error concealment of lost coefficients by spatial interpolation: (a) using pixels from four one-pixel wide boundaries, (b) using pixels from the nearest two one-pixel wide boundaries

tion does not yield accurate results, since the high-frequency DCT coefficients of prediction errors in adjacent blocks are not highly correlated. Consequently, in INTER coded blocks only the zero-frequency DC coefficient and the lowest five non-zero frequency AC coefficients are estimated from the top and bottom neighbouring blocks, while the rest of the AC coefficients are all set to zero.

4.4 Data Partitioning

To limit the effect of synchronisation loss on the decoded video quality, synch words are inserted in the video bit streams at regular fixed intervals. Unlike the core ITU-T H.263 standard which places synch words at the beginning of a frame or GOB, MPEG-4 streams are divided into a number of packets starting with a synch word and containing a regular number of bits. Figure 4.6 shows the difference between the packet structures of H.263 and MPEG-4.

Similarly to block-based video coders, the effects of errors on object-oriented compressed video streams depend on the type of the corrupted video parameter and the sensitivity of this parameter to errors. However, object-based video coded streams contain shape data, hence their increased vulnerability to errors. Since video data parameters have different sensitivities to errors, as established in Section 3.7, improvements in the error robustness of MPEG-4 could be achieved by separating the video data to two parts (Talluri, 1998). The shape and motion data of each video packet (VOP) is placed in the first partition, while the less sensitive texture data (AC TCOEFF) is placed in the second partition. The two partitions are separated by a resynchronisation code which is called a motion marker in INTER coded VOPs or a DC marker in INTRA coded VOPs. This

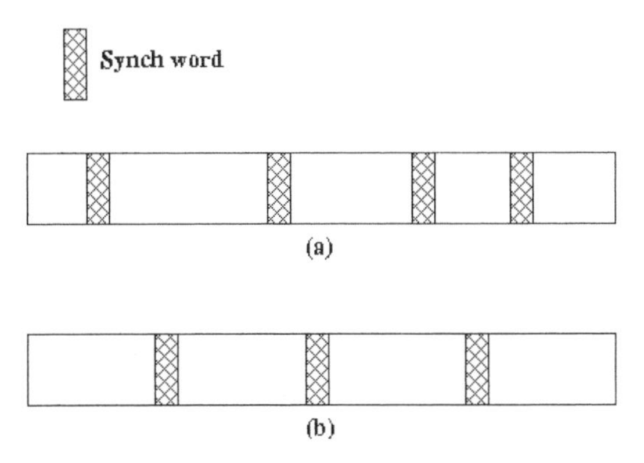

Figure 4.6 Insertion of synch words into video packets: (a) H.263, (b) MPEG-4

synchronisation code is different from the code at the beginning of a video packet. The first partition is preceded by a synch code that indicates the start of a new VOP. This MPEG-4 video data structure is illustrated in Figure 4.7. The data-partitioning scheme enables the video decoder to restore the error-free motion and shape data of a video packet when errors corrupt only the bits of the less sensitive texture data of the second partition. On the other hand, errors occurring in the second less sensitive partition can usually be successfully concealed, resulting in little visible distortion. As texture data makes up the majority of each VOP (as established in Section 3.2, Table 3.2), data partitioning allows errors to occur in a large part of the packet with relatively benign effects on video quality.

It is obvious that motion vectors are more sensitive to errors than texture data, as described in Section 3.7. However, the effect of shape data on the error robustness of an object-oriented video coder needs to be determined. The Stefan sequence is used here to analyse the error sensitivity of data in the first and second partitions of an MPEG-4 video packet. Stefan is a CIF (352 × 288) 30 frames/s fast-moving sequence that features a tennis player in the middle of a rally with two objects in the video scene, the player (foreground) and the background. The subject moves about quickly and the camera follows him by making slight multi-directional movements. 100 frames of this CIF sequence are encoded at 15 f/s to yield an average bit rate of 128 kbit/s. A packet size of 600 bits is used to limit the effect of synchronisation loss in case of errors, and an INTRA coded frame is forced once every 30 frames (1 I-frame per second). At the decoder, a simple error concealment technique sets both MVs and texture blocks of the concealed INTER

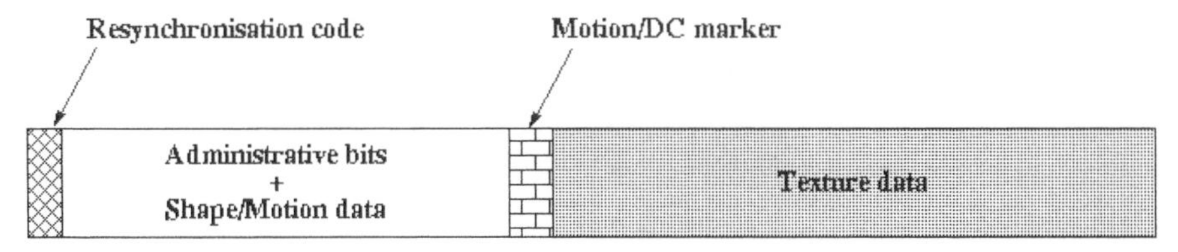

Figure 4.7 Data partitioning in MPEG-4

frames to zero, while it copies the same MB from the previous frame to the current error-concealed INTRA frames. Figure 4.8 shows that, while texture errors can be concealed with reasonable efficacy, concealment of motion and shape data results in images that contain a high degree of distortion. Since the sequence contains a great deal of motion, the video content changes significantly from one frame to another, making it difficult to conceal errors at the decoder.

The subjective results shown in Figure 4.8 confirm the error sensitivities that are demonstrated by the PSNR values of Figure 4.9. Corruption of texture produces little effect in terms of visible distortion until the bit stream is subjected to high error rates. On the other hand, shape data proves to be highly sensitive, as corruption of shape in the sequence leads to perceptually unacceptable quality.

4.4.1 Unequal error protection (UEP)

Since the video parameters of block-based and object-based video compression algorithms present different sensitivities to errors and different contributions to overall decoded quality, unequal error protection could be used for robust yet bandwidth-efficient video transmissions (Horn *et al.*, 1999). As the name implies, UEP consists of protecting video data in unequal proportions and error correction capabilities, so that the perceptual quality of video is optimised for a minimal overhead resulting from the error control paradigm. UEP was initially proposed as one of the error resilience techniques applied on MPEG-4 video data during the

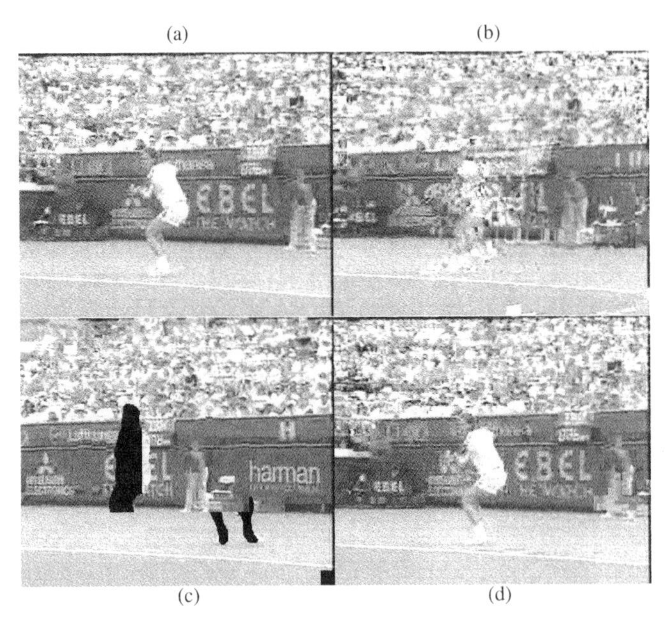

Figure 4.8 (a) Error-free Stefan sequence, (b) motion data, (c) shape data, (d) texture data, all corrupted at BER $= 10^{-3}$

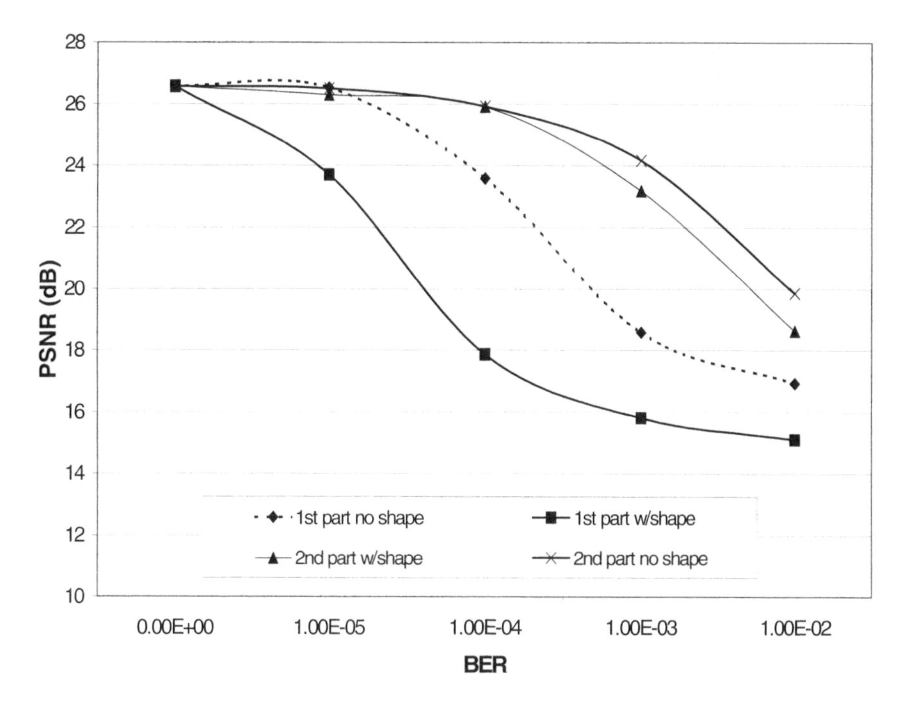

Figure 4.9 Sensitivity to errors of MPEG-4 video parameters generated by the Stefan sequence, with corruption of first and second partitions with and without shape information

development process of the standard. The UEP scheme proposed by Rabiner, Budagavi and Talluri (1998) protects fixed-length segments of data with different convolutional codes strengths, with data at the start of the packet receiving he greatest protection. However, as more motion occurs in the scene, the amount of important motion data at the beginning of each packet grows in size. This results in some of the motion information receiving less protection than required. Moreover, this UEP approach is tailored to H.324 circuit-switched applications and makes no provision for packet erasures caused by high bit error rates.

On the other hand, as data partitioning in MPEG-4 places critical data at the beginning of each video packet, the quality of data-partitioned video can benefit significantly from the UEP approach. As established in the previous section, the content of the first partition of an MPEG-4 video packet is much more sensitive to channel errors than that of the second partition. Therefore, more powerful error-protection schemes can be applied on data bits in the first partition, while only a small amount of redundancy is incurred by applying less powerful error control schemes on the less error-sensitive texture data of the second partition (Worrall *et al.*, 2000). Figure 4.10 shows the subjective quality improvement obtained by applying data partitioning and UEP onto the two partitions of an MPEG-4 video packet. More protection is then given to the first partition containing the picture headers and the error-sensitive motion data. UEP can also be applied on the multi-layer video streams, providing more powerful protection to the base layer

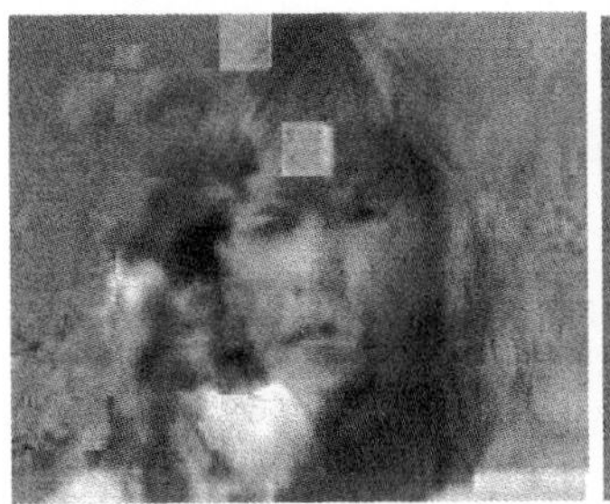

Figure 4.10 One-hundred-and-fiftieth frame of QCIF-size Suzie sequence coded with MPEG-4 at 64 kbit/s: (a) without error resilience, (b) data partitioning + UEP

stream and little protection to the less error-sensitive enhancement layer(s) stream (Lavington, Dewhurst and Ghanbari, 2000). In this case, more bandwidth is allocated to the higher-priority more heavily protected base layer to produce an acceptable end-user video quality, even when the less error-protected enhancement layer packets are lost at a rate of as high as 0.3 per cent due to network congestion caused by the TCP/IP traffic interference.

4.5 Forward Error Correction (FEC) in Video Communications

FEC techniques could also be employed to reduce the effects of errors on the decoded video quality. However, these error correction schemes inflict redundant bits on the transmitted video data. Therefore, the error detection and correction enabled by FEC techniques are carried out at the expense of bit rate overhead. In order to meet the bandwidth requirements of the network, the video source has to reduce its output rate to accommodate for more channel coding bits for error protection purposes. This process impairs the video quality in error-free or low-error conditions. The best compromise between the error performance and the error-free video quality is to make the coding rate of an FEC scheme adaptable to varying network conditions. One way of achieving this compromise is to use the rate-compatible puncture codes (RCPC) that are covered in the following subsection.

FEC techniques normally apply equal error protection (EEP) onto various video parameters. In other words, the video parameters are protected regardless of their sensitivity to errors and their contribution to overall video quality. In this case, the motion data and the transform coefficients of a block-based compressed video stream receive the same level of protection. This process makes the protection of highly sensitive data, such as motion vectors, less efficient, while leading to unnecessary waste of bandwidth by overprotecting less important data. To solve this problem, the video data parameters can be protected with unequal rates, depending on their sensitivity to errors as described in Section 4.4.1.

Due to the variable length of video parameters in a compressed bit stream, the error-protected VLC word results in another variable length code. Consequently, if the channel decoder is unable to handle the error(s) affecting a particular VLC word, the video decoder loses synchronisation, since it finds no way to identify the original size of the corrupted video parameter. In this case, the video decoder has to skip all forthcoming bits in the stream until it resynchronises on finding the next error-free synch word. This results in a huge waste of bandwidth, resulting from discarding all the error-protected parameters in the skipped video segment, thereby reducing the efficiency of the employed FEC scheme.

Because of their sensitivity to errors, motion vectors produced by block-based video coders are usually protected due to their high sensitivity to errors. In the H.263 standard for instance, the maximum length of a MV component, as indicated by the codec Huffman tables, is 13. Using a one-half convolutional coder for error protection, the length of each input codeword to the channel coder must be set to 13. If the length of a MV component turns out to be less than the maximum then the VLC word should be complemented with bits from the subsequent MV component. The half-rate convolutional coder produces a 26-bit long word that represents the protected output of this video parameter, including the padding-up section of the next MV component. If the channel decoder is, due to extremely bad channel conditions, unable to correct errors on this 26-bit word, both MV components become corrupted, creating a loss of synchronisation at the video decoder. For this reason, FEC techniques are more effective when they are used over channels with predicable BER and limited burst lengths. However, they could fail dramatically over high BER channels with long bursts of errors as the channel decoders become unable to cope with the huge number of adjacent bit errors in the coded stream, thereby leading to inefficient bandwidth utilisation and poor error protection. FEC techniques are normally applied to the fixed-length coded parameters of a video stream and used in combination with other error-resilience techniques, as will be described in Section 4.9. Figures 4.11 and 4.12 show the subjective and objective quality improvements, respectively, obtained by applying a half-rate convolutional coder to only the MV stream of an H.263 video coder.

In addition to conventional FEC techniques such as Reed–Solomon and convolutional coding, Turbo codes can also be used for protecting compressed video streams (Peng *et al.*, 1998). Despite their complexity, Turbo codes provide powerful error protection capabilities even in harsh channel conditions (Dogan, Sadka and Kondoz, 2000).

4.5.1 Rate-compatible punctured codes (RCPC)

Rate-compatible punctured convolutional codes or RCPC codes are used to provide a multi-rate channel error control (Hagenauer, 1988). The principle behind these codes is to use the same convolutional coder to provide error protection

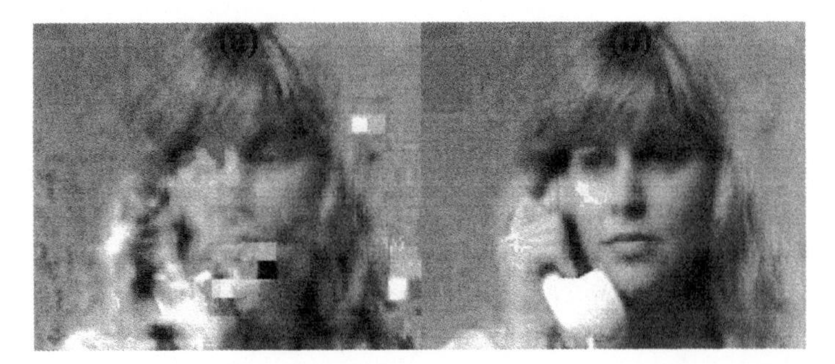

Figure 4.11 One-hundredth frame of H.263 coded Suzie sequence at 64 kbit/s with the MV stream transmitted over an AWGN channel of SNR = 12.5 dB: (a) no FEC protection, (b) MVs protected with a one-half rate convolutional coder

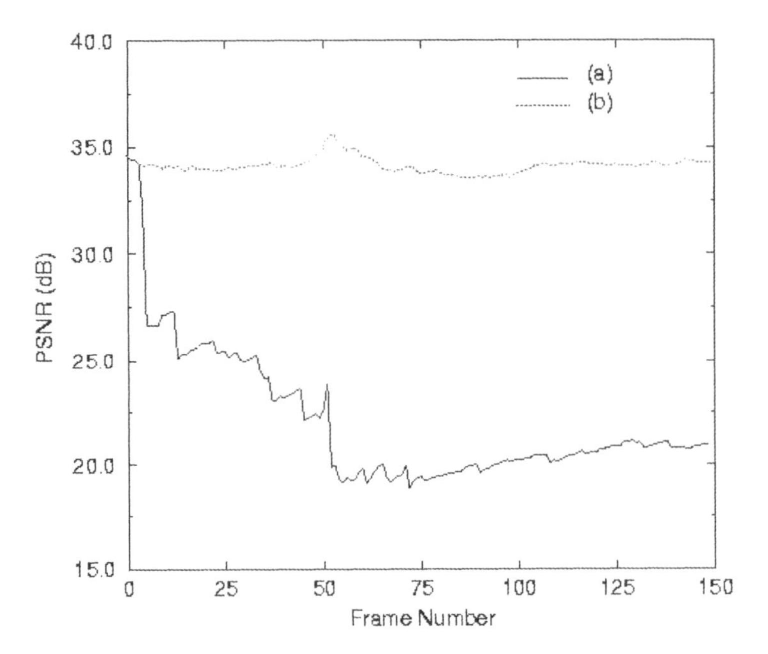

Figure 4.12 PSNR values for 150 frames of H.263 coded Suzie sequence at 64 kbit/s with MV stream sent over an AWGN channel of SNR = 12.5 dB: (a) no FEC protection, (b) MVs protected with a one-half rate convolutional coder

codes at different strengths by just eliminating some bits. When the channel conditions are time-variant, the strength of the FEC coder has to be dynamic for the optimal use of the available bandwidth. Obviously, this FEC technique must be accompanied by a very fast back channel signalling scheme that keeps the encoder updated on the status of the network. The convolutional coder starts off by sending the mother code only (with no protection bits). If the FEC decoder cannot interpret the mother code due to errors, the encoder is notified through the backward channel and consequently, the protection rate is increased accordingly. For a four-register convolutional coder, four different rates could be defined. The encoder starts with the rate set to 1 and decrements its rate when requested to do so. For degraded channel conditions, the channel coder must allocate a larger number of protection bits to the output symbols to enhance the error correction capability of the channel decoder. The rate keeps on going one further level down until the decoder is able to reconstruct the mother code bits without any detected error. When the last rate is reached while the decoder is still unable to correct the erroneous symbols, the current block is discarded and the decoder moves on to the next one. Therefore, the rate of the convolutional coder varies depending on the decoder ability to correct the corrupted bits. The higher the requested rate the more redundant bits to add to output symbols for better error protection. This multi-rate error-protection code is called a punctured code. RCPC techniques are mostly used in delay-insensitive video applications and are not particularly suited for real-time applications, due to the excessive amounts of delay that could be incurred by the feedback messages and the resulting retransmissions of damaged symbols. RCPC and back channel signalling were techniques jointly proposed for a number of experiments carried out for MPEG-4 error resilience during its standardisation process.

4.5.2 Cyclic redundancy check (CRC)

FEC data could be inserted into a video stream for a variety of reasons. One reason is to enhance the robustness of video data to channel errors, as demonstrated above. Another reason is to aid the synchronisation at the decoder by inserting synch words at the beginning of each video packet or fixed-length segment. Despite the quality improvement, the insertion of error check codes results in the bit stream being incompatible with the standard video decoder. An error control scheme that uses CRC check codes has been defined (Worrall *et al.*, 2000) that allows the insertion of channel protection data into an MPEG-4 bit stream, while still retaining compatibility with standard MPEG-4 decoders. When data partitioning is enabled in MPEG-4 as discussed in Section 4.4, the decoder identifies the number of MBs in each video packet from data in the first partition. When the last MB in the second partition is decoded, the decoder skips all the subsequent bits searching for the next synch word. Even in the case of errors, all bits following the position of error are ignored, regardless of their correctness, until the decoder resynchronises at the beginning of a new video packet. This operation could be exploited to insert user data that does not emulate a start code, at the end of the second partition, as shown in Figure 4.13, while still retaining compatibility with the standard MPEG-4 decoder.

The inserted data can therefore be located by reading backwards from the synch word at the beginning of the following video packet. For error-protection

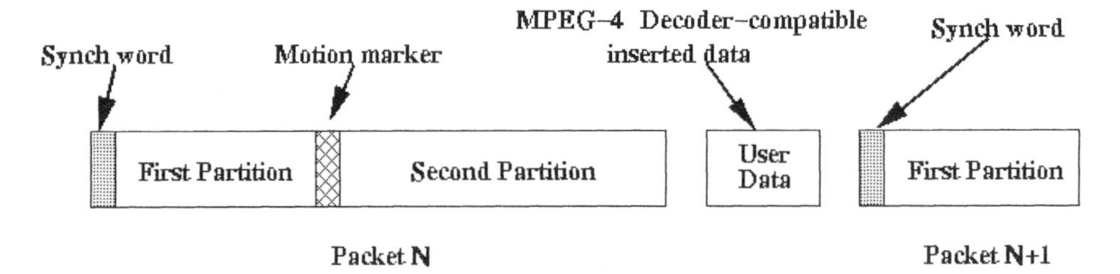

Figure 4.13 Insertion of decoder-compatible data into MPEG-4 video packet

purposes, the inserted data consists of two CRC fixed-length codes, 16-bit long each, used as a check for the first and second partitions. Therefore, the decoder-compatible inserted CRC codes are used to detect bit errors which are undetected by the standard MPEG-4 decoder. Errors detected in either one of the two CRC check codes or in the first partition of a video packet lead to a whole packet loss. However, errors that occur in the second paritition do not cause a packet loss given that no error is detected in the first partition. When a packet is dropped, error concealment is applied by replacing the corrupted MBs by their corresponding motion-compensated MBs in the previous frame (Section 4.3.1). The inserted CRC codes were found to provide a much lower variance to average objective quality, indicating that this technique provides a much more consistent video quality. Using this backward-compatible error control technique in the MPEG-4 decoder, the quality is prevented from randomly dropping to levels much lower than average. The subjective improvement of this technique is demonstrated in Figure 4.14.

4.6 Duplicate MV Information

The motion prediction in standard video compression algorithms is the main reason behind the accumulative effect of channel errors in both time and space. Motion vectors are sent in differential coordinates and predicted from candidate MVs of spatially adjacent MBs. Therefore, motion data is highly sensitive and its loss leads to a fairly fast quality degradation. To reduce the accumulative effect of errors in a video sequence, the probability of error in a MV component should be minimised. The error resilience of a video bit stream could thus be minimised by duplicating the MV data at different locations in the stream. Consequently, the probability of receiving an erroneous MV data bit can be reduced. To enable the video decoder to locate the duplicate motion data in the bit stream, a specific bit pattern is sent just prior to the start of the duplicated MVs. This specific bit pattern has to be unique and different from any combination of data bits in the video stream. This bit pattern must also be different from the synch word which

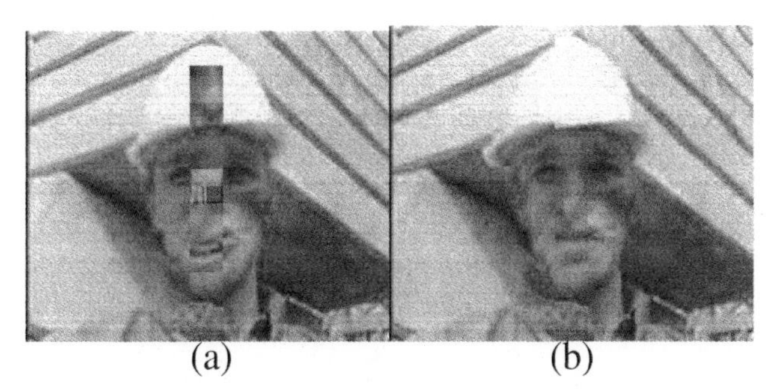

Figure 4.14 Seventy-fifth frame of the Foreman sequence encoded with MPEG-4 and sent over a mobile channel of BER = 3×10^{-4}: (a) without CRCs (variance = 3.00 dB), (b) with CRCs (variance = 0.08 dB)

normally denotes the start of a frame, a GOB or a data segment in the video stream. To reduce the likelihood that the decoder falls on a sequence of bits which resemble the unique bit pattern, the synch word could be used and followed by a five-bit word representing the decimal value 21 (10101). These five bits are normally reserved, according to the syntax of H.263, to code the sequence number of a GOB within a video frame. This five-bit word takes the decimal value 31 when the corresponding frame or GOB is the last one in the sequence. Since there is a smaller number of GOBs per frame than this five-bit word could actually indicate, one of its unused values could be used to designate the start of the duplicate-MV segment of the stream. Figure 4.15 depicts the order of transmission of an MB when the duplicate information is applied on a MB level.

Due to the variable-length coding of motion vectors, two kinds of error might arise (refer to Section 4.2). One or more bit errors hit a motion vector component in such a way that the decoder is unable to find a legitimate codeword at this position of the bit stream. In this case, the decoder assumes an error is detected, moves forward in the bit stream to locate the start of the duplicate data segment, reads the second version of the MV component and resumes decoding after returning to the position of errors and flushing the number of bits that correspond to the length of the decoded MV components. Obviously, the possibility that both copies of the same MV component are corrupted is not annihilated but the likelihood of a bit error in the same component is reduced. The second kind of error goes undetected, causing the decoder to lose synchronisation and skip the duplicate information. To avoid this scenario, a five-bit checksum representing the parity bits (Kim *et al*., 1999) of the MV components is sent. If the decoder finds no discrepancy between the calculated parity and the value of the checksum word, it

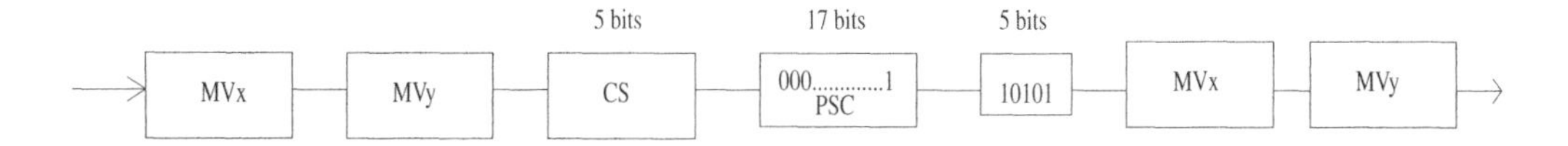

Figure 4.15 MV duplicate information applied on a MB-level

skips the duplicate motion data, assuming no error has been detected. Figure 4.16 shows the subjective improvement achieved by this technique.

The blocking artefacts in Figure 4.16(b) are due to the high level of distortion resulting from coarsely quantising the sequence, including the duplicate data and associated overhead, in order to meet the target bit rate. However, the distortion caused by increasing the quantisation parameter is highly preferable to the unpredictable effects caused by channel errors. For other conventional ITU sequences, such as Claire for instance, the quantisation distortion is less noticeable, due to the lower amount of motion and hence less duplicate MV information transmitted in the bit stream. This leads to a lower quantisation parameter and hence a better quality. A major drawback of this technique is the massive number of redundant bits added to the coded stream, making it unacceptable for very low bit rate applications. A total of 27 administrative bits are transmitted for each INTER MB apart from the MV duplicate information overhead. Moreover, this technique could completely fail when the unique bit pattern is corrupted by errors and undetected by the video decoder.

In order to reduce the bit overhead of the above mechanism, motion data duplication can be applied on a GOB level, as shown in Figure 4.17.

This GOB structure incurs a certain delay on the decoding process since the decoder has to wait, when an error is detected, until the last MB of a GOB is completely received before it can locate the unique bit pattern that is followed by the duplicated motion data. To protect the start code against channel errors and reduce the probability of failure of this technique, a Reed–Solomon (5,7) code is used to make the synch word more robust to errors. This RS code increases the overhead but enables the decoder to correct one bit-error in the start code. To reduce the overhead, however, the checksum is not transmitted in this case and

Figure 4.16 One hundred and twenty-fifth frame of the Foreman sequence encoded with H.263 at 64 kbit/s and transmitted over an AWGN channel of SNR = 12 dB: (a) ordinary H.263 stream, (b) H.263 with MB-level duplicated MV data

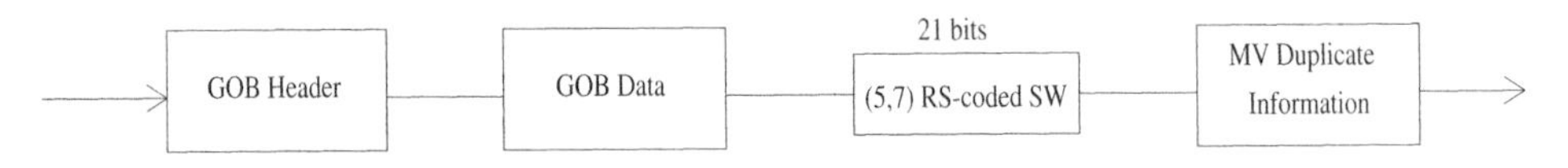

Figure 4.17 Duplicate MV data of all INTER MBs of a GOB

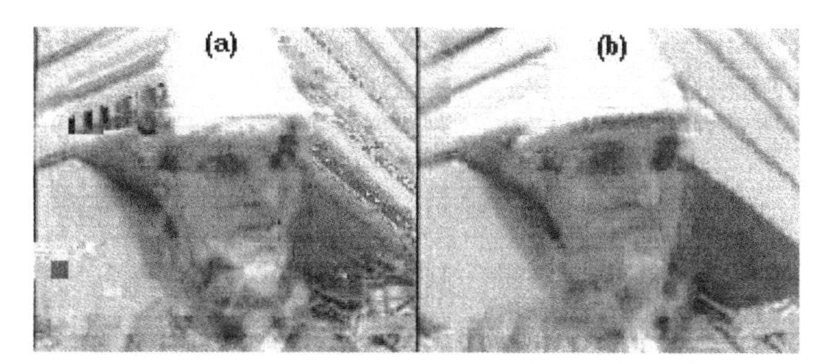

Figure 4.18 One hundred and ninety-ninth frame of H.263 coded Foreman sequence at 64 kbit/s and transmitted over a channel with random errors at BER = 10^{-5}: (a) ordinary H.263 stream, (b) H.263 stream with duplicated MV data sent on a GOB-level

error detection is left to the Huffman decoder. Therefore, the only overhead of this technique is attributed to the RS coded 21-bit start code sent once every GOB. This results in a total overhead of almost 4.7 kbit/s for a QCIF frame rate of 25 f/s. Figure 4.18 shows the subjective quality improvement achieved by this technique.

In addition to repetition of motion data, other sensitive video information could also be duplicated. In MPEG-4, the important header information that describes the video frame is repeated twice in the video packet (Talluri, 1998). When header information, such as the COD flag, temporal reference, MCBCP, CBPY, frame coding mode, timestamps, etc., is corrupted by errors, the decoder can only discard all the bits following the position of the error in the packet until it regains synchronisation at the next correctly received synchronisation word. To alleviate the error sensitivity of this important header information, a 1-bit header extension code (HEC) is sent at the beginning of each video packet. When the HEC flag is set, the header information is repeated in the video packet. If the header information at the beginning of the video packet matches the header information at the beginning of the video frame, the decoder assumes that header information has been correctly received. However, if the header data in the video frame is corrupted then the enclosed video data can still be rescued by reading the repeated header information sent within the video packet. The repetition of header information within the bit stream is very efficient in reducing the amount of discarded information, hence achieving a significant improvement in the overall video quality.

4.7 INTRA Refresh

One possible way to limit the accumulation of errors in a video sequence is to refresh the scene with an INTRA frame. INTRA frames are coded without prediction and therefore produce a low compression ratio. The number of INTRA frames should be a compromise between the error resilience of the video coder and its compression efficiency. INTRA refresh should be used in conjunction with a rate control algorithm in order to smooth out the high bit rate fluctuations produced by INTRA frames. To accommodate a large number of I-frames while keeping the bit rate below a certain limit, the quantisation parameter has to be assigned a large value. This results in coarsely quantising the DCT coefficients, thereby leading to poor video quality. If an I-frame is sent once every 20 sequence frames at a frame rate of 25 f/s, then I-frames are sent at a frequency of 1.25 f/s. Figure 4.19 shows the 200th frame of the Foreman sequence coded at 64 kbit/s and subjected to random channel errors for both the normal and increased I-frame frequency cases, while the first frame is assumed error-free.

However, if a VLC word in the first INTRA frame is hit by errors, the decoder fails to complete the reconstruction of the following part of the frame. Consequently, it becomes impossible to conceal the effect of errors until the next INTRA refresh takes place. Figure 4.20 depicts the luminance PSNR values of the Foreman sequence with increased I-frame frequency when the first frame is subject to errors. When only the first frame is INTRA coded and corrupted by channel errors, the errors propagate throughout the whole sequence time, leading to an average PSNR value of 5 dB. This marks the importance of I-frames and their contribution to overall video quality. Even with INTRA refresh, if an I-frame is hit by errors then all the following P-frames will also be damaged due to temporal prediction. The situation gets worse when the I-frame is hit by errors in early MBs, causing the decoder to discard all the forthcoming bits of the frame to restore synchronisation at the beginning of the next frame. The damaged I-frame will also entail the corruption of the next P-frames which are all temporally predicted. This is demonstrated in the low PSNR values of the first 20 frames of Figure 4.20.

Because of their importance and high contribution to perceptual video quality, I-frames must be protected against channel errors so that the INTRA refresh technique becomes successful. Since INTRADC coefficients carry a high portion of the energy of INTRA frames, they have to be made robust to channel errors.

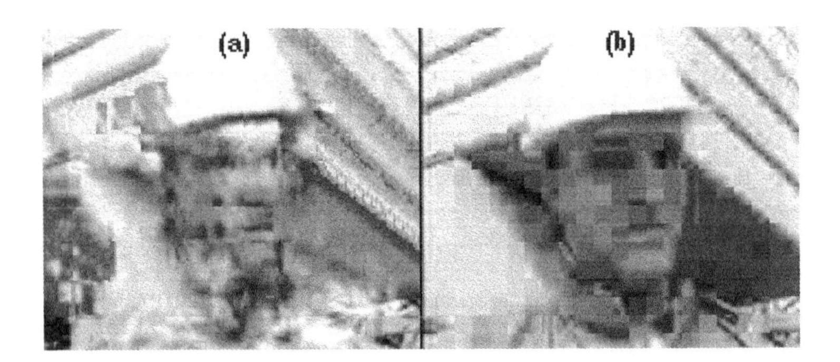

Figure 4.19 Two-hundredth frame of the Foreman sequence coded with H.263 and transmitted over a channel with random errors at BER = 10^{-4}: (a) at 64 kbit/s with only first frame coded in INTRA, (b) at 67.7 kbit/s with 1.25 I-f/s

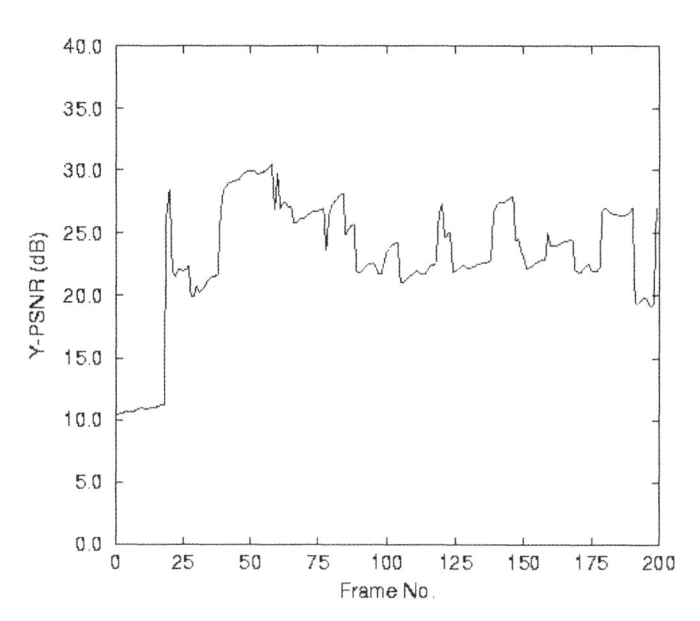

Figure 4.20 Luminance PSNR values for 200 frames of the Foreman sequence coded at 67.7 kbit/s with 1.25 I-f/s and transmitted over a channel with random errors and BER = 10^{-4}

This could be done by placing the fixed-length codes of INTRADC coefficients, with a Hamming distance of one, as close together as possible in the corresponding FLC table at both the encoder and decoder. The effect is that the most likely INTRADC codes are less sensitive to a single bit error than the less likely codes. Another possible way of protecting INTRADC coefficients is to make use of their fixed-length coding for FEC protection. In the H.263 standard, each INTRADC coefficient is eight-bit long, and therefore applying half-rate convolutional coding on each INTRADC coefficient leads to a total overhead of 5.94 kbit/s (4752 bits per QCIF I-frame) for a frame rate of 25 f/s and INTRA frame rate of 1.25 I-f/s. The remaining 63 AC coefficients of each block in an I-frame can be coded with a coarse quantiser to counter the bit rate overhead imposed by the I-frames and the FEC protection of INTRADC coefficients.

4.7.1 Adaptive INTRA refresh (AIR)

The INTRA frame refresh technique described earlier entails a large increase of the output bit rate of a video encoder. The reason for that is the low compression efficiency achieved by the INTRA coding mode and the large number of MBs to be INTRA coded. For instance, refreshing a QCIF video scene with an I-frame requires the transmission of 99 INTRA coded MBs. This process leads to the formation of undesirable spikes in the bit rate each time an I-frame is transmitted. Therefore, encoding a frame in INTRA mode produces a burst of bits that causes inevitable delays and helps build up a state of congestion in the network. Moreover, if the moving area of the image is corrupted by errors, the degradation propagates temporally, giving rise to long periods of quality deterioration until the next INTRA refresh takes place. To reduce the bit rate of INTRA coded frames while still maintaining error robustness and limiting temporal propagation of errors, a scheme known as adaptive INTRA refresh is normally used. AIR is a technique that is defined in Annex E of the MPEG-4 standard. It involves sending a limited number of INTRA MBs in each VOP, as opposed to the conventional Cyclic INTRA Refresh (CIR) where all MBs of a VOP are uniformly INTRA coded. The number of MBs to be INTRA coded in AIR is much smaller than the total number of MBs per VOP or frame. AIR selectively INTRA codes a fixed and predetermined number of MBs per frame according to a refresh map. The generation of this refresh map is achieved by marking the position of MBs which are subjected to motion, as illustrated in Figure 4.21 where the number of MBs to be INTRA coded per VOP is 2. The motion evaluation is carried out by comparing the sum of absolute differences (SAD) of a MB with a threshold value SAD_th. SAD is calculated between the MB and its spatially corresponding MB in the previous VOP and SAD_th is the average SAD value of the entire MBs in the previous VOP. If the SAD of a particular MB exceeds SAD_th, the encoder decides the MB belongs to a high motion area that is sensitive to transmission errors and thus marks the MB for INTRA coding. If the number of MBs marked for INTRA coding exceeds the number of MBs set to be INTRA coded, then the video coder moves down the frame in vertical scan order encoding INTRA MBs until the preset number of MBs have been encoded. For the next frame, the encoder starts in the same position and begins coding INTRA MBs including those marked for INTRA coding in the previous frame. The number of coded MBs is determined based on the bit rate and frame rate requirements of the video application. However, for improved robustness, the number of MBs can be made adaptive in accordance with the motion characteristics of each video frame (Worrall *et al.*, 2000). Since the moving area of the picture is frequently encoded in INTRA mode, it is possible to quickly refresh the corrupted moving area.

Obviously, increasing the number of MBs that are refreshed in each frame speeds up the recovery from errors, but results in a decrease in error-free quality at a given target bit rate. This is due to the coarser quantisation process used to achieve the target bit rate. However, AIR provides a better and more consistent objective error-free quality than the conventional INTRA refresh technique for the same target bit rate, as shown in Figure 4.22. On the other hand, AIR produces a more stable output rate, as shown in Figure 4.23, since the INTRA coded information is sent more regularly (a fixed number of MBs per frame as opposed to 99 MBs once every number of frames). Therefore, the number of MBs to be INTRA coded

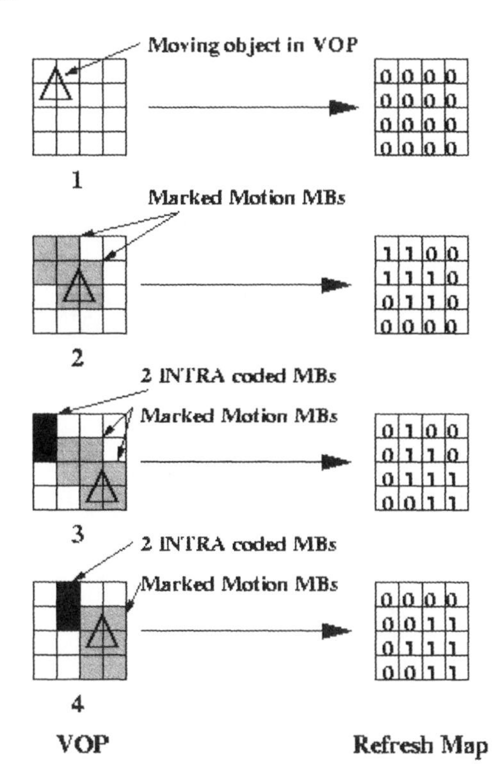

Figure 4.21 Generation of a motion map for AIR coding

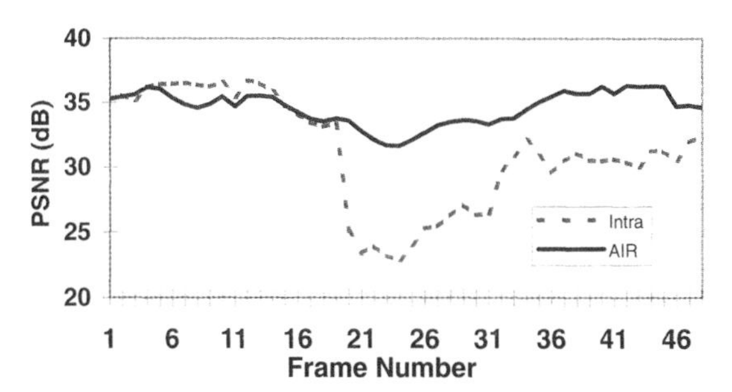

Figure 4.22 Y-PSNR values for 50 frames of Suzie sequences coded with MPEG-4 at the same target bit rate for both AIR and conventional INTRA frame refresh schemes in error-free conditions

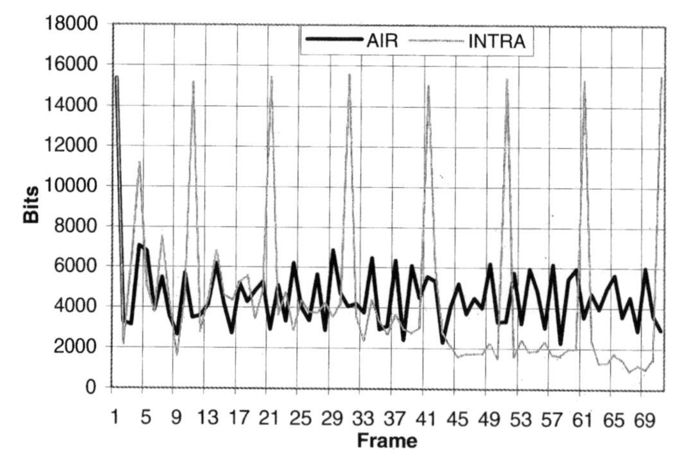

Figure 4.23 Output bit rates for various frames of Suzie sequence coded with MPEG-4 using both AIR and conventional INTRA refresh techniques

is a trade-off between the error robustness on one hand and the bit rate and error-free video quality on the other hand.

4.8 Robust I-frame

For robust video communications, the INTRA coded blocks must be optimally coded (Cote and Kossentini, 1999). Protecting the INTRADC coefficients of an I-frame with a convolutional coder, as discussed in Section 4.7, does not make an I-frame fully resilient to channel errors. A bit error that corrupts one of the variable-length codes (such as CBPY, MCBPC, TCOEFF runs and levels) of an I-frame leads to the loss of synchronisation even when INTRADC coefficients are protected with FEC techniques. This obviates the use of the convolutional coder since the added protection bits become a useless overhead when an error hits a VLC word of an I-frame. In this situation, the decoder terminates the processing of the damaged video frame and discards the next segment of the bit stream until it detects the first error-free synch word. The quality degradation persists for the next 19 P-frames until the next INTRA refreshes the scene.

To reduce the vulnerability of I-frames to channel errors that lead to loss of synchronisation, all the fixed-length INTRADC coefficients must be transmitted before the first VLC word of an I-frame appears in the stream (Sadka, Eryurtlu and Kondoz, 1997). This guarantees that the decoder receives the 594 INTRADC coefficients of a QCIF-size I-frame just before it might lose synchronisation due to an erroneous VLC word. Therefore, the transmission order of an I-frame is changed from a block level to a frame level. The I-frame consists of a fixed-length section that contains all the protected INTRADC coefficients followed by a variable-length section that consists of all the VLC words. For a QCIF-size sequence, the fixed length section of the I-frame contains 594 INTRADC coefficients, 16-bit long each (with half-rate convolutional coding). Therefore, the decoder has to read 9504 bits before it encounters the first VLC word in the I-frame. To avoid having the decoder fall on a false picture mode (INTER or INTRA), the frame mode flag is assigned a three-bit word, as opposed to just one bit in the standard H.263. For only two possible coding modes, a Hamming distance of three can be used between the two words to make the decoder tolerant to one bit error. Figure 4.24 shows the structure of both I and P frames when the robust I-frame technique is used.

If an error is still detected in the fixed-length segment of the I-frame, the decoder preserves its synchronisation and goes to the next 16-bit protected INTRADC coefficient in the I-frame. However, if an error is detected in one of the VLC words in the variable-length section, the decoder sets to zero all the AC coefficients following the position of error. This error resilience scheme gives a noticeable improvement to the subjective and objective quality of the I-frame, as indicated by Figures 4.25 and 4.26, respectively, for a bit rate increase of less than 6 kbit/s at a QCIF-size INTRA frame rate of 1.25 I-f/s and an overall frame rate of 25 f/s.

17 bits	12 bits	3 bits	9504 bits	Variable length
PSC 0........01	Pheader	Frame mode flag	594 1/2 rate convolutional coded DC coefficients	Mode + GOB No. + VLC words (CBPY,MCBPC,RUN,LEVEL,LAST)

(a)

17 bits	12 bits	3 bits	Variable length
PSC 0........01	Pheader	Frame mode flag	Group of Blocks (GOBs)

(b)

Figure 4.24 Layering structure using the robust I-frame error resilience technique: (a) I-frame, (b) P-frame

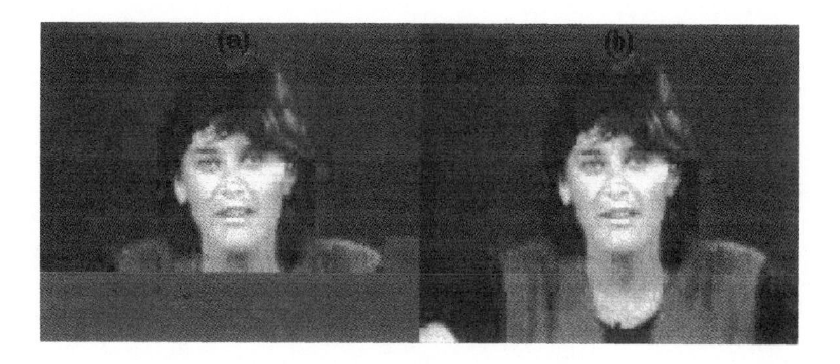

Figure 4.25 The first frame of QCIF Miss America sequence encoded with H.263 at 47 kbit/s, 1.2 I-f/s, BER = 10^{-4}: (a) ordinary H.263, (b) robust I-frame coding

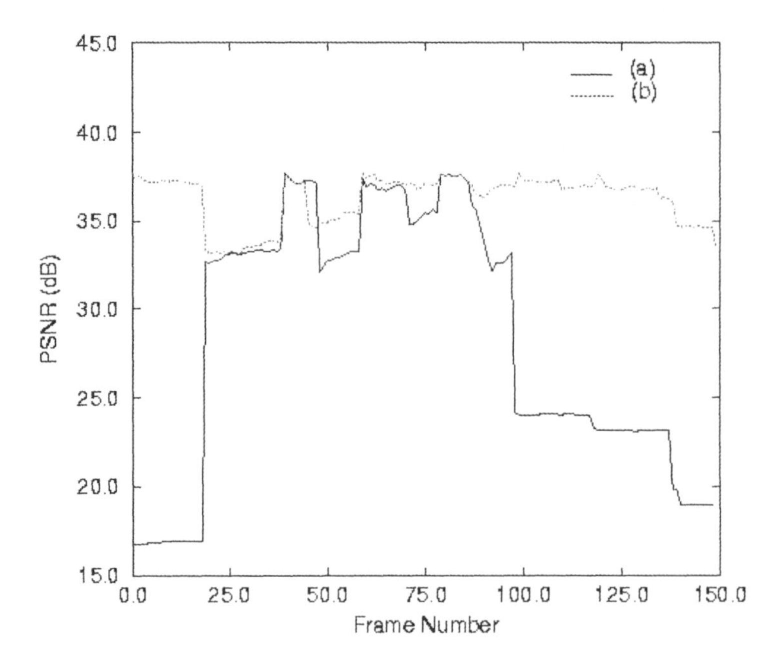

Figure 4.26 Luminance PSNR values for 150 frames of the Miss America sequence encoded with H.263 at 47 kbit/s, 1.2 I-f/s, BER = 10^{-4}: (a) ordinary H.263, (b) robust I-frame coding

4.9 Modified H.263 for Mobile Applications (H.263/M)

It has been observed in the previous section that the robust INTRA refresh scheme improves the video quality by limiting the propagation of errors within an INTRA frame. However, the P-frames remain vulnerable to errors and likely to cause severe quality degradation until the next I-frame is coded. Many standards-compliant methods (Jung, Kim and Lee, 1998; Talluri, 1998; Wenger *et al.*, 1998; Farber, Steinbach and Girod, 1996) have been proposed to improve the performance of standard video coders in error-prone environments. Some error-resilience schemes have proposed a combination of techniques to improve the error robustness of coded video streams. For example, hybrid ARQ (Automatic Repeat Request) methods were used in conjunction with feedback channel messages to help make the encoder aware of the effects of transmission errors and avoid using the corrupted information in further prediction operations (Liu and El Zarki, 1997). An H.263-compliant error resilience technique was presented by Steinbach, Farber and Girod (1997). This technique consists of a combination of ARQ, AIR and an error-concealment algorithm for improving the error robustness of H.263 streams to transmission errors in mobile environments.

To enhance the error resilience of a video coder, the factors that lead to the variability of the bit rate must be counteracted (Sadka, Eryurtlu and Kondoz, 1997). One method is to place some relatively long synch words that can notify the decoder of the start of a specific block of data in the bit stream. Apparently, this method imposes considerable overhead, and research has been conducted to optimise the length of these synch words (Lam and Reibman, 1992) for an optimal trade-off of coding efficiency and error resilience. H.263/M is a modified version of the standard H.263 coder for mobile applications. It consists of a suite of combined error resilience schemes for video communications over mobile networks.

4.9.1 Fixed length coding (FLC)

Some video parameters such as the motion vectors and control parameters, like CBPY and MCBPC, are fixed-length coded, whereas an INTRADC coefficient is already assigned eight bits by the standard H.263. Table 4.1 shows the bit allocation of various video parameters in H.263/M.

This shows that FLC is applied on a MB-level since all those parameters belong to the MB layer. However, the 63 run-length coded coefficients of a block are not fixed-length coded, otherwise the coding process becomes extremely inefficient. Undoubtedly, FLC incurs some overhead on the coded stream and reduces the coding efficiency of the compression algorithm. However, since the BERs of mobile channels are expected to be high, the distortion resulting from transmissions errors is much more detrimental than that caused by coarse quantisation. The major role of fixed-length coding is to reduce the possibility of the decoder losing its synchronisation on the occurrence of an error. If an error hits an MV component for instance, the decoder recognises the size of the corrupted MV parameter (six bits) and resumes decoding without having to skip all the forthcoming bits in search of the next error-free synch word. Another advantage of fixed-length coding is to suppress the differential coding process of motion vectors in order to improve their robustness to errors. Although these parameters are fixed-length coded, the run-length coded AC parameters are allocated variable-length codes. This implies the failure of the fixed-length coding scheme when an error hits one of the VLC words of TCOEFF. Moreover, the number of

Table 4.1 Fixed-length coded video parameters in H.263

Parameters in an MB	Number of bits per parameter
CBPY	4
MCBPC (INTRA)	4
MCBPC	5
MV-x	6
MV-y	6

parameters per MB is not fixed and this leads again to a state of confusion in the decoder on detection of a transmission error. Therefore, other techniques must be applied alongside FLC to improve the resilience of H.263/M.

4.9.2 Changed order of transmission

To make the best use of FLC, the order of transmission of video parameters has to be changed. The layering structure of H.263 described in Appendix A shows that the video parameters representing an MB are transmitted on a MB-level. Consequently, if a run-length word is corrupted by errors, the fixed-length words of upcoming MBs are either discarded or incorrectly decoded. To resolve this problem, the FLC words of a whole frame are sent at the beginning of the frame before any VLC word is transmitted (refer to Section 4.8). If an error is encountered in the variable-length section of the frame (AC coefficients), the decoder can skip the remaining bits, setting all the following AC parameters to zero. An error concealment technique could also be employed at the decoder to replace missing AC coefficients using the previously reconstructed data (refer to Section 4.3.2). Since all the fixed-length coefficients of a frame are received before the decoder could lose synchronisation on errors in the variable-length section, only minor quality degradation is perceived due to errors hitting the AC coefficients of a frame. To achieve this modification in the order of transmission from a MB-level to a frame-level, some issues need to be considered to further enhance the robustness of H.263/M.

4.9.3 COD-map coding

In INTER frame coding, the video sends a one-bit COD flag for each MB. The COD information indicates whether a MB is coded (COD = 0) or uncoded (COD = 1). Conversely, in INTRA frame coding, all MBs are necessarily coded and thus no COD information is required. Using the frame-level layering structure as described in Section 4.9.2, the COD information is no longer transmitted as a separate bit for each MB alone, but as a map of 99 (QCIF resolution) COD bits indicating the coded MBs in the whole P-frame. Since this map is highly sensitive to errors, FEC is used to protect it against transmission errors. A Reed–Solomon (3,5) channel coder is applied on each set of 9 COD bits to produce an output word of 15 bits, thereby resulting in a total of 165 bits for the protected COD map. Due to the fact that this protected COD map might contain long streams of zeros, there is a certain likelihood that the decoder would fall on a false synch word within the COD map. If an error is detected in the variable-length section of a frame, the decoder attempts to recover its synchronisation by searching for the next error-free synch word. During this search, the decoder might mistakenly fall on a sequence of bits in the COD map that accidentally resembles the bit pattern of a synch word (16 zeros followed by 1). To avoid this scenario which would cause a disastrous impact on the decoded video quality, each 15 bits of the protected COD map are XORed with a pattern of 15 alternating zeros and ones.

4.9.4 Avoiding false synch words in MV stream

Due to fixed-length coding, the motion vectors are no longer predicted and their actual values are transmitted instead. This creates another possibility that the decoder would misinterpret a certain bit pattern in the MV stream as a synch word. This could happen within the process of resynchronisation upon detection of an error. To reduce this possibility, each MV component is XORed with a sequence of six alternating zeros and ones.

4.9.5 Insertion of synch words at fixed intervals

The variable-length section of each frame contains the run-length coded AC coefficients of all blocks in the frame. Since the quantised TCOEFF might contain long runs of zeros, there is again a high likelihood that the variable-length section includes bit patterns that resemble a synch word. Therefore, when an error is detected, the decoder might fall on a false start of a new frame, thereby leading to an improper interpretation of the following segment of the stream. To resolve this problem, synch words are inserted at fixed intervals within the bit stream so that the decoder would be less likely to confuse them with similar segments of the video data stream. A fixed interval of 32 is used to maintain the byte-alignment of video frames. To reduce the appearance of long streams of zeros in the stream, the end of the frame, following the variable-length section, is stuffed with ones instead of zeros until a multiple of 32 is reached. Therefore, if the synch word is sent once every frame, the overhead induced by bit stuffing can reach a maximum of 31 bits per frame. A Hamming distance of two is allowed at the decoder for a correct synch word. If the Hamming distance is found to be greater than two, the decoder considers the sequence of 22 bits at the beginning of an interval (multiple of 32) as an ordinary data word and resumes searching for the synch word. Figure 4.27 shows the insertion of synch words at multiples of 32 bits within the bit stream and the process of one bit stuffing at the end of a frame.

4.9.6 GOB indicator coding

The mode of each MB in a P-frame is an important parameter that affects the resilience of the video stream to errors. As opposed to I-frames where all MBs are

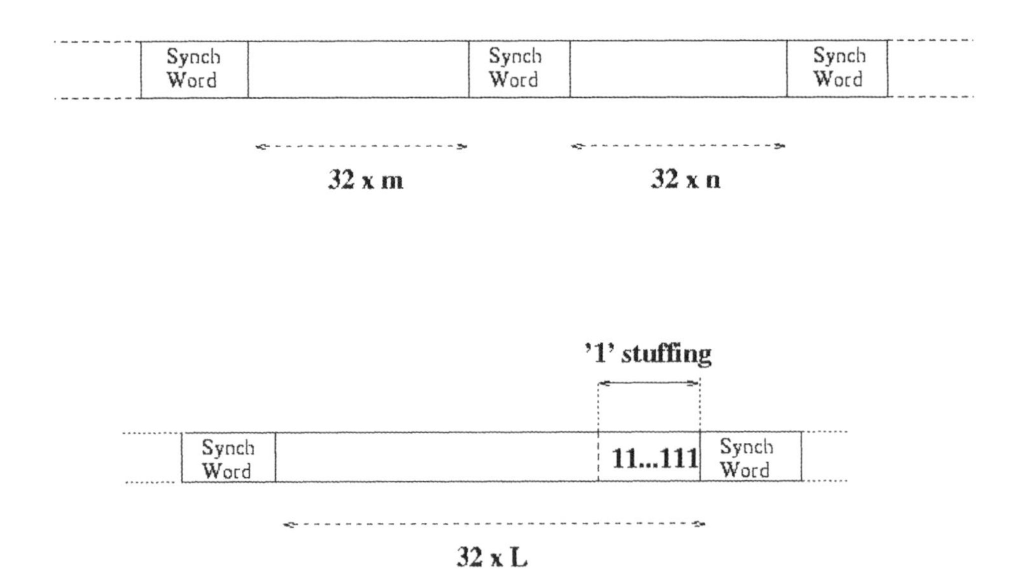

Figure 4.27 Insertion of synch words at multiples of 32 bits within the stream with one bit stuffing

INTRA coded, P-frames might include MBs of different coding modes. Therefore, it is necessary that the mode of each coded MB (COD = 0) in a P-frame is transmitted. In H.263/M, the mode information is sent on a GOB basis. If a GOB contains only INTER coded MBs, then a flag of 3 bits (010) is transmitted to instruct the decoder that all MBs of this GOB are INTER coded and therefore no mode information is required for transmission. However, if the GOB contains INTRA coded MBs, a flag of three bits (101) is transmitted prior to the mode information belonging to the enclosed coded MBs. The two values of this mode flag have a Hamming distance of three to allow for the detection and correction of one bit error in the flag. The mode information of an INTER coded MB is sent as 1 and that of an INTRA coded MB as 0. Since it is more likely to have INTER than INTRA coded MBs in a P-frame, sending 1 to indicate the mode information of an INTER coded MB reduces the possibility of detecting false synch words in case of errors. If an MB is INTER coded, a fixed-length MV (12 bits i.e. 6 bits for each component) is sent while an 8-bit INTRADC coefficient is sent for an INTRA coded MB and then followed by the AC coefficients of the variable-length section of a P-frame.

4.9.7 *Frame type indicator coding*

To indicate the frame coding mode (INTER or INTRA), a three-bit flag is sent to the decoder, as opposed to the one-bit flag of H.263 standard. A Hamming distance of three between the two values is given to allow for the correction of one bit error in the decoded flag (refer to Section 4.8). Figure 4.28 shows the layering structure of H.263/M and the modified order of transmission adopted for error resilience purposes (compare to Figure A.1 in Appendix A).

H.263/M coded streams are subject to two kinds of quality deterioration over mobile radio links. The first is due to the low compression efficiency caused by FLC which leads to a coarse quantisation at a certain target bit rate. The second and most important factor behind the quality degradation is due to transmission errors. Due to its efficient error-resilience algorithm, H.263/M provides a high level of robustness to channel errors, and thus outperforms the ordinary H.263 standard for mobile applications. In error-free conditions or at extremely low BERs, H.263/M clearly yields a slightly worse quality than H.263 for the same target low bit rate ($\leq$ 64 kbit/s). However, in mobile environments, the quality degradation caused by channel errors is much more perceptually disturbing than the distortion resulting from a poor quantisation process. Figures 4.29 and 4.30 show the subjective and objective quality improvements achieved by H.263/M, respectively. It is clear that H.263/M does not involve any retransmissions and does not require any feedback channel for interactive error control purposes. Moreover, it does not place any extra complexity on the standard algorithm, hence its suitability for mobile video applications.

4.10 Two-way Decoding and Reversible VLC

Two-way decoding is an error-resilience technique that is employed to reduce the effective error rate of a decoded video stream. As seen in Section 4.2, the bit error that is most damaging to the perceptual video quality is the one that leads to the loss of synchronisation in the decoder. In this case, the part of the stream that follows the position of error is discarded until an error-free synch word is detected. The discarded video data could, however, be received without errors and thus contribute significantly to the enhancement of perceptual quality. The number of discarded bits is a function of the distance between the position of error and the location of the next error-free synch word. This phenomenon results in an effective error rate that is much higher than the actual channel bit error ratio by orders of magnitude. In order to confine the damage to the affected area only and save the bits that are received without errors, two-way decoding also allows decoding in the reverse direction. Upon detection of an error in the forward direction, the decoder stops its operation searching for the next synch word. When the decoder restores

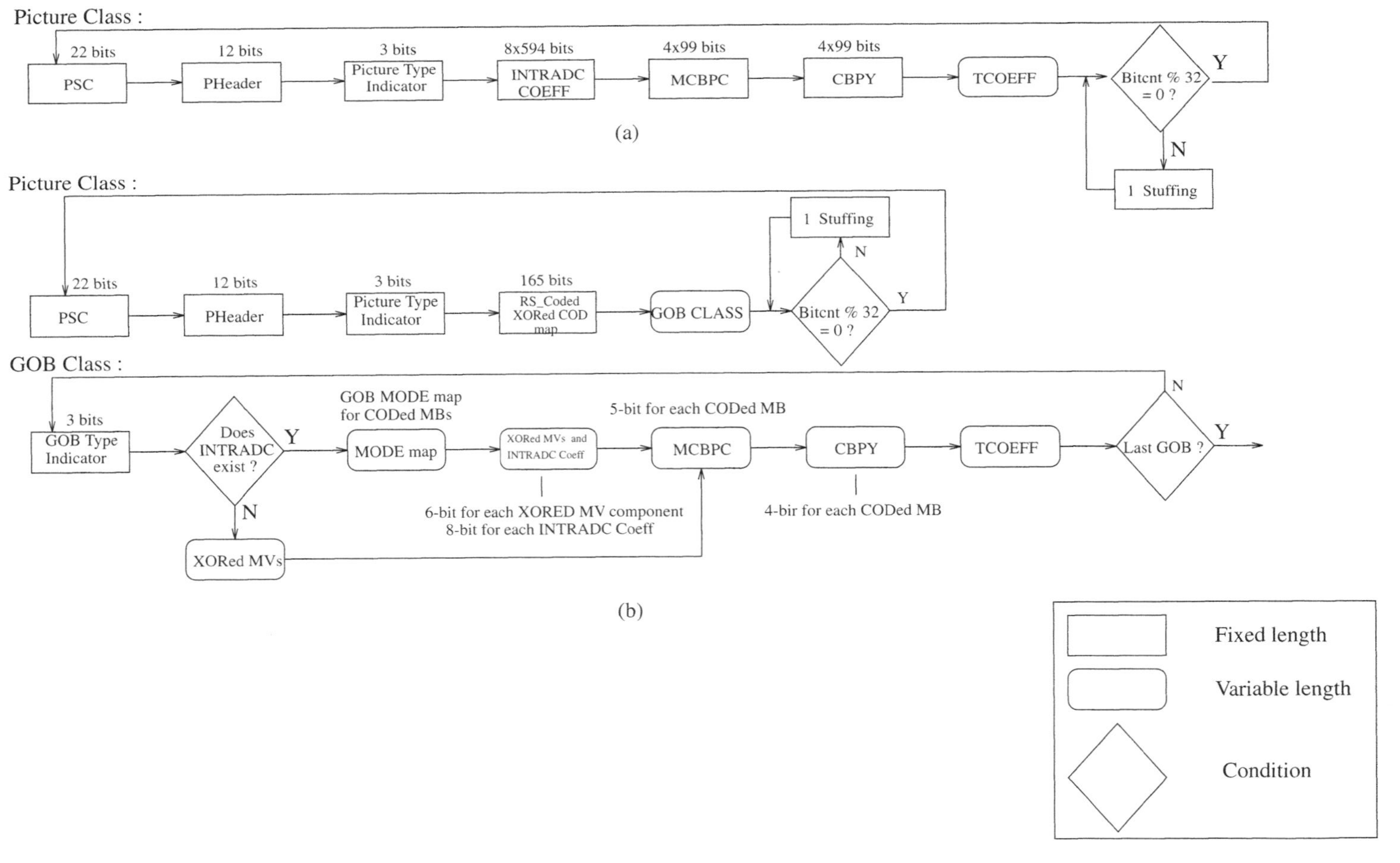

Figure 4.28 Layering structure of H.263/M for resilient video transmissions: (a) picture layer of an I-frame, (b) picture and GOB layers of a P-frame

its synchronisation at the synch word, it resumes its operation in the reverse direction, decoding the segment of the bit stream that was initially skipped in the forward direction. When both the forward and reverse decoding processes have been completed, the decoder uses one of four methods to decide about the portion of the bit stream to discard. These methods are depicted in Figure 4.31.

In Figure 4.31(a), the segment between the position of errors detected in the forward and backward directions is discarded. In (b), only MBs that are free from errors are accepted and all error-damaged MBs are discarded. In (c), MBs that are corrupted are dropped and not used in the decoding process; while in (d), only the

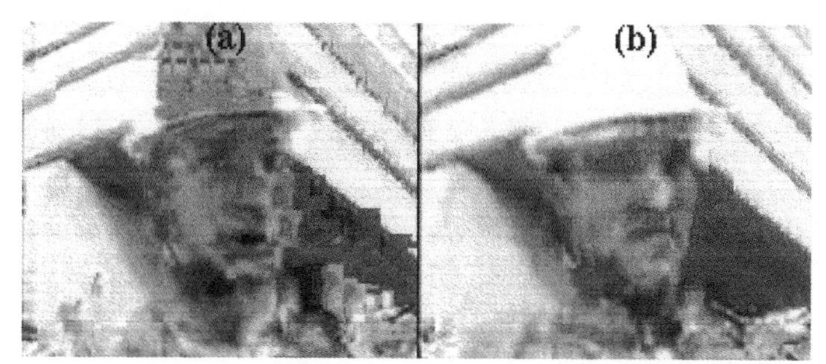

Figure 4.29 Frame 200 of the Foreman sequence encoded at 64 kbit/s and transmitted over a mobile channel with random error distribution and BER = 10^{-4}: (a) H.263, (b) H.263/M

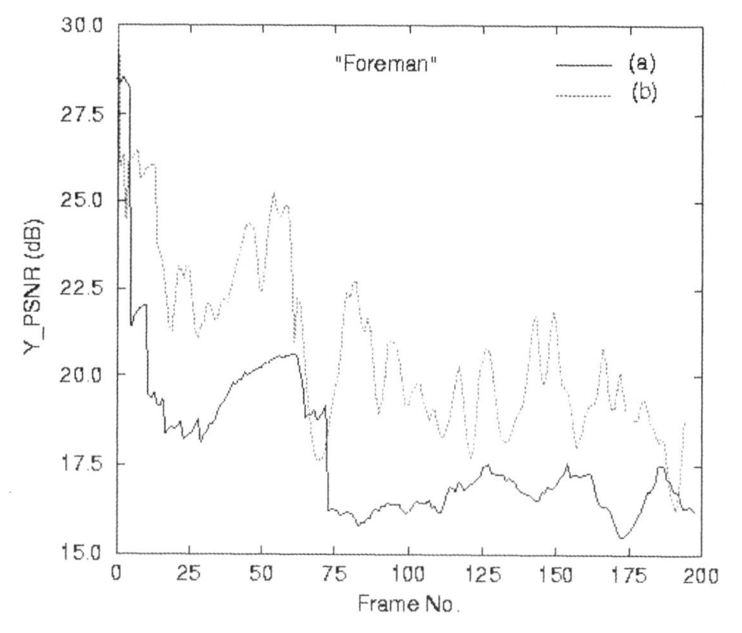

Figure 4.30 PSNR values for 200 frames of the Foreman sequence encoded at 64 kbit/s and transmitted over a mobile channel with random error distribution and BER = 10^{-4}: (a) H.263, (b) H.263/M

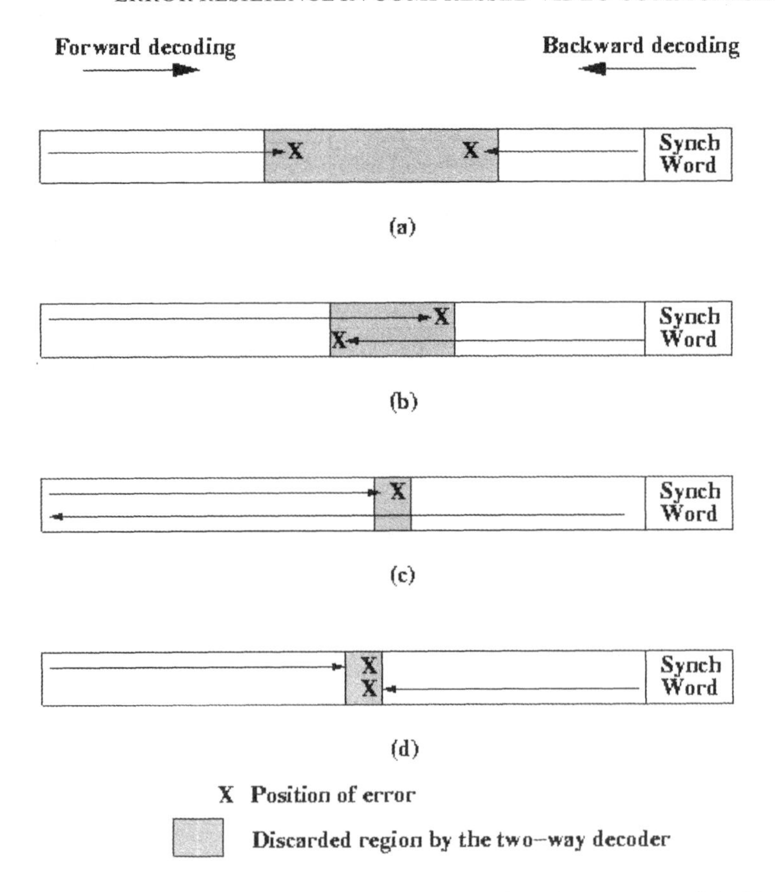

Figure 4.31 Methods used in two-way decoding to determine what segment of the bit stream to discard: (a) separated error points, (b) crossed error points, (c) error detected in only one direction, (d) errors isolated to a single MB

corrupted MB is discarded and the corresponding parameters are not used in any further reconstruction.

To enable two-way decoding, reversible VLC (RVLC) words must be used in the coded stream. As the name implies, reversible words are variable-length codes that can be decoded in both the forward and reverse directions and produce the same output. To generate a set of reversible VLC codes, one of several techniques can be used (Takashima, Wada and Murakami, 1995; Watanabe, 1996). One technique consists of keeping a constant weight (number of ones in a binary sequence) in all the generated codes. For instance, the following two steps can be followed to generate RVLCs with a weight of three, as shown in Table 4.2, where i represents the length of the RVLC code.

- Generate codes of Hamming weight 1 and length $n = 1, 2, 3, \ldots$

Table 4.2 RVLCs with a fixed number of ones per symbol

Step 1		Step 2	
n		i	
		1	0
1	1	3	111
2	01	4	1011
2	10	4	1101
3	001	5	10011
3	010	5	10101
3	100	5	11001
4	0001	6	100011
4	0010	6	100101
4	0100	6	101001
4	1000	6	110001
	⋮		⋮

- Add 1 to both prefix and suffix of each code generated in step 1. Add 0 as the shortest code.

Since the decoder knows the weight of each word, these RVLCs can then be two-way decoded. They can be identified by the number of ones, except for the first code which is the only one that starts with a zero. If the decoder detects three ones in a single word that is not a legitimate RVLC code, it interprets that as an occurrence of error. To produce a table of RVLC words with a lower average length than in Table 4.2, each of the codes is extended with m bits attached to its prefix or suffix. The decoder is then aware that each codeword consists of an RVLC plus a fixed-length prefix or suffix, and therefore is able to decode the newly formed words. For instance, using the extension method of designing an RVLC table, the output RVLC code 111 of Table 4.2 with a 2-bit extension attached to its suffix will yield four other codes, namely 11100, 11101, 11110 and 11111.

Another method to generate a constant weight RVLC table is to use a fixed number of the first symbol of a code. In other words, if the decoder starts a codeword with a zero, it searches a fixed number of zeros in the codeword, and *vice versa* for one. To design an RVLC table with a constant number (3) of the first symbol in a code as in Table 4.3, the following steps are followed:

- Generate codes of Hamming weight of 1 and length $n = 1, 2, 3, \ldots$
- Place 1 on both prefix and suffix of each code generated in step 1. Furthermore, add 0 as the shortest code.
- Add bit-inverse codes. For instance, the bit-inverse code for 010 is 101.

Table 4.3 RVLCs with a fixed number of the starting symbol (either 0 or 1)

Step 1		Step 2	Step 3	
n			i	
			1	0
			3	000
1	1	111	3	111
			4	0100
2	01	1011	4	1011
			4	0010
2	10	1101	4	1101
			5	01100
3	001	10011	5	10011
			5	01010
3	010	10101	5	10101
	⋮	⋮		⋮

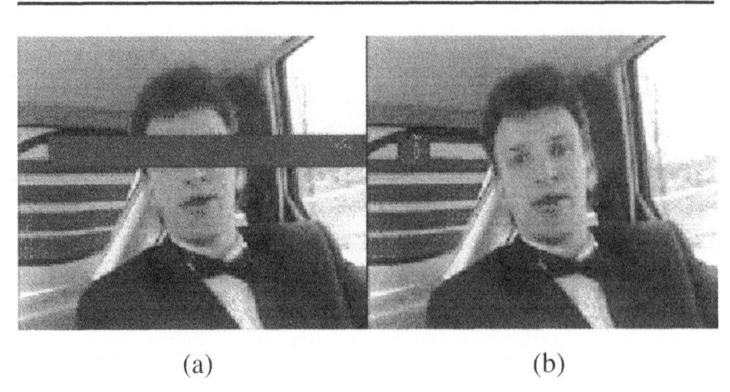

(a) (b)

Figure 4.32 One frame of the Carphone sequence coded with H.263 at 28 kbit/s and 12.5 f/s using: (a) ordinary decoding, (b) two-way decoding

Since the number of the first detected symbol in each code can be identified, these codewords can be two-way decoded. In addition to the constant weight strategy, another methodology utilised in generating RVLC table consists of assigning a fixed number of zeros and ones in each codeword. In this case, a possible set of RVLC codes consists of 01, 10, 0011, 1100, 001011, 000111, 110100, etc. Two-way decodable codes typically result in 2–3 per cent increase in the total bit rate, as compared to optimal Huffman codes, but lead to a substantial improvement in the quality of video when it is subjected to transmission errors. Figure 4.32 shows the subjective quality improvement achieved by two-way decoding applied to Carphone sequence coded stream. A synch word is transmitted once every GOB to allow the decoder to restore synchronisation, upon detection of an error, at the beginning of the next GOB. Using two-way decoding, the decoder is able to recover the bits situated between the corrupted MB and the start code of the following GOB. Figure 4.33 depicts the average PSNR values for the Carphone

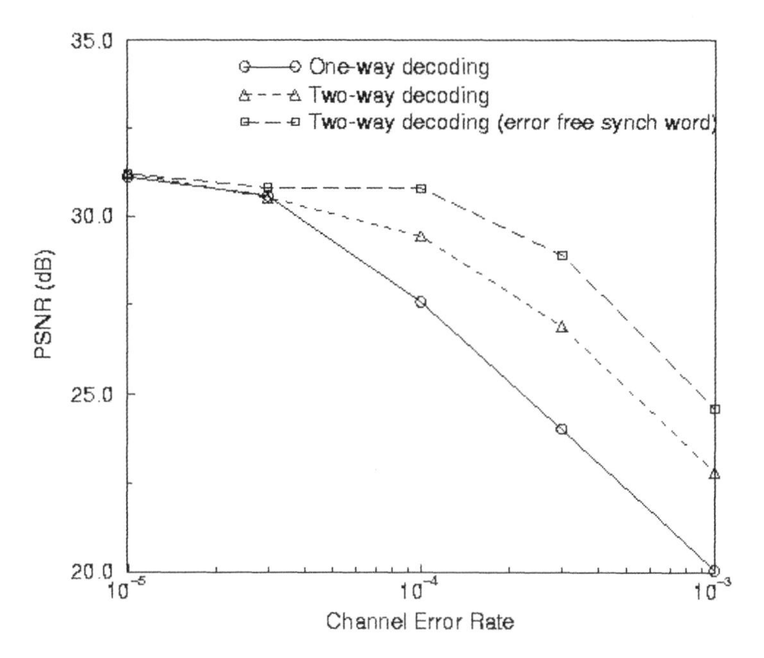

Figure 4.33 Average PSNR values for the Carphone sequence coded at 28 kbit/s and 12.5 f/s using one-way and reversible codes

sequence and demonstrates that the error robustness can be improved significantly when two-way decodable codes are employed. Protecting the synch words also improves the objective quality since they represent the position in the bit stream where the decoder restores synchronisation in case of errors.

As discussed in Section 4.2, the effect of errors on a video bit stream can be bursty when the decoder loses synchronisation, as a single bit error can lead to discarding a large number of bits in the stream. Therefore, the effective error rate is likely to be higher than the actual bit error ratio of the communication link (Eryurtlu *et al.*, 1997). If the bit length of a block of data between two successive synch words is L, then L_{av} could be determined by:

$$L_{av} = \frac{r}{fs} - l_s \tag{4.1}$$

where r is the overall bit rate of the video coder, f is the frame rate, s is the number of synch words per frame and l_s is the length of the synch word. If a synch word is inserted into the bit stream once every GOB, then L_{av} is the average length of a GOB (that corresponds to a row of MBs in QCIF format).

If e_{ch} is the channel BER, then the effective BER e_{eff} in the one-way decoding case, as shown in Figure 4.34, can be derived as follows:

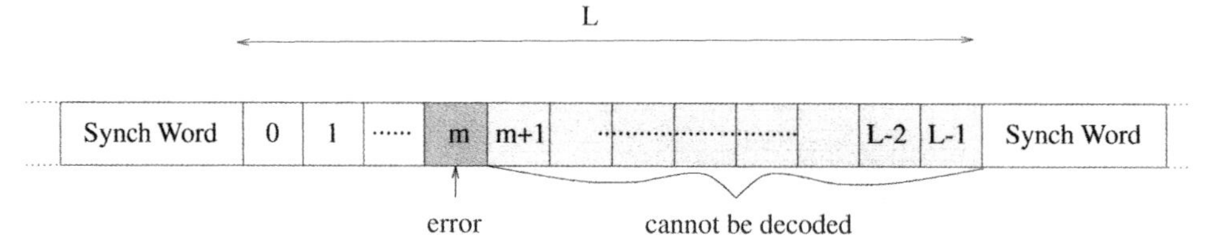

Figure 4.34 One-way decoding of variable-length codes

$$e_{eff} = \sum_{m=0}^{L=1} p_m \frac{L-m}{L} \tag{4.2}$$

where p_m is the probability of having the first corrupted bit at the bit position m and is a function of e_{ch} and m given by:

$$p_m = (1 - e_{ch})^m e_{ch} \tag{4.3}$$

Then Equation 4.2 can be rewritten as:

$$e_{eff} = \frac{e_{ch}}{L} \sum_{m=0}^{L-1} (1 - e_{ch})^m (L - m) \tag{4.4}$$

Figure 4.35 shows the effective error rates produced by the one-way decoding

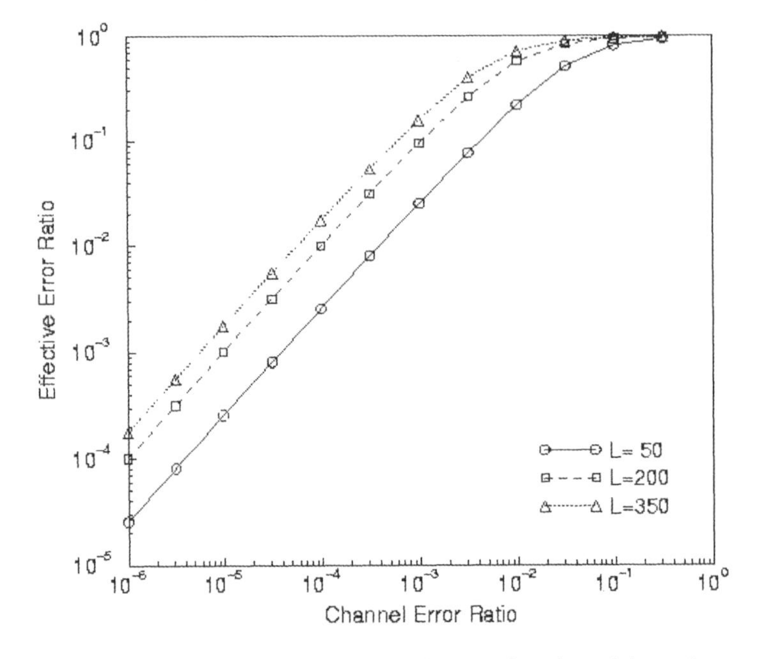

Figure 4.35 Effective error ratio in one-way decoding as a function of channel error ratio and block length for the Carphone sequence coded at 28 kbit/s and 12.5 f/s

algorithm depicted in Figure 4.34 for a range of e_{ch} values and for different values of the block length L. The graphs in Figure 4.35 demonstrate that the effective error ratio produced by the one-way decoder is larger than the channel error ratio. For a given e_{ch}, the effective error ratio increases with larger block sizes, as the number of bits dropped by the decoder in case of errors becomes larger.

However, when two-way decoding is applied, if an error is detected in the forward direction, the decoder searches for the next synch word and then resumes decoding in the reverse direction, as shown in Figure 4.36. Using this technique, more information can be correctly decoded. If no error is detected in the reverse direction then the damage is confined to the corrupted bit only. If an error is detected at position n in the reverse direction, then the two-way decoder discards all bits between positions m and n.

With reference to Figure 4.36, the effective error ratio for two-way decoding can be calculated as follows:

$$e_{eff} = \sum_{m=0}^{L-1} (1 - e_{ch})^m e_{ch} \left((1 - e_{ch})^{L-m-1} + \sum_{n=0}^{L-m-2} (1 - e_{ch})^n e_{ch} \frac{L - m - n}{L} \right) \quad (4.5)$$

$$e_{eff} = (1 - e_{ch})^{L-1} e_{ch} + \frac{e_{ch}^2}{L} \sum_{m=0}^{L-1} \sum_{n=0}^{L-m-2} (1 - e_{ch})^{m+n} (L - (m + n)) \quad (4.6)$$

Figure 4.37 shows that the effective bit error ratio in the two-way decoding case is equal to the channel bit error ratio for a wide range of BER values. However, in the one-way decoding case, the effective error rates are always larger than their corresponding e_{ch} values.

4.11 Error-resilient Entropy Coding (EREC)

Although the two-way decoding algorithm proves efficient in combating the effect of synchronisation loss, the associated reversible VLC words place some unwanted overhead in the coded bit stream. On the other hand, the changed order of transmission adopted in H.263/M does not present compliance with the existing standard algorithm. The error-resilient entropy code is an error-resilience tech-

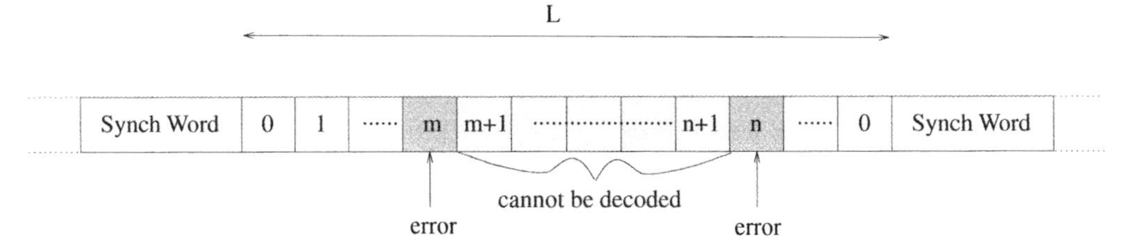

Figure 4.36 Two-way decoding of reversible variable-length codes

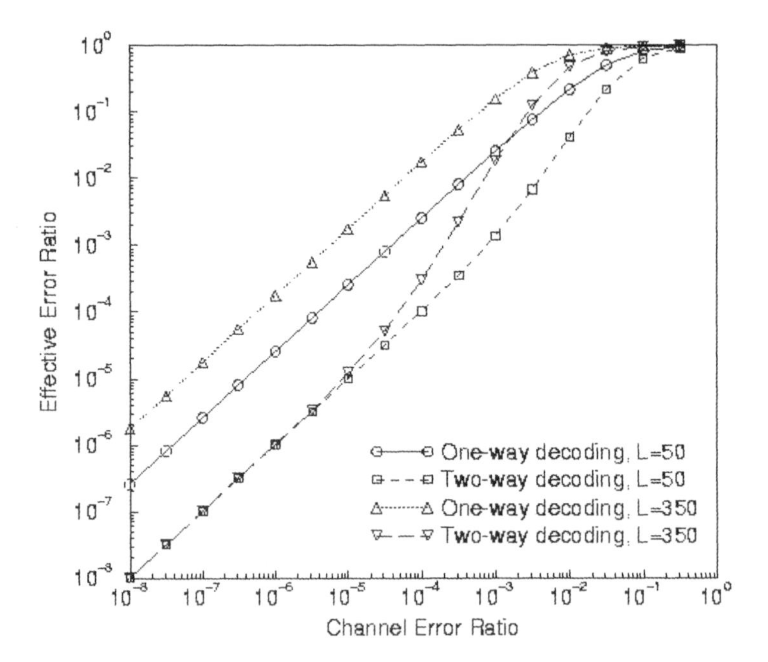

Figure 4.37 Effective bit error ratio in one and two-way decoding as a function of channel error ratio and block length L

nique (Redmill and Kingsbury, 1996) that is capable of offering significant improvement to the error resilience of coded video streams by providing an alternative to the sequential transmission of various video parameters.

The most destructive effect of errors in video communications is due to the loss of synchronisation at the decoder. This is due to the variable-length coding of video parameters and the variable number of parameters between two consecutive synch words. EREC attempts to rearrange the variable-length codes using a fixed-length slotting structure. The variable-length codes are then organised in such a way that the start of a VLC word always coincides with the start of a fixed-length slot. The size of the fixed-length slots is determined by the encoder and sent to the decoder at the beginning of each block of video prior to transmission of the slots. If the decoder detects a bit error in one of the VLC words, the synchronisation can always be recovered at the beginning of the next slot, thereby limiting the damage to the bits contained in the corrupted slot.

In order to arrange VLC words into fixed-length slots, the encoder has first to estimate the total number of bits in a video block. This can be determined by evaluating the sum of the lengths b_i of N VLC words in a video block. The encoder can then select a total data size T that is sufficient to code the N VLC words in EREC format. For this reason, T has to satisfy the following condition:

$$T - \sum_{i=1}^{N} b_i \geq 0 \tag{4.7}$$

The total data size T is then subdivided into N fixed-size slots of length s each. Therefore, the size s of each fixed-length slot is determined by:

$$s = \frac{T}{N} \tag{4.8}$$

EREC is an algorithm that consists of N number of stages whereby a VLC word is placed in each of the fixed-length slots. VLC codes with length equal to the size of a slot ($b_i = s$) will be fully coded from the first stage. VLC codes with length smaller than the actual size of a slot ($b_i < s$) will also be fully coded from the first stage, leaving a number of empty bits in the corresponding slot. VLC codes with length exceeding the size of a slot ($b_i > s$) require two or more stages to be fully coded by splitting their bits into two or more slots. For instance, if the block of VLC words consists of four MV component values then EREC will involve a four-stage procedure. If the four MV components have corresponding lengths b_i equal to 11, 7, 2, and 4 bits respectively, then EREC will produce four equal length slots of six bits each. The reorganisation process of the four MV components is depicted in Figure 4.38.

At the decoder, the original VLC words are recovered using a similar N-stage algorithm that extracts bits from slots and associate them with their corresponding VLC words. Therefore, in error-free conditions, the decoder is able to correctly reconstruct all the VLC words from the received slots. However, in the case of transmission errors, the decoder could lose synchronisation on decoding a corrupted fixed slot. In an attempt to recover synchronisation at the beginning of the next slot, the decoder skips all the embedded bits in the slot, including those bits that might belong to other VLC words in the video block. The affected bits will therefore belong to the VLC word that was placed in that particular slot at stage 1 of the reorganisation process and other bits that were added to the slot at later stages. This implies that data placed at later stages of the reorganisation process will suffer from error propagation more than data placed in the earlier stages. However, EREC provides a considerable improvement to perceptual video quality by enabling the decoder to maintain its synchronisation upon detection of a bit error. This is demonstrated in Figure 4.39 which shows the subjective quality improvement achieved by the EREC error resilience technique when applied to the VLC words of an H.263 coded stream. It is evident that EREC eliminates the quality degradation caused by a block loss of synchronisation.

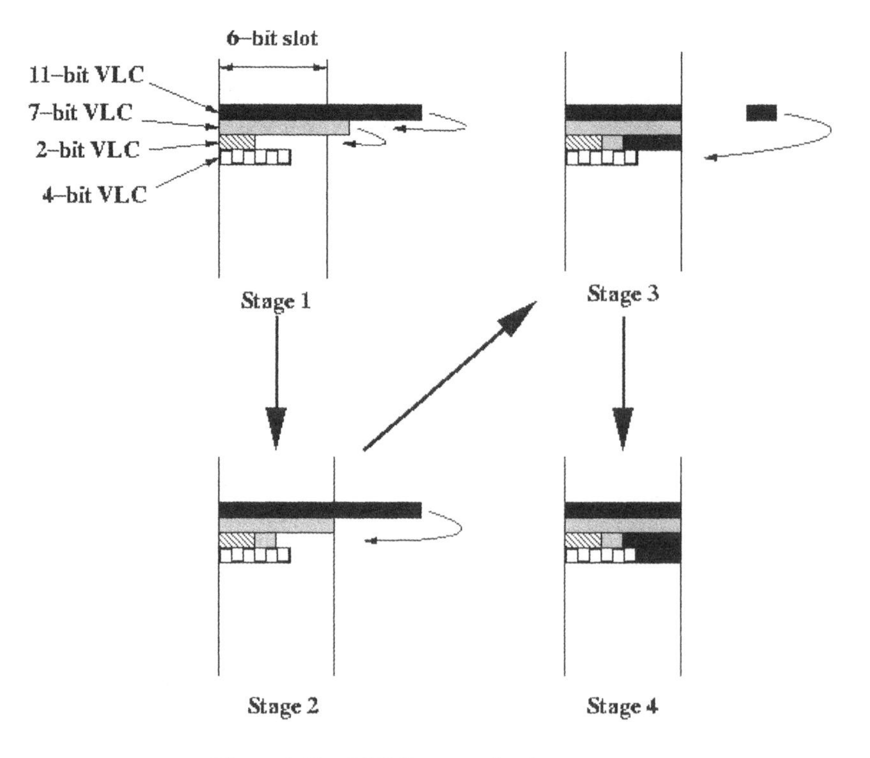

Figure 4.38 EREC reorganisation process

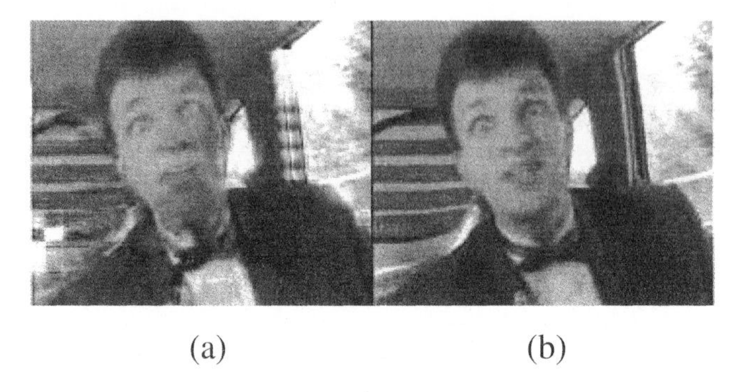

(a) (b)

Figure 4.39 Frame 155 of the Carphone sequence coded with H.263 at 80 kbit/s and 25 f/s and transmitted over a channel with BER = 0.1 per cent: (a) without error resilience, (b) with EREC applied on VLC words in the bit stream

4.12 Combined Error Resilience Schemes

In addition to the error resilience techniques presented individually above, a combination of techniques could also be used in a single video coding algorithm for optimal error resilience results. For instance, using the data partitioning scheme of MPEG-4 that produces the video packet structure described in Section

4.4 and illustrated in Figure 4.7, a variety of options arise for the resilience technique we ought to use for a particular segment of the packet in order to obtain an optimal video quality. For instance, the first partition in the packet consists of all the encoded shape and motion data of the contained MBs in addition to the corresponding administrative data such as the COD flags and MCBPC. The second partition contains the variable-length DCT data and some standard-defined control data such as CBPY and the differential quantisation step size value (DQUANT) (Talluri, 1998). To allow the decoder to automatically resynchronise on the occurrence of an error in the more important first partition, EREC is used to place the encoded motion vectors in fixed-length slots, as described in Section 4.11. The important header data of each video frame is duplicated in the video packet using the HEC (Header Extension Code) flag, as described in Section 4.6, to reduce the number of discarded frames in the video sequence. Furthermore, the TCOEFF coefficients (DCT data) of the second partition are coded using RVLC words to allow backward decoding and reduce the number of discarded DCT data resulting from loss of synchronisation. The motion vectors of the first partition could also be coded using two-way decodable codes, but EREC is preferable to cancel out the overhead introduced by reversible VLC words. Figure 4.40 shows the subjective quality improvement achieved by using a combination of error-resilience techniques on the MPEG-4 coded Suzie sequence at 110 kbit/s and 25 f/s. Firstly, data partitioning is employed to separate the coded motion vectors (first partition) of a frame from its texture data (second partition). EREC is applied to the motion data of the first partition while the DCT data is coded with reversible VLC words to allow two-way decoding of DCT coefficients when the decoder flags an error in the bit stream of the second partition. Moreover, a one-half rate Turbo coder is used to protect the header data of the first partition. Figure 4.41 shows the comparison between the objective quality that can be obtained with the MPEG-4 error-resilience tools enabled and the non-error-

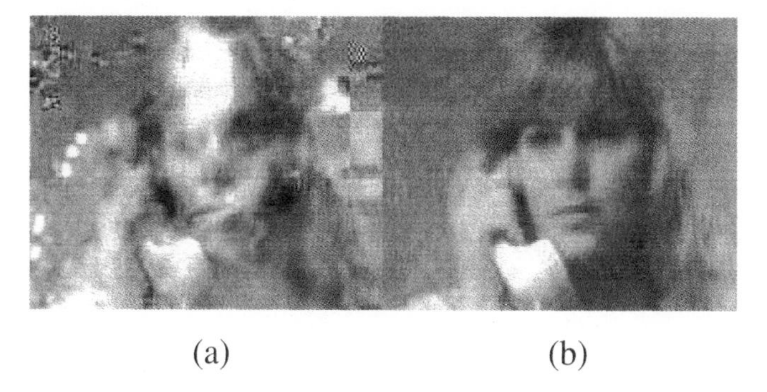

(a) (b)

Figure 4.40 Frame 124 of the Suzie sequence coded at 110 kbit/s and 25 f/s transmitted over a channel with BER $= 10^{-4}$: (a) non-error-resilient, (b) data partitioning + one-half rate Turbo coded header data of first partition + EREC on MV in first partition + two-way decoding and RVLC of DCT data in second partition

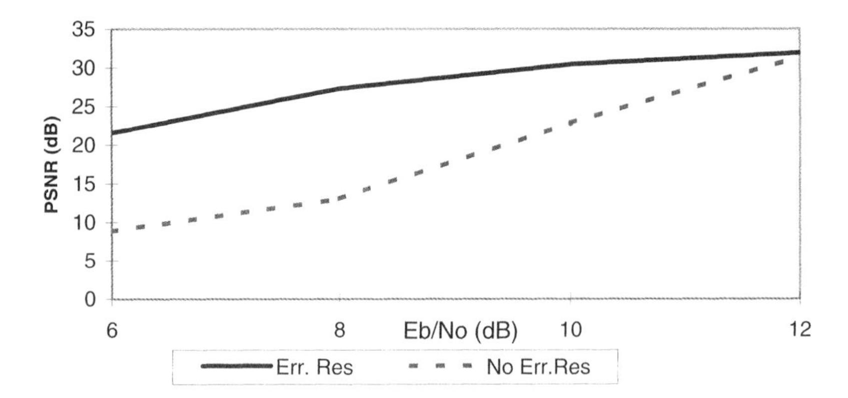

Figure 4.41 Average PSNR values for the Suzie sequence encoded with MPEG-4 at 32 kbit/s and 10 f/s with and without the combined error-resilience tools of the MPEG-4 standard

resilient video quality. It can be seen that under all but the very best channel conditions, the error-resilience tools provide a dramatic improvement, often making an otherwise unintelligible scene acceptable for most videoconferencing applications.

On the other hand, H.263+ specifies a set of error resilience algorithms that could also be employed in combinations to optimise the error resilience capabilities of compressed video for a given network and a given application (Wenger *et al.*, 1998). These error-resilience techniques are described in the annexes that are associated with the definition of the core H.263 standard. Annex H, for instance, consists of the forward error correction (FEC) mode in which the coded video stream is subdivided into a set of fixed-length (492 bits) FEC frames, each of which is protected by a 19-bit BCH checksum. These checksums can detect two bit errors and correct single bit errors in each FEC frame for an overhead penalty of 4 per cent increase in bit rate. Annex K enables the slice structure mode in which the GOB structure adopted by the baseline coder is replaced by the slice structure in processing the content of each video frame. A slice contains a number of MBs that are coded independently of the content of other slices. This implies that no motion prediction of MVs is then allowed across the slice boundaries. Although the independent slice structure impairs the error-free quality due to increased overhead (coding of a slice header at the beginning of each independent slice) and inefficient motion prediction, particularly for vertical motion (no dependencies between the video content of adjacent slices), this mode significantly enhances the video quality under error conditions. Annex R allows the independent segment decoding in which each picture is divided into a number of independently processed segments with predefined boundaries. A segment can be used to represent a detected visual object in the spatial domain of a video frame. In this case, the functionality of this mode becomes similar to that of the object-oriented VOP structure in the MPEG-4 standard. Therefore, an error-damaged segment could then be isolated from the rest of the image to avoid the spatial propagation of errors between the predefined segments of a video sequence. Furthermore, H.263+ adopts Annex N on reference picture selection to enhance the error resilience of the compressed stream by enabling more than just one reference frame in the motion compensation process, as described in the next section. On the other hand, H.263++ specifies a set of three optional error resilience techniques in Annexes U, V and W of the H.263 video coding standard (Sullivan, 2000). Annex U is the enhanced version of Annex N on reference picture selection, and is described in the next section, while Annex V specifies the optional data partitioned slice mode and Annex W describes the optional additional supplemental enhancement information. Annex V is a combination of the data partitioning scheme described in Section 4.4 and the slice structure mode specified by annex K of H.263+, as described above. The header information, motion data and DCT coefficients of each slice are sent in partitions consisting of an integer number of MBs and separated by markers which allow for resynchronisation at the end of the partition in which the error occurs. In this mode, the header data consists of RVLC words that combine both COD and MCBPC information of all the corresponding MBs in the partition. The motion data partition consists of RVLC codewords that encode the difference between the motion vectors and the motion vector predictors. Since the slice consists of a variable number of MBs, a motion vector predictor can no longer be the median value of three MVs of neighbouring MBs. However, the motion vector prediction is carried out using a single prediction thread for all the MVs in the slice, as illustrated in Figure 4.42.

The third partition of the slice contains CBPY, DQUANT and DCT coefficients coded using reversible VLC codewords that correspond to data pertaining to the embodied MBs. On the other hand, Annex W allows the encoder to send some additional information to the decoder for enhanced performance. This additional information includes new frame type (FTYPE) values to indicate the transmission of fixed-point inverse DCT (IDCT) coefficients and/or picture messages. The fixed-point IDCT indicates that a particular IDCT approximation is used in the construction of the bit stream. To control the accumulation of errors due to mismatched IDCT values at the encoder and decoder, some MBs are forced to INTRA coding mode (forced updating), at least once every 132 times, when coefficients are transmitted. However, if the decoder is capable of the fixed-point IDCT and the encoder indicates the fixed-point IDCT function type in the bit stream, then the forced updating requirement is removed and the frequency of

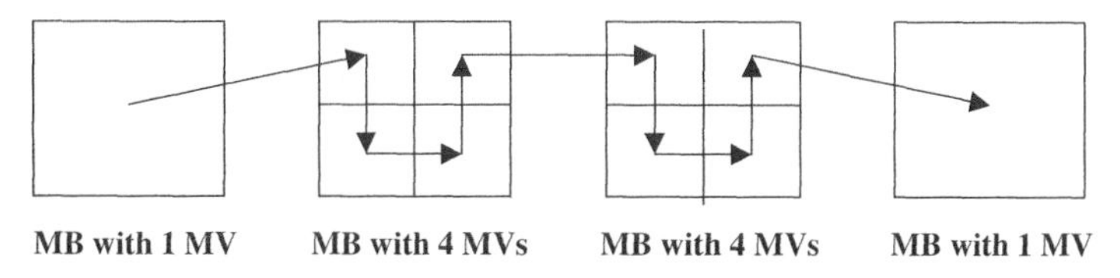

Figure 4.42 Single thread motion vector prediction in H.263++

INTRA coding is unregulated. The picture message is indicated by the message type (MTYPE) that identifies a set of optional fields such as the video copyright owner, the video description text, the current and previous pictures header repetition, the picture number, the spare reference pictures and others.

4.13 Error Resilience Based on Reference Picture Selection

The reference picture selection (RPS) mode of H.263+ is an error-resilience technique that is similar to the NEWPRED approach suggested as a part of the MPEG-4 suite of error-resilience strategies. Instead of coding the corrupted regions of a video scene in INTRA mode, RPS helps stop the propagation of errors by allowing the encoder to select one or a combination of several error-free previously decoded frames as a reference picture for prediction. It is also possible to apply the reference picture selection to individual segments within the frame rather than full pictures. However, to do that the decoder must be informed of the reference picture that was selected to predict a particular segment of frame. Therefore, the temporal reference (TR) of the reference picture to be used is conveyed in the picture or segment header to indicate which of several reference pictures the decoder ought to use to decode a particular segment or frame, respectively. RPS could be used in conjunction with (Farber and Girod, 1999; Wada, 1989) or without a feedback channel (Wenger, 1997). Feedback channel messages contain positive or negative (or sometimes both) acknowledgments as to the decoding of a particular frame as well as the corresponding TR. Using this information, the encoder can keep track of the last correctly decoded segment and/or frame, and make the selection of the reference picture accordingly. The frame that is reported to have been in error is then discarded from any further prediction process. If the encoder learns that the last correctly decoded picture is not in the encoder's buffer then it has to react by sending a fully INTRA coded frame. For this reason, both the encoder and decoder have to buffer a number of the most recent reference pictures in order to keep their picture memories matched up. During error-free transmission, the operation of the encoder is not altered and the previously decoded image is used as a reference picture.

Therefore, when a feedback channel is used, RPS can be operated in two different modes. In the ACK mode, the feedback messages only contain information about correctly received image content. In this case, the decoder acknowledges the receipt of correctly decoded pictures and the encoder uses only acknowledged image content as a reference for prediction, as shown in Figure 4.43 for a buffer size of three frames. If the round trip delay is longer than the encoded picture processing time interval, the encoder has to use a reference frame that is further back in time (Figure 4.43 shows an overall delay of two frames). Note in Figure 4.43 that the last ACKed picture is never removed from the encoder frame buffer, while the decoder always keeps the last referenced frame in its buffer.

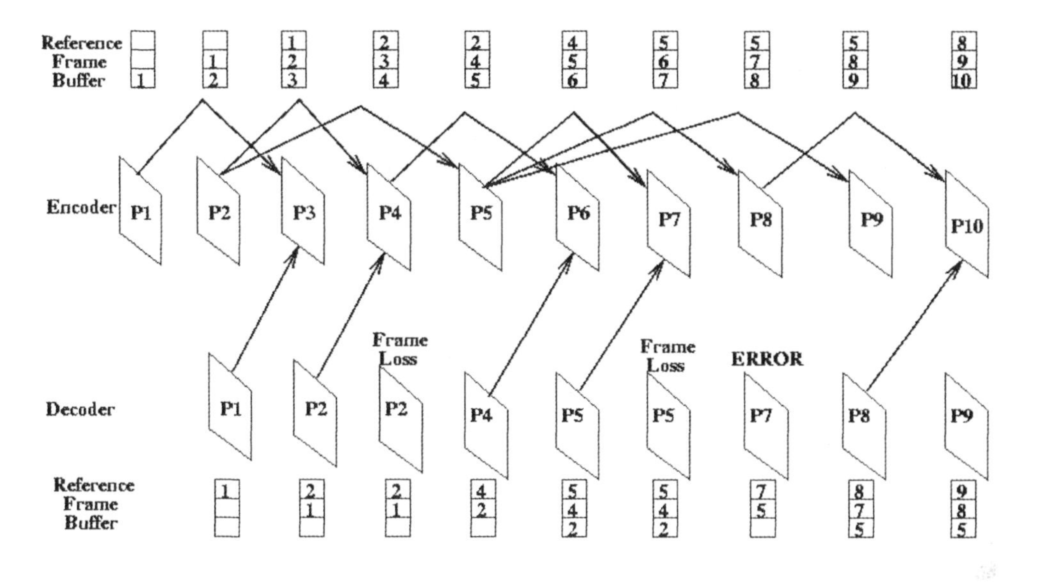

Figure 4.43 RPS ACK mode with buffer size of three and an overall delay of two frames

Obviously, the RPS error-resilience scheme leads to reduced error-free coding efficiency, especially in heavy motion scenes that result in longer motion vectors and possibly large amount of disparity between non-adjacent frames. However, in the case of transmission errors, the picture quality suffers from small fluctuations due to RPS processing and round-trip transmission latency. Figures 4.44 and 4.45 show the subjective and objective quality improvements, respectively, accomplished by applying the RPS technique operating in the ACK mode with a feedback channel of a two-frame delay (80 ms). The sharp sudden drop of PSNR values in Figure 4.45 is due to full frame losses. For a two-frame delay, the quality degradation persists for only a short interval of time before it is restored back to its error-free level after updating the motion-compensated picture memory with a successfully decoded reference picture. However, in the non-error-resilient case, the quality deterioration is noticeable for the remaining sequence time until the scene is refreshed with a fully INTRA coded frame. The RPS technique might sometimes involve a slight increase in the bit rate and hence a reduction in the compression efficiency due to the sub-optimal motion prediction resulting from estimating motion with respect to older video frames.

The second mode is called the NACK mode, and it consists of the decoder sending onto the feedback channel negative acknowledgements only for erroneously decoded pictures. NACKs report the temporal and spatial location of the video content that could not be decoded successfully. Based on the information of a NACK message, the damaged video information (frame or segment) is no longer

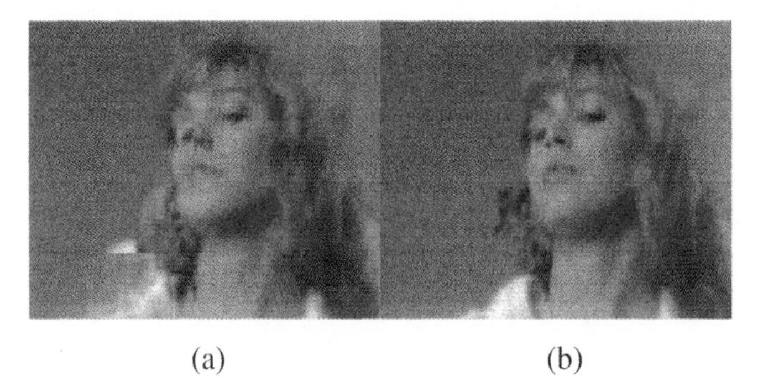

Figure 4.44 Fifty-sixth frame of the Suzie sequence coded at 25 f/s, after frame 55 was fully dropped during transmission: (a) non-error-resilient case at 57 kbit/s, (b) RPS ACK mode with a delay of 2 frames at 60.5 kbit/s

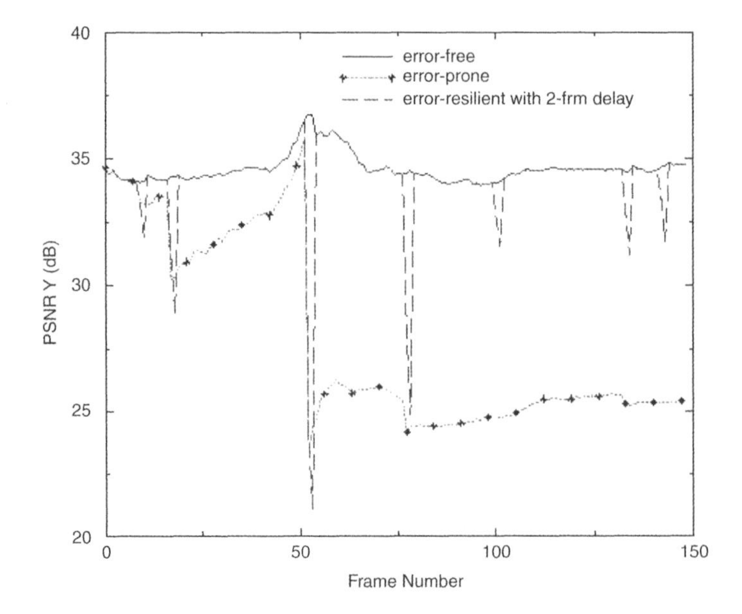

Figure 4.45 PSNR values for 150 frames of the Suzie sequence for the error-free, non-error-resilient and error-resilient transmission using RPS ACK mode with a two-frame delay back channel

selected for prediction and dropped off the reference frame buffer in both the encoder and decoder. In some cases, mode switching between ACK and NACK modes is employed for superior transmission performance (Fukunaga, Nakai and Inoue, 1996; Tomita, Kimura and Ichikawa, 1997). In addition to that, H.263++ specifies in Annex U (Sullivan, 2000) an enhanced optional version of RPS mode, namely ERPS, to provide a better coding efficiency and improved error-resilience performance. This mode extends the motion compensation at the encoder through prediction from multiple frames. The multi-frame motion-compensation process of this annex improves the coding efficiency of the video encoder and renders the coded stream more robust to random transmission errors. Therefore, the selection of a reference picture is performed on an MB basis, as opposed to a frame basis in the ordinary RPS mode. ERPS allows a reduction of the memory required to store the multiple reference pictures of RPS by partitioning each reference picture into smaller rectangular units called sub-pictures. ERPS provides the video encoder not only with the ability to use multiple references for forward prediction, but also to use more than one reference picture for backward prediction (B-frames) as well. Similarly to RPS, ERPS makes use of backward channel messages sent by the decoder to inform the encoder which pictures or parts of the pictures have been correctly or incorrectly decoded.

Therefore, the backward channel RPS techniques provide long-term memory motion-compensated prediction and thus achieve a useful advance in error-resilience schemes (Wiegand *et al.*, 2000; Wiegand *et al.*, 2001), especially in transmission environments that have low propagation delays and transit times and for point-to-point communication scenarios. Furthermore, the perceptual quality of decoded video in error-prone environments could also be further enhanced when RPS is combined with some decoder-based error concealment algorithms, as described in Section 4.3. However, in multicast environments such as the multicast backbone over the Internet where a huge number of sending and receiving hosts could all be active at the same time, feedback messages incur a significant overhead and could take the network to a state of 'network implosion'. In order to circumvent this problem, the reference picture selection is then used without a feedback channel in a mode known as video redundancy coding (VRC) (Wenger *et al.*, 1998; Wenger, 1997).

VRC divides the coded video stream into a number (two or more) of separate streams known as threads. Each original video frame is assigned to one thread in a round robin sequence, and frames that belong to a thread are coded independently of frames in the other threads. The frame rate in each thread which is then reduced to a fraction of the overall frame rate, decreases with the increasing number of threads. The frame rate in one thread is then equal to the overall frame rate divided by the number of threads defined by VRC. Obviously, the error-free video quality deteriorates due to the coding penalty incurred by the longer motion vectors and the larger changes between the adjacent predicted frames in each resulting thread. After a predefined number of frames per thread, all threads converge to a single sync frame that constitutes the starting element of a new chain of threads in the sequence. Figure 4.46 delineates VRC with three threads and two pictures per thread.

When one of the threads is subjected to transmission errors or information loss, the other threads remain intact because each thread is independently predicted. In this case, the damaged thread can still be possibly decoded for a graceful degradation of the perceptual quality until the next sync frame. Alternatively, the decoding of the corrupted thread is ceased and the next sync frame is predicted only from

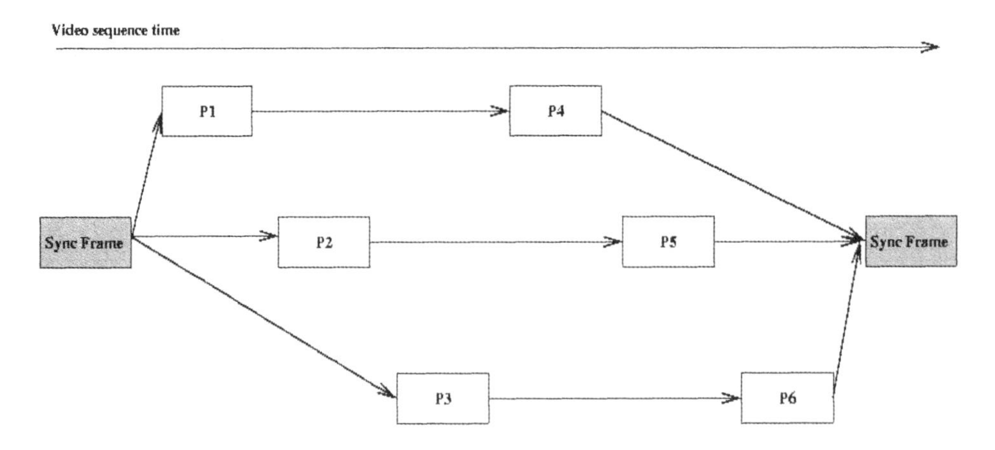

Figure 4.46 Video redundancy coding with three threads and two frames per thread

error-free threads, thereby preventing the propagation of errors at the expense of a lower frame rate. For short threads, the effects of errors (either graceful degradation of quality with same frame rate or slower frame rate with good spatial quality) would only persist for reasonably short times until a new chain of threads is started. The worst case occurs when all the threads between two consecutive sync frames are damaged, in which case the picture quality would suffer from annoying artefacts until then next error-free INTRA frame is correctly decoded, as has been the case in INTRA refresh schemes described in Section 4.7. Thereafter, a new error-free chain of threads can be predicted from the last INTRA coded frame.

4.14 Conclusions

Error control strategies are used to mitigate the effects of transmission errors on the decoded video quality. A wide variety of techniques are used for this purpose and some of them have already been incorporated in some levels and profiles of standard video compression techniques such as MPEG-4 and H.263+. The effectiveness of these techniques in terms of the achieved quality improvement is the most obvious criterion to use for assessing their performance. However, in addition to that, there are a number of factors that are equally important for the comparison and fair performance evaluation of error control schemes. A critical factor of the evaluation is the delay that an error control scheme incurs on the two-way and multipoint video communications. Furthermore, the bit rate overhead imposed by error-resilience techniques on both the application and transport levels is an important factor of evaluation in both error-free and error-prone conditions. The bit rate overhead does not apply on the post-processing error concealment techniques which operate at the decoder without relying on additional information from the encoder. Therefore, they are referred to as zero-redundancy error control schemes. The last criterion of assessment is the processing complexity which is a design issue for any system implementation. The priority of these criteria depends on the underlying application and the networking platform on which it runs.

Channel errors have a very detrimental effect on the perceptual video quality. The effects of channel errors on the video quality can take various forms. However, the most disastrous channel errors are those which lead to a loss of synchronisation at the decoder. Due to the temporal and spatial predictions employed in video coding algorithms, errors can propagate very quickly within the video sequence in both time and space. In order to limit the propagation of errors and mitigate their effects, error control schemes must be used to fulfil the user's satisfaction and meet his expectations. The simplest error control techniques are those which attempt to conceal the annoying artefacts that appear in the video content when subjected to transmission errors. Those techniques are decoder-based and thus impose no overhead on the compressed video stream. Another class of error control techniques comprises the schemes which attempt to halt the accumulation of errors in both space and time. The most prominent technique of this class is the INTRA refresh which sends an INTRA frame at regular intervals of time. An improvement of this technique is the adaptive INTRA refresh (AIR) that sends a predefined number of motion active MBs in INTRA mode once every frame. AIR achieves a more consistent video quality than the INTRA refresh since it updates the video content at more regular intervals by refreshing the most active part of the video scene.

One further class of error control schemes endeavours to regain synchronisation upon detection of errors. Examples of this class of error-resilience scheme are EREC and two-way decoding. EREC places the variable-length codes of the video bit stream into predefined fixed-length slots in such a way that the video decoder could then resynchronise at the beginning of the next slot when it fails to complete the decoding of the current VLC word. This leads to reducing the number of discarded bits caused by the loss of synchronisation. Two-way decoding has proved to be a very efficient algorithm for minimising the effective error rates of coded video streams when they are hit by transmission errors. Two-way decoding enables the decoder to decode the bit stream in the reverse direction to salvage all or part of the bit stream which is skipped in the forward direction after an error is detected. To enable the decoder to decode in the reverse direction, reversible codewords are used in the bit stream. Several ways have been discussed in this chapter in which reversible codewords could be generated, the most widely used of which are those that have the same number of the starting symbol (either 0 or 1).

An effective standard-compliant error-resilience technique used in H.263+, namely reference picture selection, also provides satisfactory results when used either with or without a reverse channel. An enhanced version of this technique is also specified in annex U of H.263++. When it is used with backward channel signalling, RPS can operate in either ACK or NACK mode. Without a back channel, VRC is used to send the video frames in a number of independent threads and robust sync frames. When one of the threads is damaged by errors, the sync frames could still be predicted with or without the damaged thread. Including the corrupted thread in the prediction of the next sync frame leads to a graceful degradation of quality, whereas excluding it results in no spatial quality degradation but leads to dropping the frame rate. Only when all threads are damaged would the quality suffer from annoying artefacts, until the next error-free INTRA frame is decoded. Throughout this chapter, a wide variety of error-resilience and concealment techniques have been discussed and analysed. Their performance evaluation has been illustrated with the support of both subjective and objective methods. Moreover, a combination of these error-resilience schemes has been examined, where appropriate (such as in H.263/M for instance), for the optimal error-resilience performance of video compression algorithms operating in error-prone communication environments.

4.15 References

Chhu, W. J., and Leou, J. J., Detection and concealment of transmission errors in H.261 images, *IEEE Transactions on Circuits, Systems, Video Technology*, **8**, No. 1, 74–84, 1998.

Cote, G., and Kossentini, F., Optimal Intra coding of blocks for robust H.263 video communication over the Internet, *EUROSIP Image Communication*, Special issue on real-time video over the Internet, 1999.

Dogan, S., Sadka, A. H., and Kondoz, A. M., Video transmission over mobile satellite systems, *International Journal of Satellite Communications*, **18**, No. 3, 185–205, June 2000.

Eryurtlu, F., Sadka, A. H., Kondoz, A. M., and Evans, B. G., Error robustness improvement of video codecs with two-way decodable codes, *IEE Electronic Letters*, **33**, No. 1, 41–43, Jan. 1997.

Farber, N., Steinbach, E., and Girod, B., Robust H.263 compatible video transmission over wireless channels, *Proc. PCS*, 575–578, Melbourne, Australia, Mar. 1996.

Farber, N., and Girod, B., Feedback-based error control for mobile video transmission, *Proc. of the IEEE*, Special Issue on Video for Mobile Multimedia, 1999.

Fukunaga, S., Nakai, T., and Inoue, H., Error resilient video coding by dynamic replacing of reference pictures, *Proc. GLOBECOM'96*, Nov. 1996.

Guillemot, C., Christ, P., Wesner, S., and Klemets, A., RTP payload format for MPEG-4 with scaleable and flexible error resiliency, IETF Internet Draft draft-guillemot-genrtp-01.txt, June 25 1999.

Hagenauer, J., Rate-compatible punctured convolutional codes (RCPC Codes) and their applications, *IEEE Transactions on Communications*, **36**, No. 4, 389–400, Apr. 1988.

Horn, U., Stuhlmuller, K. W., Link, M., and Girod, B., Robust internet video transmission based on scalable coding and unequal error protection, *Image Communications*, Special Issue on Real-time Video over the Internet, 1999.

ISO/IEC/JTC/SC29/WG11, E4: Core experiments on error resilient methods based on back channel signalling and FEC, July 1996.

ISO/IEC JTC1/SC29/WG11, E2: Error resilient core experiments on hierarchical structures, NTT's experiments on unequal error protection MPEG97, Mar. 1997.

Jung, H. S., Kim, R. C., and Lee, S. U., On the robust transmission technique for H.263 video data stream over wireless networks, *IEEE International Conference on Image Processing*, 463–466, 1998.

Kieu, L. H., and Ngan, K. N., Cell-loss concealment techniques for layered video codecs in an ATM network, *IEEE Trans. Image Processing*, **3**, 666–677, Sept. 1994.

Kim, C. S., *et al.*, Robust transmission of video sequence over noisy channel using parity-check motion vector, *IEEE Transactions on Circuits and Systems for Video Technology*, Oct. 1999.

Lam, W. M., and Reibman, A. R., An error concealment algorithm for images subject to channel errors, *IEEE Trans. on Image Processing*, **4**, No. 5, 533–542, May 1995.

Lam, W. M., and Reibman, A. R., Self-synchronisation variable-length codes for image transmission, *Proc. of the International Conferences on Acoustics, Speech and Signal Processing ICASSP*, **III**, 470–480, 1992.

Lavington, S., Dewhurst, N., and Ghanbari, M., The performance of layered video over an IP network, *Proc. of the Packet Video Workshop 2000*, Sardinia, Italy, May 2000.

Liu, H., and El Zarki, M., Performance of H.263 video transmission over wireless channels using hybrid ARQ, *IEEE Journal on Selected Areas in Communications*, **15**, No. 9, Dec. 1997.

Narula, A., and Lim, J. S., Error concealment techniques for an all-digital high-definition television system, *Proc. SPIE Conference on Visual Communication Image Processing*, Cambridge, MA, 304–315, 1993.

Parthsarathy, V., Modestino, J. W., and Vastola, K. S., Design of a transport coding scheme for high-quality video over ATM networks, *IEEE Trans. Circuits and Systems for Video Technology*, **7**, 358–376, Apr. 1997.

Peng, Z., *et al.*, On the trade-off between source and channel coding rates for image transmission, *Proc. of the IEEE International Conference on Image Processing*, 118–121, 1998.

Rabiner, W., Budagavi, M., and Talluri, R., Proposed extensions to DMIF for supporting unequal error protection of MPEG-4 video over H.324 mobile networks, ISO/IEC JTC1/SC29/WG11, m4135, Atlantic City, Oct. 1998.

Redmill, D. W., *et al.*, Error-resilient image and video coding for wireless communication system, *IEE Electronics and Communication Engineering Journal*, 181–190, Aug. 1998.

Redmill, D. W., and Kingsbury, N. G., The EREC: an error-resilient technique for coding variable-length blocks of data, *IEEE Transactions on Image Processing*, **5**, No. 4, 565–574, Apr. 1996.

Sadka, A. H., Eryurtlu, F., and Kondoz, A. M., Improved performance H.263 under erroneous transmission conditions, *IEE Electronics Letters*, **33**, No. 2, 122–123, Jan. 1997.

Sadka, A. H., Eryurtlu, F., and Kondoz, A. M., Error resilience improvement for block-transform video coders, *IEE Proceedings on Visual Image Signal Processing*, **144**, No. 6, Dec. 1997.

Salama, P., Shroff, N. B., and Delp, E. J., Error concealment in encoded video streams, *Signal recovery techniques for image and video compression and transmission*, edited by Galatsanos and Katsaggelos, Kluwer Academic Publishers, 199–233, 1998.

Shirani, S., Kossentini, F., and Ward, R., A concealment method for video communications in an error prone environment, *IEEE Journal on Selected Areas in Communications*, 1998.

Shirani, S., Erol, B., and Kossentini, F., Error concealment for MPEG-4 video communication in an error prone environment, *Proc. of ICASSP2000*, **4**, 2107–2110, Istanbul, Turkey, June 2000.

Soares, L. D., and Pereira, F., An alternative to the MPEG-4 object-based error resilient video syntax, *Proc. of the IEEE International Conference on Image Processing*, 467–471, 1998.

Steinbach, E., Färber, N., and Girod, B., Standard compatible extension of H.263 for robust video transmission in mobile environments, *IEEE Trans. Circuits and Systems for Video Technology*, **7**, 872–881, Dec. 1997.

Sullivan, G., Rapporteur for Q.15/16 – Draft for H.263++, Annexes U, V and W to Recommendation H.263, ITU Telecommunication Standardisation Sector, Nov. 2000.

Takashima, Y., Wada, M., and Murakami, H., Reversible variable length codes, *IEEE Trans. on Communications*, **43**, No. 2, Part 3, 158–162, Feb./Mar./Apr. 1995.

Talluri, R., Error resilient video coding in the MPEG-4 standard, *IEEE Communications Magazine*, 112–119, June 1998.

Tomita, Y., Kimura, T., and Ichikawa, T., Error resilient modified Inter-Frame coding system for limited reference picture memories, *Proc. of the Picture Coding Symposium*, 743–748, Berlin, Germany, Sept. 1997.

Wada, M., Selective recovery of video packet loss using error concealment, *IEEE Journal on Selected areas in Communications*, **7**, No. 5, 807–814, June 1989.

Wang, Y., and Zhu, Q. F., Error control and concealment for video communication: a review, *Proc. of the IEEE*, **86**, No. 5, 974–997, May 1998.

Watanabe, T., Designing process for reversible variable-length code (RVLC), Technical report from Toshiba, Oct. 1996.

Wenger, S., Knorr, G., Ott, J., and Kossentini, F., Error resilience support in H.263+, *IEEE Transactions on Circuit and Systems for Video Technology*, **8**, No. 7, Nov. 1998.

Wenger, S., Video redundancy coding in H.263+, *Proc. AVSPN 97*, Aberdeen, UK, 1997.

Wiegand, T., Farber, N., Stuhlmuller, K., and Girod, B., Error-resilient video transmission using long-term memory motion-compensated prediction, *IEEE Journal on Selected Areas in Communications*, **18**, No. 6, 1050–1062, June 2000.

Wiegand, T., Farber, N., Stuhlmuller, K., and Girod, B., Long-term memory motion-compensated prediction for robust video transmission, *Proc. of the 2000 International Conference on Image Processing*, **2**, 152–155, Jan. 2001.

Worrall, S., Fabri, S., Sadka, A. H., and Kondoz, A. M., Prioritisation of data partitioned MPEG-4 video over mobile networks, *Proc. of the Packet Video Workshop 2000*, Sardinia, Italy, May 2000.

Worrall, S., Sadka, A. H., Sweeney, P., and Kondoz, A. M., Backward compatible user defined data insertion into MPEG-4 bit streams, *IEE Electronics Letters*, **36**, No. 12, 1036–1037, June 2000.

Worrall, S., Sadka, A. H., Sweeney, P., and Kondoz, A. M., Motion adaptive INTRA refresh for MPEG-4, *IEE Electronics Letters*, **36**, No. 23, 1924–1925, Nov. 2000.

Yoo, K. Y., Adaptive resynchronisation marker positioning method for error resilient video transmission, *IEE Electronics Letters*, **34**, No. 22, 2084–2085, Oct. 1998.

5

Video Communications Over Mobile IP Networks

5.1 Introduction

The near future will witness the universal deployment of the third-generation mobile access networks that are expected to revolutionise the world of telecommunications. In addition to conventional voice communication services provided by the second-generation GSM networks, the third-generation mobile networks will support a greatly enhanced range of services due to the higher throughput made available by embracing a number of new access technologies. These include TDMA and a variety of CDMA radio access families such as the direct sequence Wideband-CDMA (WCDMA) and multi-carrier CDMA. Consequently, the most prominent development brought forward by the third-generation family of standards and protocols, namely IMT-2000, compared to second-generation GSM systems, is the provision of high data rates that will enable the support of a wide range of real-time mobile multimedia services including combinations of video, speech/audio and data/text traffic streams with QoS control (Third-generation Partnership project). This chapter examines the issues involved in the provision of video services over the 2.5G and 3G mobile networks, and evaluates the perceived service quality resulting from video transmissions over these networks under various operating conditions. The focus will also be on describing and analysing the performance of a number of tools specifically designed to improve the perceptual video quality over the new mobile access networks.

5.2 Evolution of 3G Mobile Networks

The second-generation GSM technology has resulted in a major success for the delivery of telephony and low bit rate data services to mobile end users. On the other hand, the tremendous growth of the Internet has given rise to a new range of multimedia applications that have penetrated the global market at an explosive pace. The aim of the third-generation mobile networks is to combine the multimedia services of the Internet and the digital cellular concept of mobile radio networks in order to support the provision of multimedia services over mobile wireless platforms.

In order to accommodate a new range of services with much higher data rates than those provided by GSM, the most fundamental improvement that is required from the third-generation mobile systems is to embrace a number of new access technologies that will allow for a high-throughput access and true real-time multimedia services. The fundamental voice communication services provided by the 2G GSM will be preserved by the new mobile systems, while assuring an improved audio quality across the network along with improved call management and multiparty communication. In addition to conventional voice services, the mobile users will have the ability to connect to the Internet remotely while retaining access to all its facilities, such as e-mail and Web browsing sessions. Mobile terminals will be enabled to access remote websites and multimedia-rich databases with the use of multimedia plug-ins embedded into the Web browsers of these terminals. The conversational video communications over 3G networks will also support multi-user capabilities such as multi-party videoconferencing among various fixed and mobile users. The ubiquity of connection that is allowed by portable mobile terminals will significantly enhance the functionalities of such devices, especially in scenarios involving e-commerce and e-business applications. This will be made possible through the implementation of mobile work environments and virtual offices. Last but not least, the next generation of mobile networks will also support the selective and on-demand coverage of live events such as breaking news and sports in the form of streaming audiovisual content. This will also be accompanied by the on-demand access to archived media such as high-quality highlights of TV scenes and remote audiovisual clips.

The support to all the mobile multimedia services mentioned above will have its implications for the design of the end-to-end mobile network architecture. Firstly, the quality of service (QoS) offered to client applications will be a function of different connection parameters such as throughput, end-to-end delays, error rates and frame dropping rates. Therefore, each mobile terminal will have access to a number of bearer channels, each offering a different QoS to the various services being used. On the other hand, the standardised protocols that were adopted for the Internet Protocol and have consequently led to the widespread success of the Internet have allowed an extremely diverse range of terminals and devices to communicate with each other. Moreover, the accepted application-layer standards such as the HyperText Transfer Protocol (HTTP) have also allowed multimedia applications to be deployed and to proliferate. The combination and interoperability of these universally accepted application and network-layer standards will certainly constitute the core of the architecture of 3G systems, and will identify the mechanism of operation of multimedia services over these mobile platforms. This chapter will focus on the real-time transmission of compressed

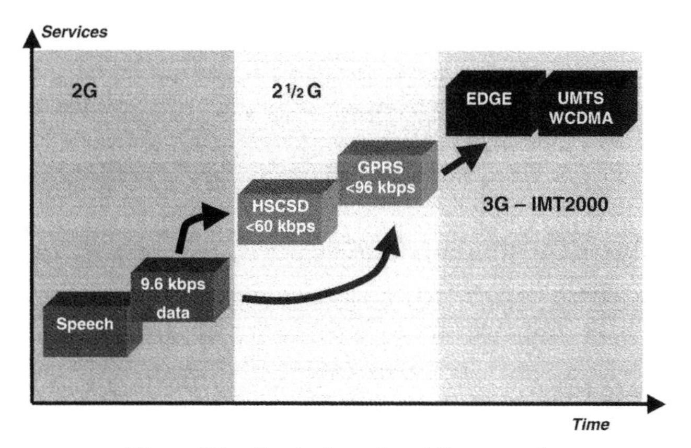

Figure 5.1 Evolution of mobile networks

video data encapsulated in IP packets over the future mobile networks. Figure 5.1 illustrates the time evolution of mobile networks as a function of their provided services. This evolution was consolidated by the remarkable migration from the second-generation GSM network to the third-generation EDGE (Enhanced Data rate GSM Evolution) and UMTS (Universal Mobile Telecommunication System) networks through the 2.5G packet-switching GPRS (General Packet Radio Service) and circuit-switching HSCSD (High Speed Circuit Switched Data) systems.

5.3 Video Communications from a Network Perspective

One of the main design trends of multimedia networks is to achieve a connection between two or more users by bringing digital content, such as video, to their desktops. Video telephony, videoconferencing, telemedicine and distance learning are all examples of multimedia applications that aim at providing video (along with voice) services in a networking environment. Beyond the desktop, multimedia technology relies on high-capacity digital networks to carry video content and support real-time services such as messaging, conversation, live and on-demand streaming, etc. In video telephony and conferencing for instance, users are geographically far from each other and therefore the video streams must be transmitted in real time over a communication network. In video on-demand applications, the storage medium is remote, and thus video must be retrieved and streamed over a network for being delivered to the requesting user. In distance learning applications, video is captured and then transmitted to remote learners using a shared communication medium. In all these cases, a communication network is obviously required.

Since the users are located far from each other, multimedia services must be offered in the presence of a telecommunication system that performs the routing of multimedia traffic across a network. On the other hand, a multimedia service might involve more than two users at the same time (such as videoconferencing). This requires the presence of a sophisticated network infrastructure with an integral communication protocol for the end-to-end routing, transport and delivery of multimedia traffic. Without the development of corporate networks to route the video traffic among various users, little chance exists to commercialise multimedia and broaden its applications from the PC-based software and hardware to multi-sharing services on a worldwide basis.

5.3.1 Why packet video?

The time synchronisation between the sender and receiver is a key issue in any communication session. To achieve synchronisation, either one of two approaches is adopted, namely synchronous and asynchronous transmissions. Asynchronous communication consists of sending the stream of data in the form of symbols, each represented by a pre-defined number of bits. Each symbol is preceded by a start bit and followed by a parity bit, thereby leading to an overhead of two bits per symbol. With synchronous transmission, characters are transmitted without any start and end indicators. However, to enable the receiver to determine the beginning and end of a block of data (set of characters), each block of data begins with a preamble bit pattern and ends with a post-amble bit pattern, as is the case in asynchronous communication systems. This block of data is referred to as a packet. The packet can be of fixed length such as the ATM cell (53 bytes), or variable length as for IP packets.

Unlike data streams, coded video has a very low tolerance to delay, and therefore dropped video information cannot be retransmitted. Alternatively, compressed video data has to be fitted into a certain structure that enables error control to be applied in case of information loss and bit errors. This structure is called a packet and consists of a video payload and a protocol header. The process of fitting the video payload into this packet structure is called packetisation, and the part of the communication system where packetisation is performed is known as the packetiser. Figure 5.2 is a block diagram of a typical packetiser with one input video source.

A number of advantages are obtained from packetising a compressed video stream before transmission.

It is intended that a number of applications would be running between two end-points at the same time. Moreover, the traffic flow between these two end-points may consist of a number of various traffic types. Therefore, the successful end-to-end control and delivery of routed multimedia information would be impossible if the information bits were not sent in packet format. The traffic type of the payload is then identified by the content of the type field in each packet header. Using the packet structure, it would be possible to multiplex various streams of

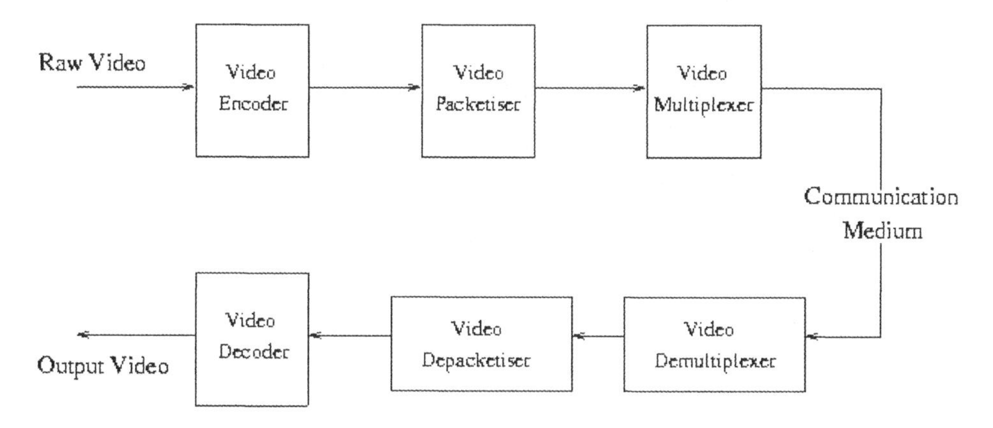

Figure 5.2 Block diagram of a video packetiser/depacketiser system

data onto the same bearer since the depacketiser would then be able to identify the source of each packet from the content of its type field. Once the source is known, the payload is then delivered to the corresponding decoder. Consequently, the packet structure enables the multiplexing of various streams of data, thereby resulting in an efficient sharing of the available bandwidth.

Due to excessive delays and interference, the video data is subject to information loss and bit errors, respectively. As examined in Chapter 4, a single bit error could lead to a disastrous degradation of the decoded video quality. If a packetisation scheme is employed, the effect of bit errors and information loss could be confined to a single packet since the video decoder would then resynchronise at the beginning of the following error-free packet. Moreover, the MBs contained in a video packet can be predicted independently of the MBs in other packets (Independent Segment Decoding in Annex R of H.263+ described in Section 4.12), thereby improving the error robustness of video data.

The packet structure enables the datagram or connectionless service of the network layer routing protocol. As opposed to the virtual circuit connection, the connectionless routing strategy shows a high flexibility in the selection of the path between source and destination at any instant of time. It also results in a much higher channel utilisation, since it does not require any prior bandwidth allocation, as is the case for virtual circuit connections. To prevent out-of-sequence arrival of packets, resulting from multipath fading and varying network conditions, the depacketiser can re-order the received packets in accordance with their sequence numbers before passing their payload up to the video decoder.

One further advantage of packet transmission is the ability of the decoder to acknowledge the receipt of error-free packets. In many situations, it is paramount that the video encoder is aware of the network conditions so that it adapts its output rate and error protection mechanism accordingly. The acknowledgement of correct delivery can be periodically sent to the encoder in the form of feedback reports that update the encoder on the latest status of the network. This mechanism can be used for various purposes such as flow control, as described in Chapter 3, and error resilience, as described in Section 4.13 on the reference picture selection (RPS) technique.

The packet structure also enables the prioritisation of video data in accordance with its sensitivity to errors and contribution to overall video quality. Some levels of priority can then be assigned to video packets depending on their payload (the prioritised information loss of Section 3.7). In case of reported network congestion, the video encoder drops low-priority packets, hence reducing its output rate for graceful quality degradation.

5.4 Description of Future Mobile Networks

The second-generation mobile cellular networks, namely GSM, do not provide sufficient capabilities for the routing of packet data. In order to support packet data transmission and allow the operator to offer efficient radio access to external IP-based networks such as the Internet and corporate Intranets, GPRS (General Packet Radio Service) has been developed by ETSI (European Telecommunication Standards Institute) and added to GSM. GPRS is an end-to-end mobile packet radio communication system that makes use of the same radio architecture as GSM (Brasche and Walke, 1997). GPRS permits packet mode data transmission and reception, on both the radio interface and the network infrastructure, without employing circuit switched resources. Although GPRS was initially designed for the provision of non delay-critical data services, this packet-switched system can be a suitable medium for video communications due to two main reasons. Firstly, the throughput capability of a single GPRS terminal can be increased using the multi-slotting feature of the GPRS system simply by allocating more timeslots or PDTCH (Packet Data Traffic Channels) to a single terminal. Another important feature of GPRS is its IP support, and this allows for accessing and interworking with the video applications of the Internet.

The network infrastructure for implementing the GPRS service is based on IP technology. For data packet transmission in the GPRS network, the mobile terminal is identified by an IP address assigned to it either permanently or dynamically at the time the session is set up. The routing of IP packets is performed by a logical network entity that is referred to as the GPRS Support Node (GSN). The Serving GPRS Support Node (SGSN) that is connected to the access network is the node that serves the GPRS mobile terminal, retaining its location information and performing operations related to security and access control. The Gateway GPRS Support Node (GGSN) is seen from outside as the access port to the GPRS network and acts as an interworking unit for the external packet-switched networks. Within the network, GGSN and SGSN are connected by means of an IP-based transport network. The IP packets and all relevant overlying transport protocol headers are forwarded to the Subnetwork Dependent Convergence (SNDC) protocol layer which formats the network packets for transmission over the GPRS network. The SNDC protocol carries out header compression and the multiplexing of data from different sources. The Logical Link Control (LLC) layer operates above the Radio Link Control (RLC) layer to provide highly reliable logical links between the mobile station and the Serving GPRS Support Node (SGSN). Its main functions are specifically designed to maintain a reliable link. If the network packet size does not exceed the maximum LLC frame size (1520 octets), each IP packet is mapped onto a single LLC frame. The LLC frames are then passed onto the RLC/MAC (Medium Access Control) layer where they are segmented into fixed-length RLC/MAC blocks. At the MAC layer, multiple mobile stations are allowed to share a common transmission medium. GPRS allows each time slot to be multiplexed between up to eight users, and allows each user to use up to eight timeslots, thereby achieving great flexibility in the resource allocation mechanism. The RLC blocks are arranged into GSM bursts for transmission across the radio interface where the physical link layer is responsible for forward error protection, as described in Section 5.5.2. In the physical link layer, interleaving of radio blocks is performed and methods to detect link congestion are also employed. Figure 5.3 depicts the logical architecture of a GPRS network connection involving a Mobile Station (MS) and a Base Station Subsystem (BSS).

The GPRS service introduced in the GSM system is an intermediate step towards the third-generation UMTS network. EGPRS (Enhanced GPRS) is an enhanced version of GPRS that allows for a considerable increase in throughput availability to a single user given enough traffic availability from active sources and benign interference conditions. This implies that EGPRS can provide video services with higher data rates than is possible with GPRS. EGPRS uses the same

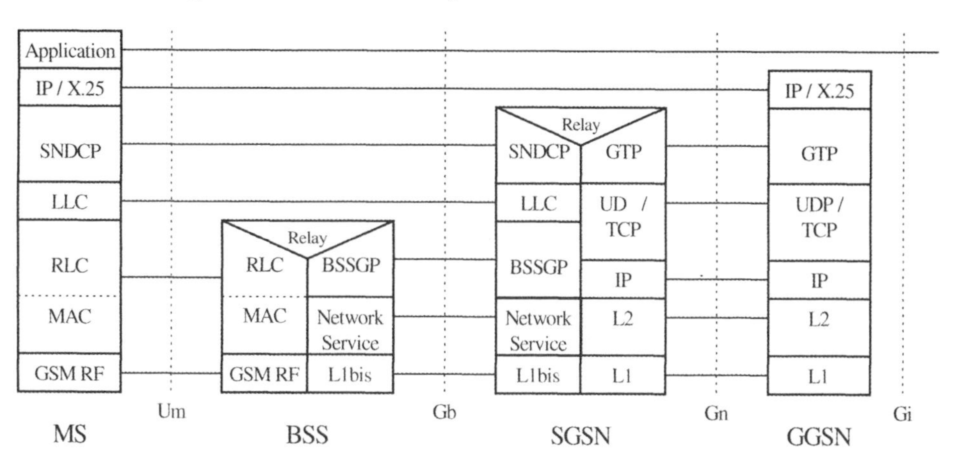

Figure 5.3 GPRS logical protocol architecture

protocol architecture of GPRS described above, with improvements of the modulation scheme employed in the EDGE (Enhanced Data rate GSM Evolution) radio interface that lead to the increase in throughput availability. Similarly, UMTS uses an innovative radio access approach to increase the available capacity of the radio interface. The UMTS infrastructure is integrated with GSM so that the UMTS core network can perform both the circuit- and packet-switching functions. However, the major technological innovations of UMTS are incorporated in the packet-switched IP nodes. The structure of the packet switched part of the UMTS core network is similar to that of the GPRS described above, where the BSS access segment is replaced by the UTRAN (Universal Terrestrial Radio Access Network) access network that is based on W-CDMA (Wideband Code Division Multiple Access) technologies. The connection between the UMTS core network and UTRAN access network is guaranteed by a new interface called I_u, which specialises in managing both the packet-switched and the circuit-switched components. The main improvements achieved by UMTS compared to GPRS are in the IP mobility management and the quality of service control. UMTS offers a range of QoS levels that are suitable for real-time video communications, namely those specified in the conversational and streaming classes. The main feature that defines the capability of a QoS class to accommodate a real-time video service is its sensitivity to delay. The conversational class allows videoconferencing sessions in which the delay factor must be minimised and the temporal relationship between various streams (voice and video for instance) must be maintained stationary. In the streaming class that allows for real-time streaming of multimedia data, the requirement for low transfer delay is not stringent but the various stream components must be kept temporally aligned. In addition to the conversational and streaming classes, UMTS offers the interactive QoS class which enables the mobile user to interact with a remote device on the network such as a video database or a website. The main requirements of this class are a limited round-trip delay and data integrity represented by low bit error rates.

5.5 QoS Issues for Packet Video over Mobile Networks

In real life, transmitted video packets are subject to loss and the contained information is susceptible to bit errors. When packets are corrupted, any one of three possible kinds of error might result. If the sequence number of the packet is affected, the decoder becomes unable to figure out the correct order of packet transmission. As a result, the depacketiser fails to merge the information of consecutive video packets in order to properly reconstruct the video sequence. This has a damaging effect on the video quality regardless of whether or not the data bits of affected packets have arrived intact. The second kind of error arises when some of the payload of a certain video packet is hit by errors in such a way that the resulting sequence pattern resembles a packet delimiter (start or end code). The latter would then be misinterpreted by the video depacketiser as the end of the current packet and the start of a new one with a different sequence number. Consequently, the depacketiser carries out an incorrect split of video data, thereby causing loss of synchronisation and a number of subsequent false merges and splits of video packets. The third kind of error affects the payload of a packet while the headers remain error-free. This type of error is more frequent than the first two since the payload constitutes the higher proportion of the packet length. In this case, the bit errors result in the same effects that have been examined in Chapter 4. However, in packet video networks, quality degradation could also be due to network congestion and link overflows. These network problems result in completely discarding the video packets that have been subject to excessive amounts of delay. In order to mitigate the effect of packet loss, some intelligent content-based packetisation schemes must be employed.

5.5.1 *Packetisation schemes*

The structure of a packet depends on the layer at which the packet is defined and the networking platform upon which the packets are transmitted. As described in Section 4.4, MPEG-4 defines an application layer packet structure where each packet consists of two main partitions. The first partition contains the more error-sensitive shape and motion data, while the second partition consists of the more error-tolerant texture data. This packetisation scheme allows the video decoder to successfully reconstruct (with minor quality degradation) the MBs contained in a packet using their motion and shape data (first partition) when errors hit only the texture data (second partition) of the packet. This application layer MPEG-4 packet differs from the transport layer packet in which the MPEG-4 packets are encapsulated. The latter has additional protocol headers which reduce the overall throughput available to the video source. The overhead imposed by the packetisation scheme depends on the transport mechanism employed for the transmission of video packets. For instance, packing coded video streams in RTP (Schulzrinne *et al.*, 1996) packets for real-time video transmission over IP networks has different implications from packing the same video data into ATM cells for transport over the B-ISDN networks (Broadband Integrated Service Digital Network).

The layering structure of video coding standards requires that some information should be specified in the video packet at each level of the hierarchy. For instance, at the frame level, information such as temporal reference and picture header is contained in the output stream. At the GOB level, the GOB number and the quantiser level for the entire GOB are indicated. At the MB level, both coded and non-coded MBs are identified and an optional quantiser is specified, as well as information about the coded blocks such as MVs. This structure requires that the frame header should be first decoded to decode the GOBs, and so should be the information contained in the GOB header to decode the MBs. Therefore, the logical sequence of the frame components implies that all packets containing a certain picture must be received before the picture components are successfully reconstructed. To overcome this problem when no restriction on the packet size is imposed, each video frame can be packed into a single packet. However, a frame or even a GOB can sometimes be too large to fit into a single packet. Moreover, the loss of a video packet would in this case lead to the loss of a whole video frame, thereby leading to poor error performance. In this case, the packetisation scheme has to adopt the MB as the unit of fragmentation, thus causing packets to start and end on an MB boundary. Consequently, an MB would not be split across multiple packets, and then a number of MBs could be packed into a single packet when they fit within the maximal packet size allowed. Since the MBs belonging to the same video frame may not necessarily be embodied in the same packet, the loss of a video packet would result in damage of the corresponding frame, even when adjacent packets are correctly received. In order to limit the propagation of errors between various packets, each packet could contain an independent segment of a video frame and each segment could be coded separately from others, as is the case in the independent segment decoding mode (Annex R of H.263+) described in Section 4.12. Moreover, to enable the decoder to resynchronise on the occurrence of a packet loss, each packet should contain the picture header and the GOB header that indicate to which frame and GOB the contained video payload belongs, respectively.

On the other hand, when the packet has a fixed size, as is the case for ATM cells, for instance, the packetisation conditions become more stringent. An ATM cell has an overall size of 53 bytes, 5 bytes of which are occupied by the cell header. In the 48-byte payload, the coded video can be packed using one of two different approaches (Ghanbari and Hughes, 1993), as illustrated in Figure 5.4.

In the close packing scheme, video data is packed continuously in the payload field until the ATM cell is completely full. This leads to the possibility that some MBs can be split between two adjacent cells. In the second approach, i.e. the loose packing, each ATM cell contains an integral number of MBs. In both methods, an eight-bit field is assigned to the cell sequence number and a five-bit one to the picture number. Moreover, in both methods, the first complete MB inside the ATM cell is absolutely addressed with reference to the picture information, while all the following MBs in the cell are relatively addressed. The use of absolute addressing is useful in eliminating the effect of cell loss propagation into the forthcoming correctly received cells. A unique bit pattern is used in the close packing methodology to designate the end of the variable-length section of data belonging to the previous cell. This unique bit pattern must be different from the GSC (GOB Start Code) so that the depacketiser will not fall on a false start of a GOB. The shorter this bit pattern, the higher the probability of falsely detecting it due to combinations of other codewords in the ATM cell. However, it is a

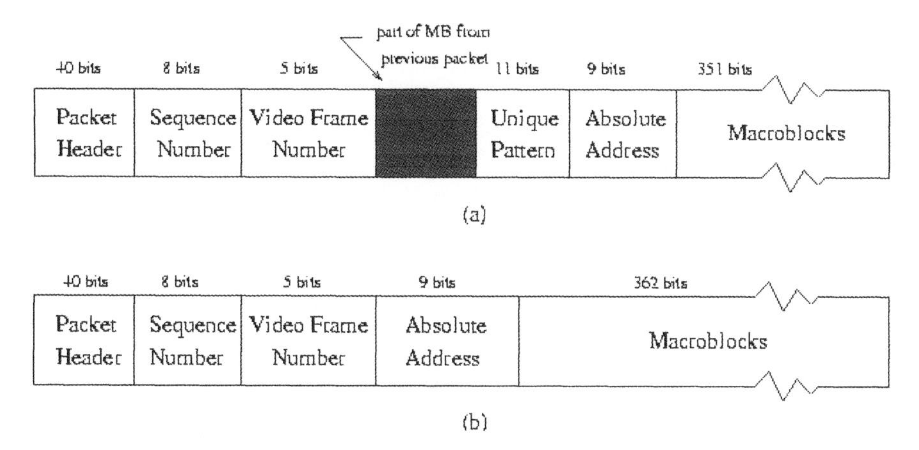

Figure 5.4 Packing video in ATM cells: (a) close packing, (b) loose packing

requirement to reduce the size of this unique bit pattern in order to minimise the amount of overhead imposed by the close packetisation scheme. As a trade-off between throughput and error robustness, the size of the unique bit pattern is set to 11 bits. Therefore, the total overhead of the close packing scheme is 4.125 bytes, whereas it is only 2.75 bytes for the loose packing technique. However, the loose packing scheme results in a less efficient use of bandwidth, especially when ATM cells carry the traffic of multiple video sources.

Apart from bandwidth utilisation, the packetisation scheme also has an effect on the error performance of the packet video application. In the ATM cell close packing technique, the loss of a cell affects not only the MBs of the discarded cell, but those in adjacent cells as well. The loss of a cell entails the loss of all the MBs within the cell in addition to portions of two more MBs shared with both the previous and next cells. Exceptions exist only when the end of the lost cell coincides with the end of its last MB, or when the start of the cell coincides with the start of its first MB. However, when a loose packing ATM cell is lost, only the enclosed MBs are lost, thereby leading to an improved error performance as compared to that of the close packing technique. In variable-size packets, the size of the lost packet is an important metric in assessing the error performance of the packetisation technique. Longer packets lead to improved throughput resulting from lower overheads, but yield a lower tolerance to loss which would then hit a larger segment of video payload. Eventually, the damage to video quality resulting from a packet loss is further exacerbated by the predictive video coding technique and the temporal/spatial dependencies of video data contained in different packets. As a result of the prediction used in the INTER coding mode, the loss of a packet would also cause disastrous damage to the forthcoming video data that is predicted from the lost information in both time and space. The effects of packetisation on the service quality of real-time video transmissions over IP-based mobile networks will be analysed in Subsection 5.6.1.

5.5.2 Throughput and channel coding schemes

In addition to the packet structure, the quality of service of video communications over the future mobile networks depends on a number of other parameters, namely the available throughput and the employed channel coding schemes. For example, the GPRS data is transmitted over the Packet Data Traffic CHannel (PDTCH) after being error-protected using one of four possible channel protection schemes, namely CS-1, CS-2, CS-3 and CS-4. The first three coding schemes use convolutional codes and block check sequences of different strengths to produce different protection rates. CS-2 and CS-3 use punctured versions of the CS-1 code, thereby allowing for a greater user payload at the expense of reduced performance in error-prone environments. However, CS-4 only provides error detection functionality and is therefore not suitable for video transmission purposes. For video applications, it has been experimentally proved that only CS-1 and CS-2 could achieve acceptable video quality. Table 5.1 shows the data rates provided per timeslot for each one of these GPRS channel coding schemes.

As can be observed in Table 5.1, the payload available in a GPRS radio block depends on the channel coding scheme used. The rate of the RLC/MAC data payload, i.e. the rate presented to the LLC layer, varies from 8 kbit/s for CS-1 to 20.35 kbit/s for CS-4. Depending on the multislotting capabilities of the mobile GPRS terminal, the throughput available to the terminal is a multiple of these data rates. These data rates represent only the throughput at which LLC PDUs (Packet Datagram Unit) are transmitted across the radio interface. However, when considering the GPRS protocol stack illustrated in Figure 5.3, it can be seen that the RLC/MAC data payload will contain header and other related signalling overheads from the LLC, SNDC, IP, UDP and RTP layers. The presence of these overheads will reduce the true throughput presented to the application layer, i.e. the video source coder. The protocol overheads constitute approximately 10 per cent to 15 per cent of the total throughput at the RLC layer for QCIF video transmissions at frame rates of 5 to 10 f/s when no header compression is applied. For this reason, the total throughput, as seen by the application layer in the GPRS protocol stack, for all combinations of timeslots (TS) and channel coding schemes (CS) allowed by GPRS, is depicted in Table 5.2.

Table 5.1 GPRS data rates per timeslot for each of the four channel protection schemes

Scheme	Code rate	Data payload	Radio block size (headers + data)	Data rate (kbit/s)
CS-1	1/2	160	181	8.0
CS-2	$\approx 2/3$	247	268	12.35
CS-3	$\approx 3/4$	291	312	14.55
CS-4	1	407	428	20.35

Table 5.2 Video source throughput in kbit/s for all GPRS timeslot/CS combinations

Scheme	1 TS	2 TS	3 TS	4 TS	5 TS	6 TS	7 TS	8 TS
CS-1	6.8	13.6	20.4	27.2	34	40.8	47.6	54.4
CS-2	10.5	21	31.5	42	52.5	63	73.5	84
CS-3	12.2	24.4	36.6	48.8	61	73.2	85.4	97.6
CS-4	17.2	34.4	51.6	68.8	86	103.2	120.4	137.6

Table 5.3 EGPRS data rates allowed per timeslot for each of the nine channel protection schemes

Scheme	Code rate	Header code rate	Radio block size (headers + data)	Data rate (kbit/s)
MCS-1	0.53	0.53	176	8.8
MCS-2	0.66	0.53	224	11.2
MCS-3	0.8	0.53	296	14.8
MCS-4	1.0	0.53	352	17.6
MCS-5	0.37	1/3	448	22.4
MCS-6	0.49	1/3	592	29.6
MCS-7	0.76	0.36	2 × 448	44.8
MCS-8	0.92	0.36	2 × 544	54.4
MCS-9	1.0	0.36	2 × 592	59.2

Like GPRS, EGPRS supports its own nine joint modulation-coding schemes which are referred to as MCS-1 to MCS-9. One vital difference between the coding schemes used in EGPRS and those employed by the GPRS PDTCHs is that the radio block headers are encoded separately from the data payload. One further difference is that schemes MCS-7, MCS-8 and MCS-9 allow the insertion of two RLC/MAC blocks into a single radio block, while in GPRS only one-to-one block mapping is allowed for all schemes. The data rates allowed per timeslot and presented to the LLC layer for each of the MCS schemes employed by EGPRS are depicted in Table 5.3.

As in GPRS, due to the overheads imposed by the protocols overlying the RLC/MAC layer, some protocol efficiency has to be compromised. Similarly, in EGPRS, a protocol efficiency of 85 per cent can be achieved for QCIF frame rate of 5 f/s, assuming an overall header size of 44 bytes in each RLC/MAC block. Consequently, the throughput presented to video sources at the application layer is less than that available at the RLC/MAC layer and can vary with the employed MCS scheme. Using a single timeslot at the radio interface, it is possible to provide the 5 f/s video coder at the application layer of an EGPRS terminal with a source throughput varying from 7.5 kbit/s for MCS-1 to 50 kbit/s for MCS-9. Using the multislotting capabilities of the radio interface, the video source can have multiples of these data rates, as shown in Table 5.4. This reflects the large spread in the values of available throughput for video services over EGPRS. The choice of a suitable CS-TS combination for video services over mobile networks depends

Table 5.4 Video source throughput in kbit/s for all EGPRS TS/MCS combinations

Scheme	1 TS	2 TS	3 TS	4 TS	5 TS	6 TS	7 TS	8 TS
MCS-1	7.5	15	22.5	30	37.5	45	52.5	60
MCS-2	9.6	19.2	28.8	38.4	48	57.6	67.2	76.8
MCS-3	12.6	25.2	37.8	50.4	63	75.6	88.2	100.8
MCS-4	15	30	45	60	75	90	105	120
MCS-5	19	38	57	76	95	114	133	152
MCS-6	25.2	50.4	75.6	100.8	126	151.2	176.4	201.6
MCS-7	38	76	114	152	190	228	266	304
MCS-8	46.2	92.4	138.6	184.8	231	277.2	323.4	369.6
MCS-9	50.31	100.6	150.9	201.2	251.5	301.8	352.1	402.4

highly on the activity of the video source and error characteristics of the radio network.

5.6 Real-time Video Transmissions over Mobile IP Networks

The main objective of transmitting video over mobile networks is to provide interactive and conversational services. This implies that all the video services offered over GPRS for instance must run in real time with one-way delay not exceeding 200 ms per service. In order to meet these delay requirements, it is not possible to use retransmissions or repeat-request systems (ARQ). Alternatively, the RLC (Radio Link Control) layer operates in its unacknowledged mode of operations, which does not include any retransmissions. On the end-to-end level, the transport layer protocol employed is the User Datagram Protocol (UDP), as opposed to TCP, which overlays IP and does not make use of repeat-request systems.

On the other hand, IP networks do not provide any guarantee for the delivery of packets due to the best-effort service of the IP protocol. Furthermore, they do not have any guarantee on the packet time arrival. Consequently, the inter-arrival time of packets would vary, hence giving rise to the jittering effect of video frames. The packets could also be delivered out-of-sequence. This implies that in order to provide real-time services with acceptable quality of service, some transport-layer mechanism must be employed in order to provide some reliable timing information from which streamed video could be properly reconstructed. The most popular transport-layer protocol used for such purposes is the IETF (Internet Engineering Task Force) Real-time Transport Protocol (RTP) (Schulzrinne *et al.*, 1996). RTP provides end-to-end network transport functions suitable for real-time data transmissions. These functions include payload type identification, sequence numbering, timestamping and delivery monitoring. Typically, real-time applica-

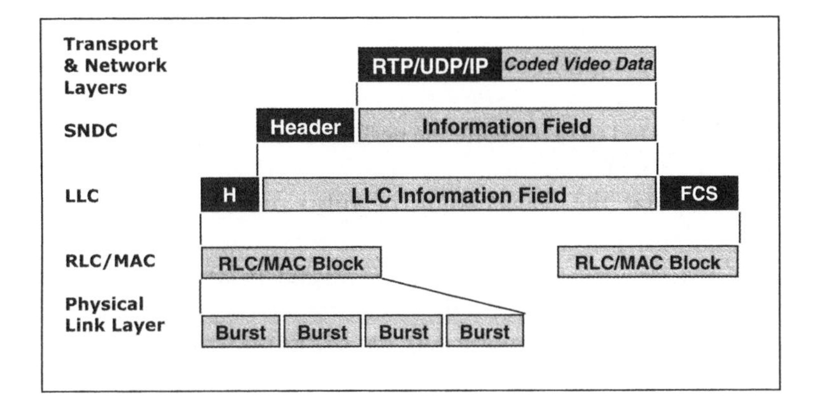

Figure 5.5 Protocol architecture for real-time video transmission over IP-based mobile radio network

tions run RTP over UDP rather than TCP, since the latter imposes huge delays resulting from data retransmissions that are not suitable for real-time applications. Therefore, video frames are segmented and encapsulated into RTP packets, which are then embodied in the packet structure of the underlying protocols, namely UDP and IP as shown in Figure 5.5.

5.6.1 Packetisation of data partitioned MPEG-4 video using RTP/UDP/IP

The careful packetisation of video data is necessary to ensure the optimal trade-off between the channel utilisation and error robustness. Several researchers (Basso, Varakliotis and Castagno, 2000) have attempted to develop optimal techniques in order to pack compressed video data into RTP packets for real-time transmission over IP networks. The main focus of their work has been on the ability to synchronise MPEG-4 streams with other RTP payloads, the monitoring of MPEG-4 delivery performance through the use of the RTP control protocol, namely RTCP (Real Time Control Protocol), on the reverse channel, and also the combination of MPEG-4 with other real-time data streams into a set of consolidated streams by means of RTP mixers. However, these packetisation techniques did not focus on the error-resilience issues of packet video over mobile networks. The size of the video payload and the sequence of video data within each packet do have a direct influence on the error robustness and channel utilisation of the video application. Therefore, in order to achieve the best quality of service, the error-resilience aspects of the packetisation scheme have to be considered.

On the other hand, due to the time-varying nature of the mobile channel conditions, the packetisation techniques ought to be adaptive in order to maintain an optimal trade-off between throughput and error resilience at any instant of time. The adaptation of the video source rate and error resilience to the channel conditions has been comprehensively examined in Chapters 3 and 4. However, in addition to the application-layer link adaptation schemes previously described, researchers have started to investigate the performance of adaptive transport-layer packetisation schemes (Worrall *et al.*, 2001). Their work was mainly motivated by the ramifications that the transmission of video over RTP has on both the throughput and error performance of the video service in mobile networks. The adaptation of the payload size in each RTP packet is based on both the error conditions of the network and the motion activity of the video content.

As described in Chapter 4 (Section 4.4), the MPEG-4 stream is usually broken up into a sequence of independently decodable video packets of regular length, with each packet starting with a resynchronisation word. However, these packets are created at the application layer and considered a part of the MPEG-4 video compression algorithm. Therefore, they should be separated from the packets created by the underlying layers such as IP, UDP and RTP. In each MPEG-4 packet, video data is split into two major partitions, where the first partition contains header and motion data, and the second partition consists solely of texture data. The corruption of the first partition leads to the loss of the whole MPEG-4 packet since the second partition can only be decoded when the first partition is properly reconstructed. If the second partition is corrupted while the first is error-free, then only the video data following the position of errors is discarded. Apart from the video data sensitivity to errors, the corruption of the synchronisation word, i.e. the MPEG-4 packet header, results in the loss of the whole packet since the decoder could only then resynchronise at the beginning of the next packet. Similarly, the corruption of any part of the RTP/UDP/IP header results in the loss of the whole RTP packet. Consequently, selecting a long packet size implies that more data is lost after each RTP/UDP/IP packet corruption, but it also implies that headers would occupy a smaller proportion of the packet, thereby reducing the likelihood of header corruption. This is in addition to the fact that the header size directly affects the channel utilisation because it reflects the proportion of overhead in the packet. Therefore, for a fixed header size (40 bytes for IP/UDP/RTP), the video payload size is a paramount factor that controls the trade-off between error robustness and throughput achieved by a given packetisation scheme.

Two packetisation schemes could be employed to encapsulate MPEG-4 video data into RTP packets. In the first scheme each MPEG-4 packet is encapsulated within a single RTP packet, whereas in the second scheme each RTP packet contains a video frame (a number of MPEG-4 packets), as shown in Figures 5.6 and 5.7, respectively.

The eight-bit Cyclic-redundancy Check (CRC) codes are inserted at the end of each MPEG-4 packet to aid with error concealment in the video packet data while retaining backward compatibility with the standard MPEG-4 decoder. In Figure 5.7, e_{eff} represents the effective error rate that is due to corruption of the corre-

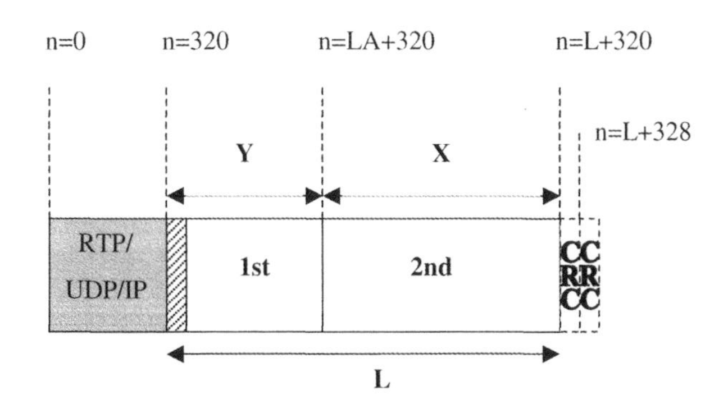

Figure 5.6 One MPEG-4 packet per RTP packet

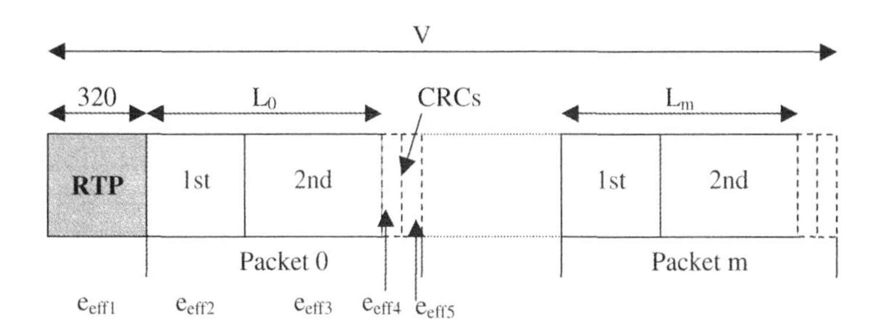

Figure 5.7 Several MPEG-4 packets per RTP packet

sponding segment of the packet. Using a similar mathematical analysis to that established in Section 4.10, the effective error rates for both packetisation schemes could be evaluated. The objective performance evaluation of both packetisation schemes shows that, for a given channel bit error rate and an optimally selected packet length, the second scheme with one frame per RTP packet produces slightly lower effective error rates and hence higher PSNR values than the other scheme. This could be justified mainly by the larger proportion of the RTP/UDP/IP header (40 bytes) with respect to the overall packet size set by the first packetisation scheme as compared to the second scheme. The RTP/UDP/IP packet header and the error-sensitive first MPEG-4 partitions in each RTP packet constitute a much larger portion of the RTP packet than they do in the second scheme. This implies that they are more likely to be hit by errors, thereby leading to the loss of the whole RTP packet. Furthermore, the corruption of any first-partition in the second packetisation scheme of Figure 5.7 leads to the loss of the video packet only and not the whole RTP packet. However, for both packetisation schemes, the performance is observed to be always optimal at a certain RTP packet size and for a given bit error rate of the channel, as shown in Figure 5.8.

It can be seen that for a given C/I ratio, the perceptual video quality is optimal at a certain RTP packet size. When the packet size is 200 bits, the extra overhead caused by larger header proportions in the bit stream is more damaging to video

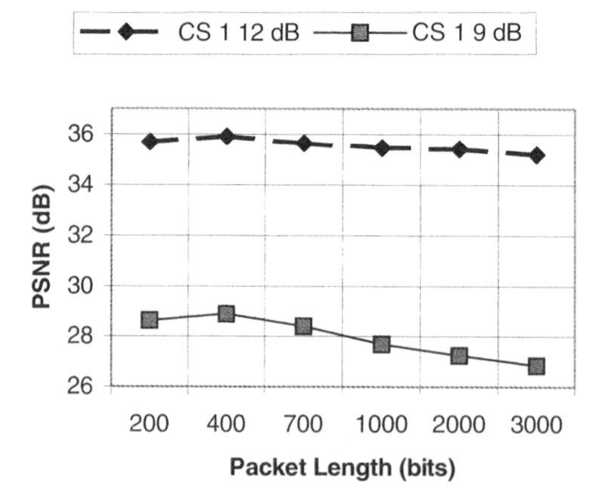

Figure 5.8 Average PSNR values for the Suzie sequence coded with MPEG-4 at 64 kbit/s, 10 f/s, and sent over a GPRS using CS1 at two different C/I ratios and for various RTP packet lengths

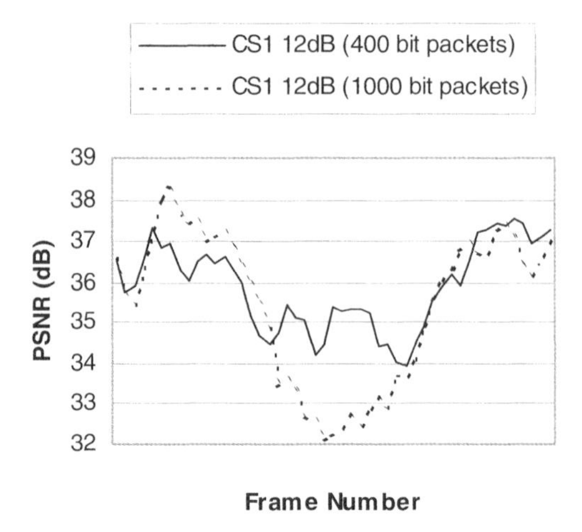

Figure 5.9 PSNR values for 50 frames of the Suzie sequence coded with MPEG-4 at 64 kbit/s and 10 f/s, and sent over a GPRS channel using CS-1 at C/I = 12 dB with two different RTP packet lengths

quality than the effect of GPRS channel errors. This is even more obvious for lower values of C/I as is noted in the 9 dB case. A more degraded error performance could certainly be expected when other GPRS channel coding schemes (CS-2, CS-3 or CS-4) are employed at C/I ratios that are less than 12 dB. Therefore, the RTP packet length must be changed adaptively to achieve optimal video quality in time-varying mobile channels. Figure 5.9 shows the frame-by-frame PSNR values for 50 frames of the Suzie sequence coded with MPEG-4 and transmitted over a GPRS channel with C/I of 12 dB, using the CS-1 channel coding scheme.

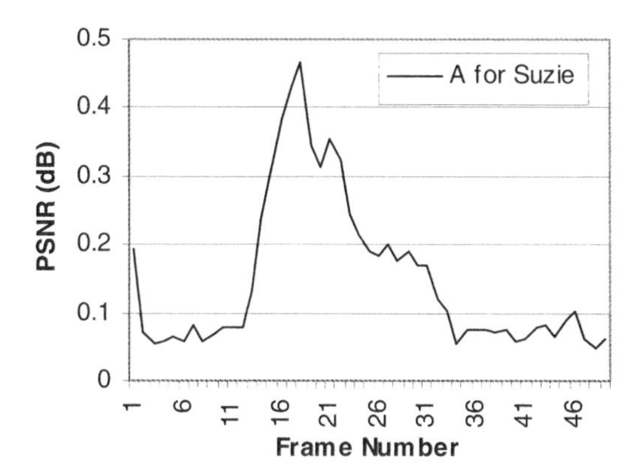

Figure 5.10 The variation of A for the first 50 frames of the Suzie sequence

In the first few frames of the Suzie sequence, where a small amount of motion is detected, it can be seen that a larger packet size produces a better performance. However, with the head shake in the sequence when more activity is detected in the video scene, a smaller packet size results in higher PSNR values. This is indicative of the direct effect of the motion activity of the video sequence on the choice of the RTP packet length for optimal video quality. In compressed video, it is conventional that a high motion activity in the sequence is equivalent to an increase in the output bit rate of the video coder. The extra bits are mainly due to the transmission of more motion vectors (in addition to the transmission of more residual video data that represents the motion compensated frames). Since the motion data is embodied in the first partition of an MPEG-4 packet, the proportion of the MPEG-4 packet occupied by the first partition is a good indication of the amount of motion in the video scene. Consequently, the PSNR values shown in Figure 5.9 indicate that a scheme that varies the RTP packet size with the first partition size should provide quality improvements, especially for video sequences featuring sudden bursts of motion. If A is a variable defined as the proportion of the packet size occupied by the first partition then A is related to the amount of motion, and can be expressed in reference to Figure 5.6 as:

$$A = \frac{\overline{Y_{MB}}}{\overline{X_{MB}} + \overline{Y_{MB}}} \tag{5.1}$$

where $\overline{Y_{MB}}$ is the average number of bits per MB in the first partition, and $\overline{X_{MB}}$ is the average number of bits per MB in the second partition. The time-varying nature of the variable A is illustrated in Figure 5.10 for the first 50 frames of the Suzie sequence. It is obvious that the variation in A is consistent with the quality improvement achieved by the variation of the RTP packet length shown in Figure 5.9. Therefore, for optimal video quality, the size of the RTP packet can be

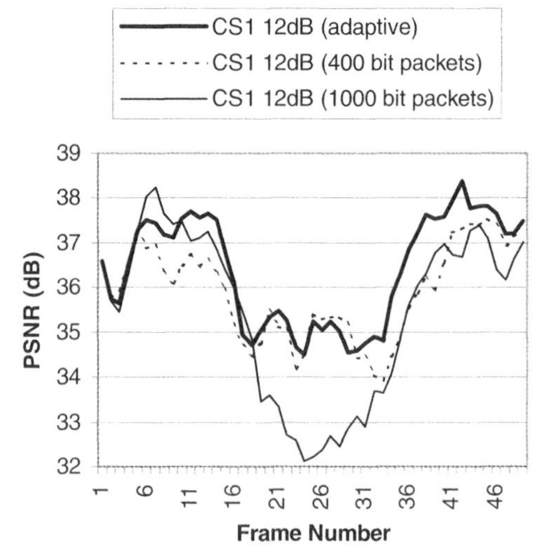

Figure 5.11 Objective performance of adaptive RTP packetisation against fixed RTP packetisation for real-time video over GPRS using CS-1 with C/I = 12 dB

adaptively selected in accordance with the value of the variable A, as shown in Figure 5.11.

5.7 Quality Optimisation for Video Transmissions over Mobile Networks

It is of paramount importance to ensure that the video content is delivered to the end-user with an acceptable level of quality. Several mechanisms have been adopted to optimise the performance of a video service in mobile radio networks. In addition to the transport and application layer packetisation schemes described above, error and flow control plays an important role in the quality optimisation process of video transmissions over mobile networks. A variety of rate-control and error-resilience techniques have been examined in Chapters 3 and 4, respectively. In mobile IP networks, the time-varying nature of the channel necessitates the use of adaptive quality control tools to deliver to the end-user a consistent and optimal quality of service. The adaptation of the quality control tools is performed in accordance with the mobile network conditions and the type of the video application. A recently completed annex (Annex X) of H.263++ specifies a number of profiles and levels for the most efficient use of its optional features and combinations of them in accordance with the overlying video application, the networking platform and the conditions of operation. These specifications were proposed in order to minimise the limitations placed on the encoder in selecting the mode combinations as opposed to the baseline mode of operation (Profile 0) that consists of the core standard recommendation without any optional mode of operation.

5.7.1 Enhanced video quality using advanced error protection

Profile 3 in Annex X of H.263 specifies a set of optional modes of operation for enhanced coding efficiency and improved error resilience in the delivery of interactive and streaming video services over wireless platforms. This profile consists of the baseline design (Profile 0) of the standard, in addition to four additional modes of operation, namely the advanced INTRA coding (Annex I), the de-blocking filter (Annex J), the slice structure mode (Annex K) with arbitrary slice ordering (ASO), and the modified quantisation (Annex T). Only one of these four options is directly related to the error-resilience capability of the coder, namely Annex K described in Section 4.12. The slice structure mode is included in this profile due to its enhanced ability to provide resynchronisation points within the video bit stream to enable the decoder to recover from the effect of erroneous or lost data. The ASO sub-mode of the slice structure mode is also included to provide compatibility for cross-network communication with packet-based network systems. Support for the rectangular slice (RS) sub-mode of the slice structure mode is not included in this profile, in order to limit the complexity requirements of the decoder. The additional computational burden imposed by the slice structure mode is minimal, limited primarily to bit stream generation and parsing. Profile 4 of Annex X is also an improved version of Profile 3 for enhanced coding efficiency and error resilience in the delivery of interactive and streaming video services over wireless platforms. This profile provides several enhancements for the support of wireless video transmissions and consists of Profile 3 in addition to two additional modes of operation. Firstly, Profile 4 enables the data partitioned slice mode of Annex V with arbitrary slice ordering sub-mode. This feature enhances error-resilience performance by separating motion vector data from DCT coefficient data within the slices (similar to data partitioning in the MPEG-4 video packet), and protects the motion vector information (the most important part of the MB data) by using reversible variable length coding (RVLC). Support of the ASO sub-mode is included so that Profile 4 will always indicate enhanced performance relative to that provided by Profile 3. Furthermore, Profile 4 enables the picture header repetition and supplemental enhancement information defined in Annex W of H.263++ (Section 4.12). This feature allows the decoder to recover the header information from a previous frame in case of data loss or corruption. In addition to interactive and streaming video applications, Annex X specifies a set of optional modes in Profile 6 for conversational video services over IP networks. In conversational video applications, the delay factor is of prime importance and excessively delayed packets have to be discarded by the network in order to achieve real-time operation. Profile 6 consists of the optional modes enabled by Profile 5 in addition to several enhancements used for improved coding efficiency. The error-resilience capability of a video compression algorithm has a great impact on the performance of video services over mobile networks. This will be demonstrated later in the chapter (Section 5.9) in the course of the performance evaluation study of video transmissions over GPRS and UMTS.

In addition to the efficient packetisation schemes and the standard-compliant error-resilience techniques, the quality of video services in mobile networks can be improved by enhancing the error detection and correction capabilities of the video decoder. This enhancement has to be carried out while making no alteration to the standard coding algorithm or retaining backward compatibility with the standard video decoder. An example of this kind of error control mechanism in mobile video communications is the CRC code inserted at the end of each MPEG-4 video packet, as discussed in Section 4.5.2. The effectiveness of inserting these CRC codes into each video packet is in the provision of additional error checks to both partitions of the packet enclosure. The detection of an error triggers the decoder to activate its error concealment mechanism by replacing the MBs of the affected video packet by their corresponding motion-compensated MBs. This enhances the quality of the mobile video service by providing consistency and preventing sharp quality degradations.

5.7.2 Content-based adaptive quality control for mobile video transmissions

The error-prone nature of mobile channels implies that the provided video service must be fairly robust to severe network conditions. An optimal packetisation scheme can be combined with an efficient error control tool to set up an adaptive quality control mechanism for mobile video communications. In Section 5.6.1, it was observed that modifying the RTP packet length with respect to the varying motion activity in the video scene resulted in optimal quality performance. Since the motion data is highly sensitive to errors, the RTP packet becomes less tolerant to errors with the increasing amount of motion data embodied in the first partition of the packet. Therefore, this packetisation scheme attempts to set a smaller RTP packet size when the motion activity in the video content exceeds a predefined threshold. Although shorter packets imply larger overheads and hence lower error-free video quality, the smaller RTP packet size results in improved robustness over the error-prone mobile channels. This content-based adaptive transport-layer packetisation mechanism can be combined with the application-layer resilience tools in order to further improve the quality of video service.

A content-based error resilience tool is the adaptive INTRA refresh (AIR) defined in Annex H of the MPEG-4 video coding standard. AIR is an application-layer error-resilience tool that is also based upon an estimation of the amount of motion in a video scene. Once the motion MBs are identified, they are entered into a motion map and a fixed number of these motion MBs is INTRA coded in every video frame (Section 4.7.1). In order to establish the best trade-off between error resilience and throughput, the number of AIR MBs per frame could be adapted to the amount of motion in the video scene (Worrall *et al.*, 2000). As illustrated in Section 5.6.1, there is a strong correlation between the activity of the video scene and the size of the motion information required to encode it. During periods of high scene activity, a large number of AIR MBs would cause a large drop of video quality. This quality deterioration is mainly attributed to the low efficiency of the INTRA coding mode. To optimise the video quality, it is necessary to increase the MB refresh rate after periods of high motion. This could be accomplished by observing the change in the variable A as described in Figure 5.10 for the Suzie sequence. If A increases then the number of AIR MBs encoded within a frame also increases, and *vice versa*. Therefore, this adaptive scheme achieves the fast updating of the video scene following periods of high motion. On the other hand, it reduces the number of AIR MBs during periods of high activity, thereby preventing an unnecessary drop in video quality. Figures 5.12 and 5.13 show the subjective and objective quality improvement, respectively, achieved by the adaptive scheme when the MPEG-4 coded Suzie sequence is transmitted over a GPRS channel using CS-1 and a C/I of 12 dB. The Suzie sequence is encoded at a bit rate of 64 kbit/s, a frame rate of 10 f/s and then encapsulated in RTP packets for real-time transmission over GPRS using the packetisation scheme of Figure 5.7. For the fixed quality control scheme, the size of each RTP packet is set to 700 bits, while the number of AIR MBs encoded per video frame is set to 8. It can be seen that a significant improvement in quality is achieved by the adaptive content-based quality control technique, both subjectively and objectively. In Figure 5.13, the PSNR recovers more rapidly with the adaptive scheme, after the high motion section, due to the increase in AIR MBs after the motion peak.

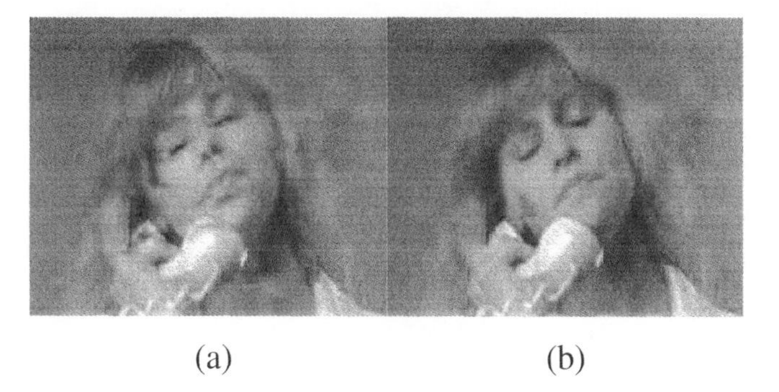

(a) (b)

Figure 5.12 Frame 28 of Suzie sequence coded with MPEG-4 at 64 kbit/s, 10 f/s, after transmission over a GPRS channel (CS-1, C/I = 12 dB): (a) fixed quality control scheme (RTP packet size = 700 bits, 8 AIR MBs/frame), (b) adaptive content-based quality control scheme

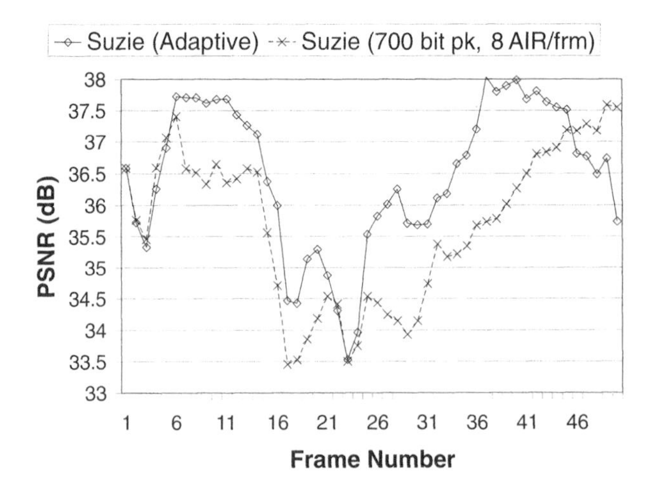

Figure 5.13 PSNR values for 50 frame of Suzie sequence coded with MPEG-4 at 64 kbit/s, 10 f/s, after transmission over a GPRS channel (CS-1, C/I = 12 dB) for both the fixed and the adaptive content-based quality control schemes

5.8 Prioritised Transport for Robust Video Transmissions over Mobile Networks

In addition to its scaleability benefits, the layered video coding discussed in Chapter 3 has inherent error-resilience benefits, particularly when the base layer can be transmitted with higher priority and the enhancement layer(s) with lower priority. The layered video coding is usually accompanied by the use of UEP (Unequal Error Protection) to enable the high-priority base layer to achieve a guaranteed service quality and the enhancement layers to produce quality refinement, as examined in Section 4.4.1. This approach is known as layered coding with transport prioritisation, and is used extensively to facilitate error resilience in video transport systems (Wang and Zhu, 1998). An improvement to conventional layered coding algorithms used for error-resilience purposes introduces the rate-distortion optimisation factor (Gallant and Kossentini, 2001) in the multi-layer video coder. In this technique, the rate-distortion optimisation of each layer is performed in both the error-free and error-prone cases in accordance with the available bandwidth and the particular network conditions. The error-free case involves determining the optimal allocation of bit rate among the source-coding elements, and the error-prone case involves the optimal allocation of bit rate between the source-coding and channel-coding elements with emphasis on the priorities assigned to the generated video layers (UEP).

A similar method for improving the quality of video transport over networks is the prioritisation of different parts of the video bit stream by sending data as two separate streams (refer to Section 3.10). This enables the video encoder to demand that the network send the data using channels with different priorities, allocating more important and error-sensitive data to more reliable and secure channels. Therefore, the stream of motion and header data for instance is assigned a higher level of error protection and then sent over a more reliable bearer than that of the texture stream. In MPEG-4 for instance, data partitioning places critical data at the beginning of each video packet, thereby suppressing the likelihood of losing the video packet when errors hit the less sensitive texture data in the second partition of the packet. This scheme enables the UEP protection of data partitions, with the first partition receiving the greatest protection (Rabiner, Budagavi and Talluri, 1998).

Consequently, the method of prioritisation can include video layering, video data partitioning, UEP and multi-bearer prioritised video transport. However, in mobile radio networks, the implementation of the prioritisation scheme at the application layer would leave all network and transport layer headers unprotected. Moreover, the UEP mechanism applied to the prioritised video information (layer, partition, stream, etc.) does not provide any protection against packet loss. In mobile networks, the high bit error rates also cause frame and packet erasures that are due to the corruption of an important part of the packet enclosure, such as the header or sensitive payload data. Furthermore, applying the prioritisation scheme to the application layer places some limitations on the interoperability of the application. This implies that the development of enhanced services would be impeded, since improving the application capability would then require the change of all the underlying network protocols. To enable the efficient use of video prioritisation schemes in mobile networks, the video transport mechanism has to be taken into consideration (Worrall *et al.*, 2001). This scheme makes use of the data partitioning technique employed by MPEG-4 in each corresponding video packet to send the partitions in two separate video streams and using different GPRS channel protection schemes. The first partition of each MPEG-4 video packet is allocated to the high-priority stream and the second partition is allocated to the low-priority video stream. The synchronisation between the two streams is achieved by means of the time-stamping feature of the MPEG-4 standard. Since the video information is arranged into different streams, UEP can then be employed to provide a guaranteed quality of service for the high-priority stream that involves the sensitive video information. Each stream is transported over a different mobile bearer channel as offered by the underlying network whereby each bearer channel meets different QoS levels set by the application, thereby optimising the performance over the access networks. This prioritised two-stream video transport mechanism is illustrated in Figure 5.14.

As discussed in Section 5.5.2 above, using the multislotting capabilities of GPRS, it is possible to provide data rates of up to 60 kbit/s. This means that the GPRS access channel will be a suitable environment for the provision of real-time video services by using RTP over IP. Therefore, the transport of the prioritised video streams can be accomplished using packet switching technology over the GPRS access network infrastructure. The four channel coding schemes supported

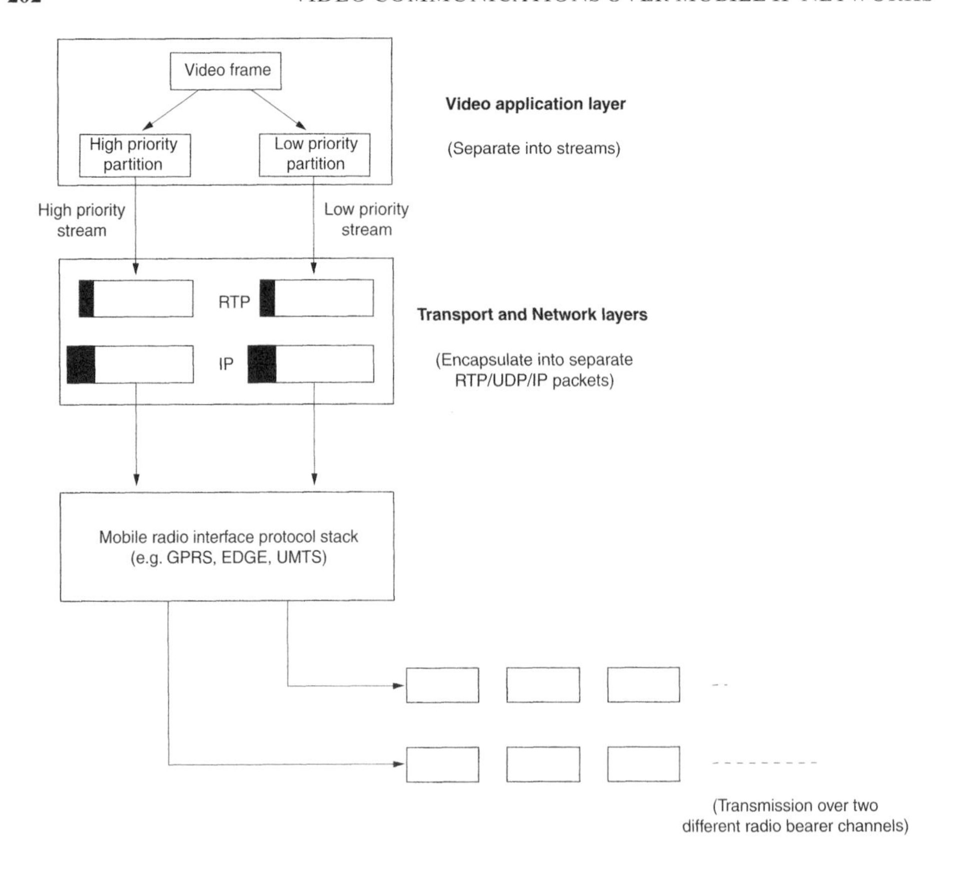

Figure 5.14 Transport of prioritised video streams over mobile access networks using different bearer channels

by GPRS can be used to offer different levels of protection to separate video streams. For real-time operation, video packets are encapsulated into RTP packets and transmitted using the UDP transport protocol. Although the packet video stream is subject to bit errors within the access network, a single bit error in the RTP/UDP/IP header for instance could lead to a whole packet erasure. Therefore, the robustness of packet headers to channel errors plays an important role in dictating the performance of the video service over mobile networks. Figure 5.15 shows the PSNR values obtained for the Suzie sequence coded with MPEG-4 and sent over GPRS using the prioritised video transport technique described above. Two data-partitioned output streams are generated by the MPEG-4 encoder and then encapsulated into RTP packets for real-time transmission. CS-1 is used to protect the high-priority video stream, whereas CS-3 is used to protect the low-priority stream for the prioritised video transport of partitioned video data. However, only CS-2 is used to protect the single stream output with no prioritisation. The PSNR values show that the prioritisation of video steams for UEP protection and transport over two GPRS radio bearer channels offers a

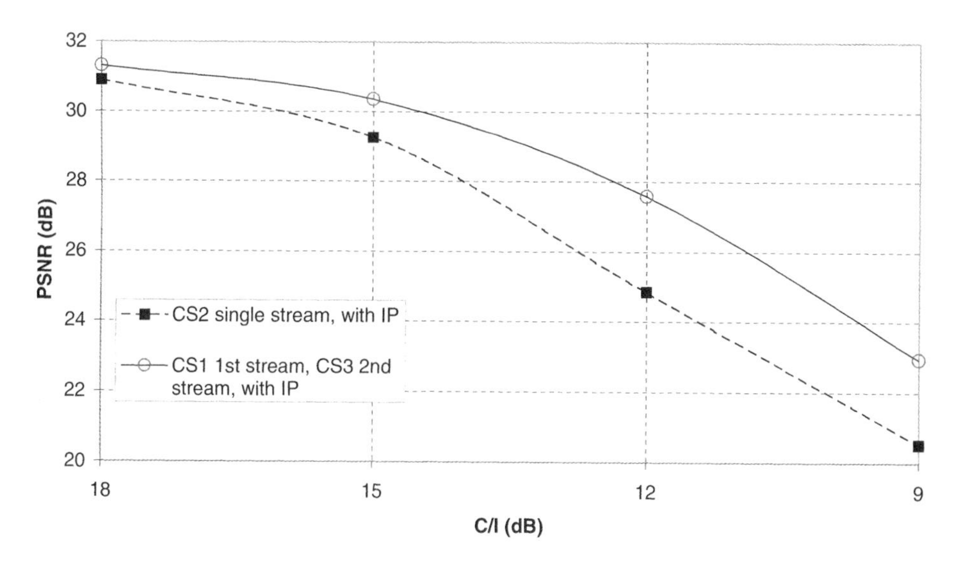

Figure 5.15 PSNR values for the Suzie sequence encoded with MPEG-4 at 64 kbit/s and 10 f/s using the data partitioning option and transmitted over a GPRS channel using both the single stream and the prioritised stream transport mechanisms

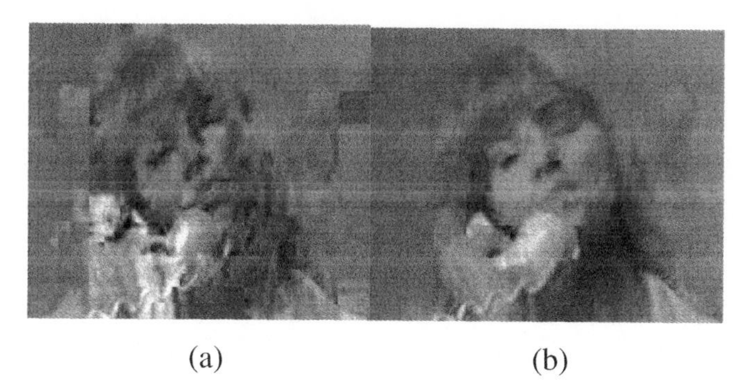

(a) (b)

Figure 5.16 Frame No. 75 of the Suzie sequence coded with MPEG-4 at 64 kbit/s and 10 f/s using the data partitioning option and transmitted over a GPRS channel with C/I = 12 dB: (a) single stream transmission, (b) prioritised stream transmission

better objective performance than the single stream case with EEP (Equal Error Protection) and no transport prioritisation. The corruption of IP packet headers also has a detrimental effect on the perceptual quality of the video sequence since a single bit error in the RTP/UDP/IP packet header leads to the loss of the whole RTP packet. Figure 5.16 depicts the subjective quality improvement achieved by the prioritised video transport over GPRS networks.

5.9 Video Transmissions over GPRS/UMTS Networks

The performance of video services over mobile networks depends on a number of factors. Firstly, the bandwidth allocated to an offered video service dictates the output rate of the video source and hence the temporal/spatial quality of the video sequence. Due to the multi-slotting capabilities of the mobile radio interface, the throughput available to a video source can be increased with various rates of error protection. Therefore, the bandwidth is directly related to the number of time slots used by the mobile radio interface, as discussed in Section 5.5.2. Obviously, the output rate of the video source is a direct consequence of the encoding parameters used such as the quantisation level (Qp), the frame rate and the spatial image resolution. These parameters have a direct impact on the performance of a given video service.

On the other hand, the data rate available at the lower layers of the mobile access network protocol architecture (GPRS/UMTS) is different from that available at the video application layer. At each layer of the protocol stack, a header is associated with the payload of the video packet. This introduces a bit rate overhead at each layer of the hierarchy and reduces the net throughput presented to the video source. The data rate available to a video source per time slot is also a function of the channel coding scheme employed. These schemes achieve different levels of protection onto video data and therefore inflict different amounts of redundancy on the payload. Consequently, the video throughput is a function of both the number of time slots used at the radio interface and the channel coding scheme employed. Table 5.5 shows the data rate per time slot available at the physical layer (PDTCH) of the GPRS protocol architecture for the supported channel coding schemes. At the video application layer, the net throughput available to the source is affected by the percentage of overhead placed by the protocol headers in relation to the video payload size in the corresponding packets. In Section 5.5.2, it was established that this overhead could reach a percentage of 10 to 15 per cent for a video frame rate of 5 to 10 f/s and a QCIF frame resolution.

Consequently, the combination of a channel coding scheme and the number of time slots employed in the video transmission is a major factor that affects the

Table 5.5 Data rates (kbit/s) per timeslot presented to the physical link layer by the four channel coding schemes of GPRS

Scheme	Code rate	Radio blocks (bits)	Data rate (kb/s)
CS-1	1/2	181	9.05
CS-2	≈2/3	268	13.4
CS-3	≈3/4	312	15.6
CS-4	1	428	21.4

performance of a mobile video service. The number of time slots used at the radio interface determines the data rate at the physical layer, whereas the employed channel coding scheme identifies the level of error protection received by the video payload and also the throughput per time slot. Figure 5.17 shows the effect of CS/TS combinations on the performance of a mobile video service. The PSNR values are obtained for transmitting a QCIF video sequence encoded at 32 kbit/s and 10 f/s over a GPRS channel (Fabri *et al.*, 2000). Obviously, different levels of video quality could be obtained for different combinations of a channel coding scheme and timeslot allocation. It can be observed that the increase in time slot allocation results in improving the decoded video quality. When two time slots are allocated to the mobile terminal, only a minimal video quality could be achieved by CS-2 and CS-3 at the rate of 5 f/s, which is half of the original frame rate of the coded video sequence. However, CS-1 cannot sustain the 5 f/s frame rate with only two time slots, since a lot of video frames would then be dropped by the rate control mechanism (of the video encoder), thereby degrading the temporal video quality. The availability of three time slots at the radio interface allows for an acceptable quality of video service at the three considered channel coding schemes.

In addition to bandwidth efficiency, the channel coding schemes affect the error performance of the video service. A scheme that results in a larger amount of redundancy and hence a less efficient bandwidth utilisation offers a higher protection rate, and *vice versa*. Therefore, the selection of the most appropriate channel coding scheme must be made in the light of the observed channel conditions to achieve an optimal trade-off between the bandwidth efficiency and the error performance of the video service. In good channel conditions, the error protection rate is reduced to increase the throughput of the video source. Conversely, in bad

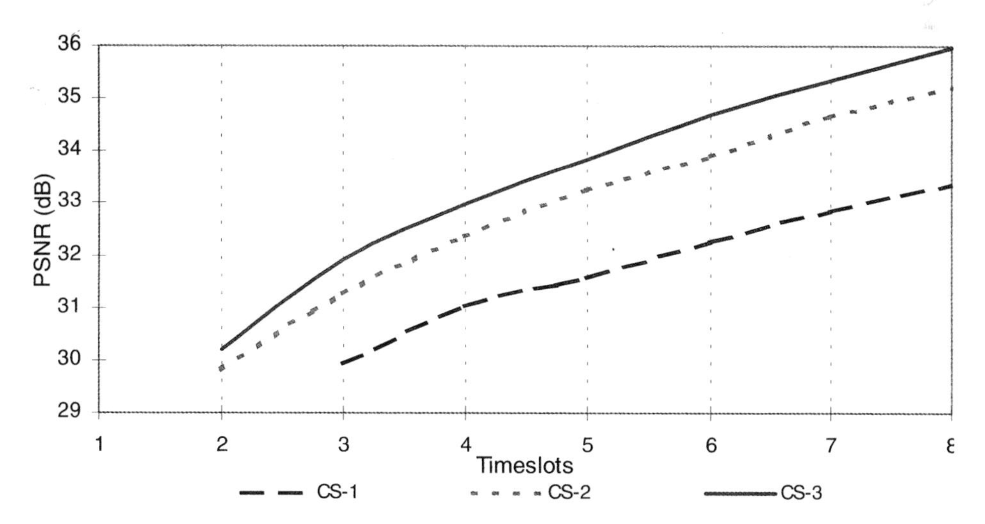

Figure 5.17 Objective quality levels obtained using different CS-TS combinations to transmit a video sequence coded at 32 kbit/s and 10 f/s over GPRS

channel conditions, the video source reduces its rate to enable a more powerful error protection, thereby achieving a guaranteed quality of service to the transmitted video data. Figure 5.18 shows the PSNR values obtained for transmitting the same QCIF video sequence as in Figure 5.17 over a GPRS channel at different Eb/No ratios. It can be seen from the PSNR graph that CS-1 offers error protection at the error-free quality level for Eb/No values of slightly in excess of 11 dB. CS-2 and CS-3 achieve the same level of protection at higher Eb/No values of 14 and 18 dB, respectively. Therefore, in order to achieve an error-free video quality, the channel coding scheme has to be chosen in accordance with the channel condition, i.e. Eb/No ratio, which is 11 dB for CS-1, 14 dB for CS-2 and 18 dB for CS-3. For Eb/No values which are lower than 11 dB, even with the CS-1 coding scheme, error-free video quality is not achievable. However, an acceptable level of quality could still be sustained for Eb/No values as low as 7 dB, representing a BER value of 1 per cent, when CS-1 is employed for error-protection purposes.

In addition to the bandwidth factor and CS-TS combinations, equally important factors that control the performance of video services in mobile networks are the error-resilience capabilities of the video encoder and the error-concealment effect of the video decoder. In Chapter 4, the effects of bit errors and packet loss were addressed and a variety of techniques for error control in video communications were presented. These error-resilience schemes are used to mitigate the effects of mobile channel errors and the resulting packet loss on the decoded video quality. Combinations of these error resilience tools can be used to optimise the error-resilience capabilities of the video encoder, as described in Section 4.12. Standard-compliant combinations of error-resilience tools for wireless video applications have been specified in Annex X of H.263++, as mentioned in Section 5.7.1. To evaluate the performance of video transmissions over GPRS and UMTS, five main combinations of error-resilience tools are selected in accordance with the

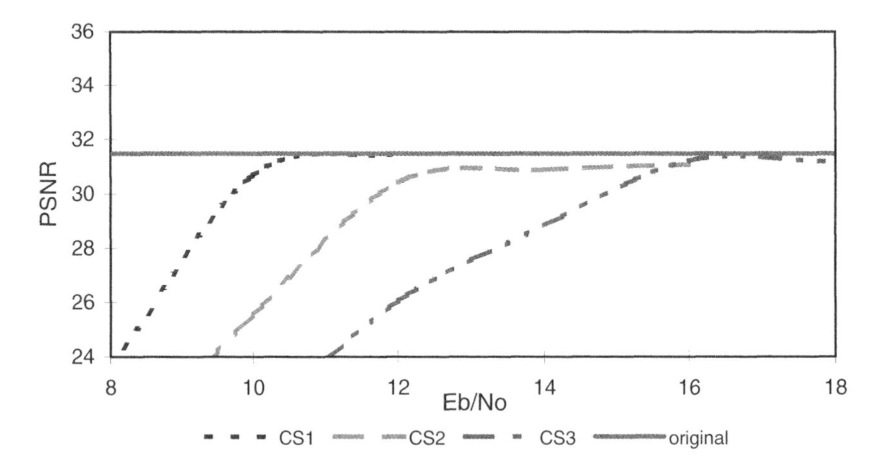

Figure 5.18 Objective quality levels obtained using different CS schemes for transmitting a video sequence coded at 32 kbit/s and 10 f/s over GPRS

profiles specified in Annex X of H.263, as outlined in Table 5.6.

Synchronisation markers refer to the packet structure of MPEG-4, where each packet is a unit of synchronisation consisting of an integer number of MBs. This error-resilience tool is similar to Annex K of H.263+ on the slice structure mode. It is used for its enhanced ability to provide resynchronisation points within the bit stream that enable the recovery from error corruption or data loss. The packet structure of MPEG-4, or alternatively Annex R of H.263+, ensures that errors inside a packet or a segment, respectively, do not propagate to other areas of the scene, and so can reduce the spatial spreading of errors. The data partitioning tool of MPEG-4 refers to splitting the packet into two parts, with the first containing motion and headers data, and the second purely consisting of texture data. It is equivalent to the data partitioned slice mode (Annex V) of H.263+. No error resilience corresponds to Profile 0 of Annex X, sync only refers to Profiles 3 and 6 of the same annex and full error resilience is equivalent to Profile 4, with the only exception of the picture header repetition enabled by Annex X for enhanced supplemental information. In addition to the error-resilience tools outlined in Table 5.6, error concealment is enabled at the video decoder by replacing the error-affected MBs by their corresponding motion-compensated ones in the previous frame. The performance of the video service over GPRS is evaluated here using a QCIF video sequence coded with MPEG-4 at a throughput of 32 kbit/s and a frame rate of 5 f/s. This represents the data rate achievable using five time slots multiplexing at CS-1 or 3 slots at CS-2. Each of the five encoded bit streams is subject to 13 different error patterns generated by a model that simulate the Typical Urban multipath environment of the GPRS channel with ideal frequency hopping and a single co-channel interferer, as shown in Table 5.7. The average of the 13 different PSNR values obtained in each of the five cases is shown in Figure 5.19.

The PSNR values in Figure 5.19 clearly demonstrate the effectiveness of the error-resilience tools in improving the quality of a video service over GPRS. The use of synchronisation codewords is seen to provide between 2 and 5 dB improvement in quality, depending upon the carrier-to-interference ratio. The resilience performance can be improved significantly by introducing data partitioning,

Table 5.6 Combinations of standards-compliant error-resilience schemes used for the performance evaluation of video transmissions over GPRS and UMTS

Error-resilience scheme	Synchronisation markers	Reversible codewords	Data partitioning
No error resilience	×	×	×
Sync only	√	×	×
Sync + RVLC	√	√	×
Sync + DP	√	×	√
Full error resilience	√	√	√

Table 5.7 Parameter settings of the GPRS physical link layer

Parameter	Settings
Channel coding scheme employed	GPRS PDTCH CS-1
Interleaving	Block rectangular over four frames for GPRS, Diagonal for HSCSD
Modulation	GMSK
Interference characteristics	Single co-channel interferer
Fading characteristics	Rayleigh fading for each path. Fading varies during one burst
Multipath characteristics	GSM typical urban
Transmission capabilities	Ideal frequency hopping
Mobile terminal velocity	3 km/h
Carrier frequency	900 MHz
Antenna characteristics	0 dB gain for both transmitter and receiver. No antenna diversity
Signal-to-noise characteristics	26 dB AWGN source at receiver
Burst recovery	Synchronisation based on the cross-correlation properties of the training sequence
Equaliser	16-state soft output equaliser
Channel decoding	Soft-decision Viterbi convolutional decoder. Fire correction and detection

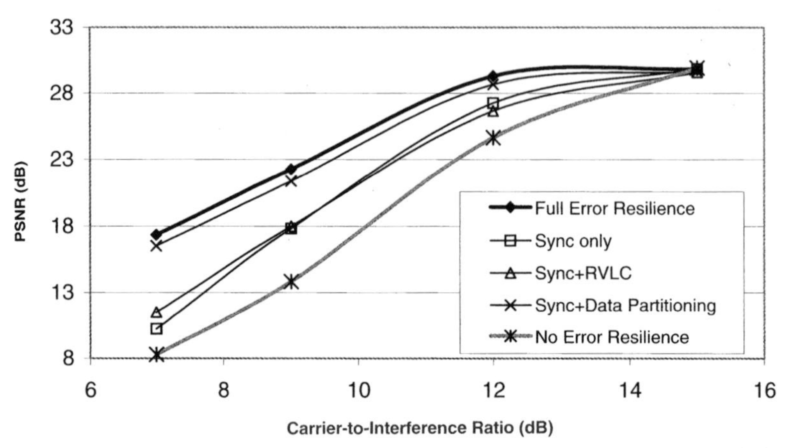

Figure 5.19 Average PSNR values showing the performance of video transmissions over GPRS at 32 kbit/s and 5 f/s using CS-1 and five time slots

which gives a further 4–1 dB gain in PSNR. Reversible codewords are not seen to provide a large improvement, although when used in conjunction with data partitioning, this gives a fairly constant 1 dB improvement in quality. This is due to the small video packet size used (700 bits). This means that whenever an error occurs, only a limited number of MBs are irreversibly corrupted. If sync markers were not used, the effect of RVLCs would be much more pronounced. The objective results clearly indicate that the best error performance is obtained when all error-resilience tools are used together, with the effect of the employed tools becoming more pronounced as the channel conditions degrade. Figure 5.20 shows the subjective video quality achieved by the transmission of MPEG-4 coded Foreman sequence over GPRS with the same physical link parameter settings of Table 5.7 without and with full error resilience.

Similarly, the performance of compressed video transmissions over UMTS is evaluated using the same MPEG-4 coded sequence above at two throughputs of 128 and 64 kbit/s and frame rates of 10 and 5 f/s, respectively. The error conditions are those expected to be found in Wideband-CDMA environments as used in UMTS Radio Access Networks. The error patterns are generated by a model that simulates the UMTS channel, with the parameters settings outlined in Table 5.8.

Figures 5.21 and 5.22 show the PSNR results obtained by transmitting MPEG-4 video data over the UMTS channel characterised in Table 5.8 for the five error-resilience cases of Table 5.6. Figure 5.23 depicts the corresponding subjective quality achieved. It is clear that the combined error-resilience tools of the MPEG-4 video coder offer an objective quality improvement of 2 dB over the no-error-resilience case under a bit error rate of 5×10^{-3}. When transmitting at 128 kbit/s, this improvement is reduced to 1.5 dB under the same error conditions. The use of synchronisation markers, RVLCs and data partitioning is seen to give varying degrees of improvement at error rates above 10^{-4}. Below this value, the difference in performance between the different combinations of schemes is marginal. In general, it is seen that the best performance is once again achieved with all error-resilience tools enabled. This is another indication of the importance of the error-resilience tools (or efficient combinations of them) in controlling the performance of the video service and defining the quality level for a given mobile channel condition.

Two trends may be observed from these results. The relative improvement obtained by employing error-resilience mechanisms decreases as the source

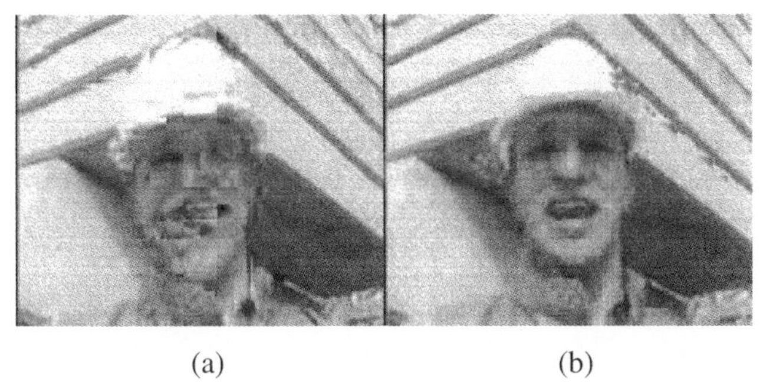

(a) (b)

Figure 5.20 A frame extracted from the Foreman sequence coded with MPEG-4 at 32 kbit/s and 5 f/s and transmitted over GPRS at C/I = 12 dB using CS-1 and five time slots: (a) no error resilience, (b) full error resilience

Table 5.8 Parameter settings of the UMTS physical link layer

Frequency	1920 MHz
Chip rate	4.096 Mcps
Transmission direction	Uplink
Multipath profile	ITU Outdoor-to-indoor A
Mobile speed	3 km/h and 50 km/h
Bit rate	64, 128, kbps
Interleaving depth	40 ms
DPDCH spreading factor	32, 16, 8, 4
RLC-PDU = frame with CRC bits. FER is calculated per RLC-PDU	640 bits
Number of RLC-PDUs per interleaving period	2–24
CRC bits	16
Coding	1/3-rate Turbo code, 4 states
DPCCH spreading factor	256
Power difference between DPCCH and DPDCH	32, 64 and 128 kbps: −6 dB 384 kbps: −9 dB
Power control step	1 dB
Power control signalling errors	0%
RX-antenna diversity	Yes (2 antennae)
Simulation length	180 s

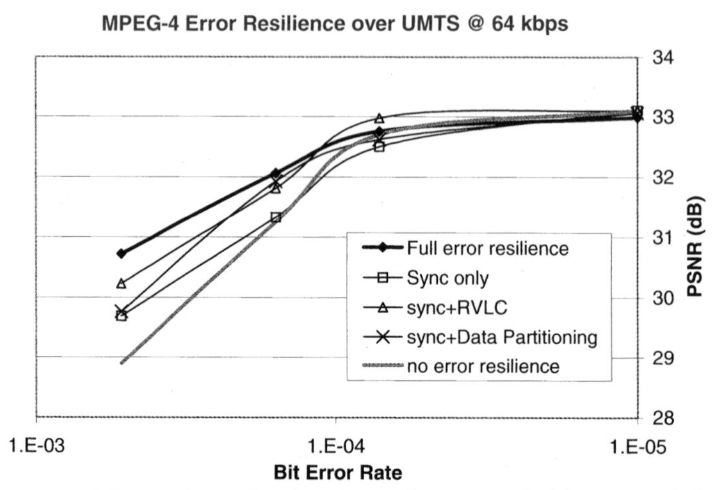

Figure 5.21 Average PSNR values, showing the performance of video transmissions over UMTS at 64 kbit/s and 5 f/s

throughput increases. This is clear from a comparison between the performance at 64 and 128 kbit/s. In addition, the use of synchronisation words reduces the throughput available to represent the actual video information. This means that at low error rates, the error-resilient bit streams are likely to produce lower-quality video output than the standard non-resilient schemes. These characteristics show

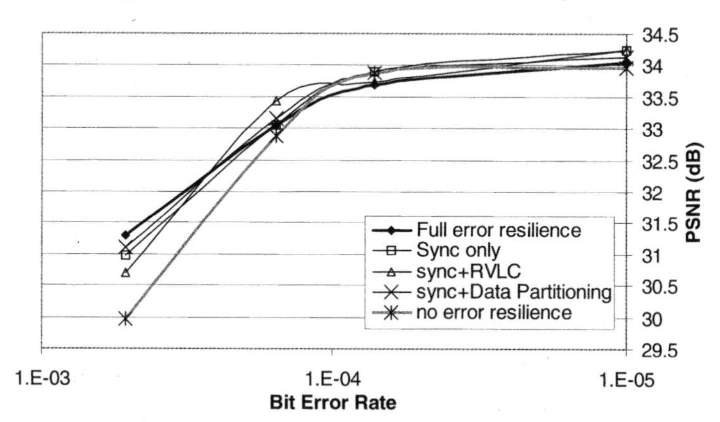

Figure 5.22 Average PSNR values, showing the performance of video transmissions over UMTS at 128 kbit/s and 10 f/s

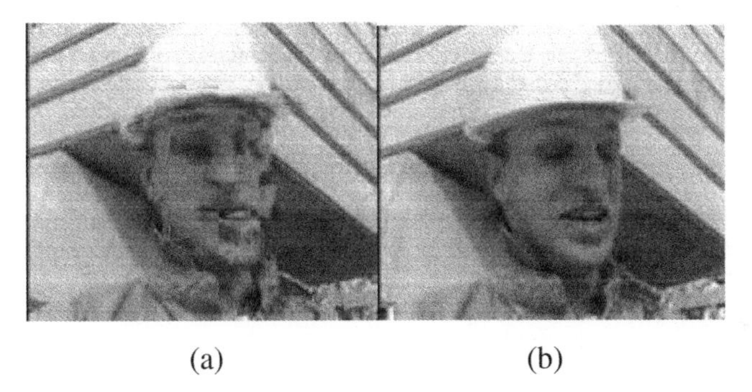

Figure 5.23 A frame extracted from the Foreman sequence coded with MPEG-4 at 128 kbit/s and 10 f/s and transmitted over UMTS WCDMA at BER $= 1.5 \times 10^{-4}$: (a) no error resilience, (b) full error resilience

that error resilience is more critical in GPRS environments, where not only are the error characteristics likely to be more demanding than in the UMTS WCDMA environments, but also the lower source bit rates available cause the compressed video to be more susceptible to errors. This can be attributed to the higher proportion of the total number of bits carrying header information at lower bit rates.

It must also be pointed out that the objective PSNR results do not always provide a reliable metric for the performance evaluation of a video communication service. Since video signals are three-dimensional, the temporal factor plays a major role in the overall quality assessment. Although the PSNR results of Figures 5.21 and 5.22 reflect slight differences between the non-error-resilient and error-resilient video transmissions for BER $\leq 10^{-5}$, the subjective evaluation performed by comparing the perceptual quality of the two video clips shows a noticeable quality enhancement brought forward by the error-resilient version of the standard, as shown in Figure 5.23, for instance, for a BER of 1.5×10^{-4}. This improvement may not in all cases be well reflected by the objective PSNR quality metric.

5.10 Conclusions

The multi-slotting capabilities of the mobile radio interface help increase the throughput available to a mobile terminal, thereby enabling the provision of video services over the new generation mobile networks. The number of time slots available to the mobile terminal and the error protection provided by the channel coding schemes are two major elements that control the video service quality. The selection of the optimal combination of time slots and channel coding scheme is a trade-off between the error performance and the throughput of the video application.

RTP is a transport-layer protocol that is used for the provision of real-time multimedia services over IP-based networks. Since video services are very delay-sensitive, retransmissions of lost or corrupted video data are not desirable. Therefore, the end-to-end error control mechanism of real-time video services is performed by the video application using the feedback messages sent in the RTCP (Real-time Transport Control Protocol) packets. One of the most efficient application-layer error control mechanisms employed in mobile video communications is error resilience. Annex X of H.263 has recommended a number of profiles that specify combinations of error-resilience tools used in various mobile video applications. The performance evaluation of mobile video services over GPRS and UMTS shows that the combined error-resilience schemes achieve a noticeable improvement in video quality, both subjectively and objectively.

In addition to the error resilience and the optimal combination of a channel coding scheme and the number of time slots available, the video quality of service is influenced by the packetisation scheme employed. In real-time mobile video services, the compressed video payload is encapsulated in RTP/UDP/IP packets. The headers of each packet do not only impair the channel utilisation but also present a very high sensitivity to errors. The corruption of the headers leads to the whole packet erasure and hence the loss of its video data enclosure. The study presented in this chapter demonstrates the importance of using error control to mitigate the effects of packet loss on the received video quality. Primarily, error concealment makes use of previously received error-free video data to compensate for the loss of data in the dropped packets. More importantly, error-resilience provides immunity to transmitted video data in order to improve the ability of the decoder to resynchronise upon detection of an error. Furthermore, error control using the channel coding schemes of mobile network protocols improves the ability of the mobile video system to detect and correct errors in the received bit stream. Finally, the packetisation scheme specifies the size and structure of the packet, thereby controlling the throughput and error performance of the overlying video application. Adaptive content-based packetisation schemes can also be used to achieve optimal error performance by adaptively modifying the packet size in accordance with the motion activity of the video scene.

5.11 References

Basso, A., Varakliotis, S., and Castagno, R., Transport of MPEG-4 over IP/RTP, Packet video workshop, May 2000, Sardinia, Italy 2000.

Brasche, G., and Walke, B., Concepts, services and protocols of the new GSM phase 2+ General Packet Radio Service, *IEEE Communications Magazine*, 94–104, 1997.

Digital Cellular Telecommunication System, GSM Radio Access Phase 3; Channel Coding, GSM 05.03, V. 6.1.

Digital Cellular Telecommunications System (Phase 2+); Radio Transmissions and Reception, GSM 05.05 V 7.0.

ETSI/SMG, GSM 03.64 1998, Overall description of the GPRS radio interface stage 2, V. 5.2.0.

Fabri, S., Worrall, S., Kondoz, A. M., and Sadka, A. H., Real time video communications over GPRS, *Proceedings of the IEE 3G Conference*, London, UK, Mar. 2000.

Gallant, M., and Kossentini, F., Rate-distortion optimised layered coding with unequal error protection for robust Internet video, *IEEE Trans. on Circuits and Systems for Video Technology*, **11**, No. 3, 357–372, Mar. 2001.

Ghanbari, M., and Hughes, C. J., Packing coded video signals into ATM cells, *IEEE/ACM Transactions on Networking*, 505–509, Oct. 1993.

ISO/IEC JTC 1/SC 29/WG11: Information technology – Generic coding of audio-visual objects – Part 2: Visual, ISO/IEC 14496-2, MPEG Vancouver meeting, July 1999.

ITU-T Recommendation H.263 (draft), Video coding for low bit rate communication, Nov. 2000.

Rabiner, W., Budagavi, M., and Talluri, R., Proposed extensions to DMIF for supporting unequal error protection of MPEG-4 video over H.324 mobile networks, Doc. M4135, MPEG Atlantic City meeting, Oct. 1998.

Schulzrinne, H., Casner, S., Frederick, R., and Jacobson, V., RTP: a transport protocol for real-time applications, Audio-Video Transport Working Group, RFC 1889, Jan. 1996.

Tdoc SMG2 EDGE 401/99, Working assumption for receiver performance requirements, EDGE Drafting Group, Aug. 1999.

Tdoc SMG2 086/00, Outcome of drafting group on MS EGPRS Rx performance, EDGE Drafting Group, Jan. 2000.

Third-generation Partnership Project, Technical Specification Group, Services and System Aspects; Service Aspects; Mobile multimedia services including mobile Intranet and Internet services, 3G TR 22.960, Version 3.0.1, April 1999.

Third-generation Partnership Project, Technical Specification Group, Services and System Aspects; Services and service capabilities. 3G TS 22.105, Version 3.6.0, October 1999.

Third-generation Partnership Project, Technical Specification Group, Services and System Aspects; QoS concept and architecture. 3G TR 23.107, Version 3.0.0, October 1999.

Wang, Y., and Zhu, Q. F., Error control and concealment for video communication: a review, *Proc. IEEE*, **86**, 974–997, May 1998.

Worrall, S., Sadka, A. H., Sweeney, P., and Kondoz, A. M., Optimal packetisation of MPEG-4 using RTP over mobile networks, *IEE Proceedings of Communications*, 2001.

Worrall, S., Sadka, A. H., Sweeney, P., and Kondoz, A. M., Motion adaptive INTRA refresh for MPEG-4, *IEE Electronics Letters*, **36**, No. 23, 1924–1925, Nov. 2000.

Worrall, S., Fabri, S., Sadka, A. H., and Kondoz, A. M., Prioritisation of data partitioned MPEG-4 video over mobile networks, *European Transactions on Telecommunications*, **12**, No. 3, 169–174, May/June 2001.

6

Video Transcoding for Inter-network Communications

S. Dogan, A. H. Sadka

6.1 Introduction

Due to the expansion and diversity of multimedia applications and the underlying networking platforms with their associated communication protocols, there has been a growing need for inter-network communications and media gateways. Eventually, these applications will encounter compatibility problems. Not only will asymmetric networks run different set of communication protocols, but they will also operate various kinds of incompatible source coding algorithms that are characterised by different target bit rates and compression techniques. Therefore, the interoperability of these source coders necessitates the presence of a control unit which acts as a media traffic gateway lying on the borders of the underlying networking platforms. This chapter is dedicated to the investigation of various methods which achieve the interoperability of compressed video streams while taking into consideration the application-driven constraints and the varying network conditions. The video transcoding algorithms are examined and analysed, and their performances are evaluated using both subjective and objective methods.

6.2 What is Transcoding?

Video transcoding comprises the necessary operations for the conversion of a compressed video stream from one syntax to another one for inter-network communications. Thus, the tool that makes use of this algorithm to perform the necessary conversions is called a video transcoder.

The original idea behind video transcoding was the scaleability of video coding techniques (Ghanbari, 1989; Radha and Chen, 1999). These techniques comprise a layered video encoder structure that provides different layers of compressed video, with each layer coded at a different bit rate. Scaleability allows the video coder to produce different video streams at different bit rates and QoS levels using only a single video source. At the time, this was necessary due to the wide deployment of video-on-demand (VoD) applications, where high-resolution high-quality video was required for delivery to network subscribers with bandwidth-limited or congested links. In such cases, the most appropriate low bit rate version of the bit stream could be chosen at the expense of smaller resolution and lower perceptual quality. Layering was accomplished with one base layer providing the minimum requirements for the reconstruction of low bit rate video and several enhancement layers (on top of the base layer) for enhanced quality resulting in increased bit rates. According to the varying network conditions, adequate bit rates were achieved by selecting either the base layer only or the base plus one or more enhancement layers. However, scaleable encoding required the use of complex scaleability techniques, leading to extra processing power requirements and additional delays resulting in complex and sub-optimal video encoder and decoder implementations.

Besides complexity, the frequent changes in network conditions and constraints require necessary actions to be taken at a different location (other than encoder and decoder) within the network. This specific location, as seen in Figure 6.1, is referred to as video proxy, that enables faster network responses. The video proxy helps the video encoders and decoders remain free of unnecessary

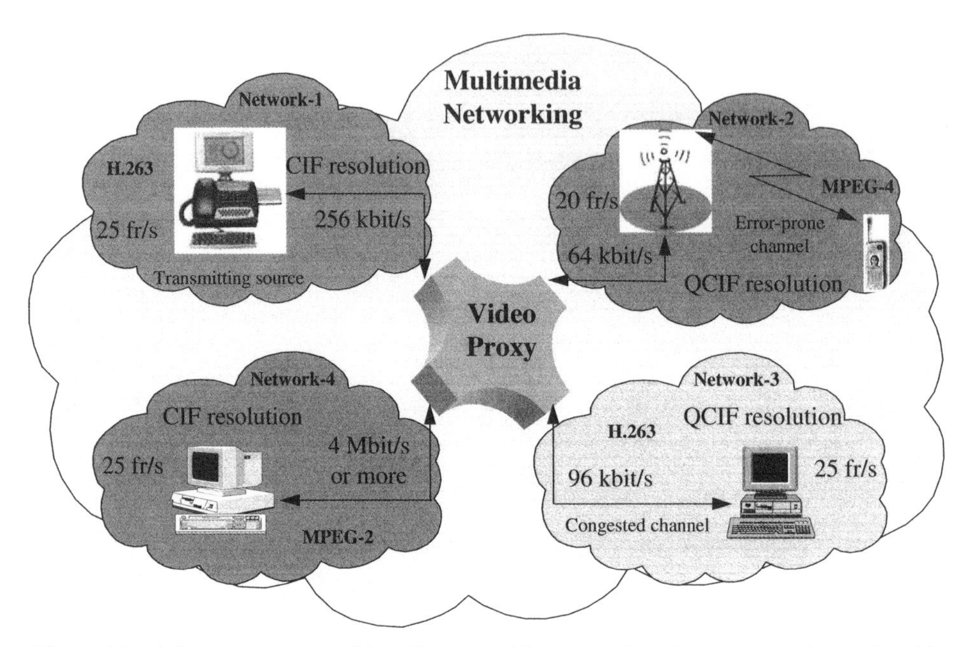

Figure 6.1 A heterogeneous multimedia networking scenario using a transcoder at the video proxy

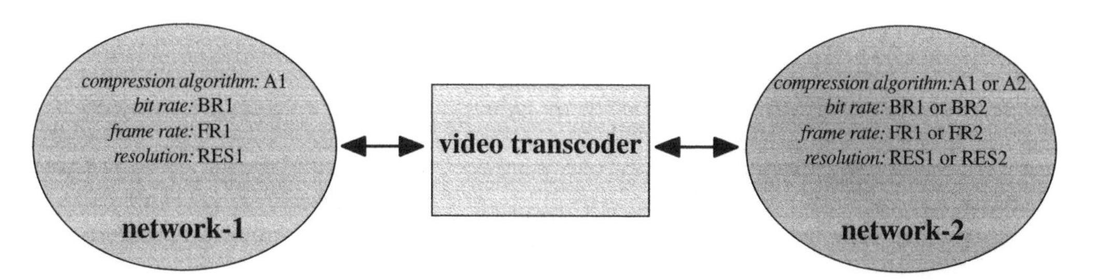

Figure 6.2 Video transcoding

complexities incurred by the scaleability algorithms. A video proxy can consist of a single or a group of video transcoders operating simultaneously.

Therefore, video transcoding is a process whereby an incoming compressed video stream is converted to a different video format, size, transmission rate or simply translated to a new syntax without the need for the full decoding/re-encoding operations, as depicted in Figure 6.2. Using transcoding, the complexity, processing power and delay incurred by the necessary conversion operations are kept minimal while achieving an improvement to the decoded video quality (Bjork and Christopoulos, 1998; Kan and Fan, 1998; Keesman *et al.*, 1996).

Four major types of video transcoding algorithms have been proposed and presented (Assuncao and Ghanbari, 1996; Kan and Fan, 1998; Keesman *et al.*, 1996; de los Reyes *et al.*, 1998; Warabino *et al.*, 2000; Youn, Sun and Xin, 1999; Youn and Sun, 2000). The most commonly discussed one is the homogeneous video transcoding that comprises bit rate, frame rate and/or resolution reduction algorithms for varying transmission conditions. Heterogeneous video transcoding has become popular as diverse multimedia networks have emerged and become operational. Moreover, the third and fourth types are gaining increasing attention for error resilience applications and multimedia traffic planning purposes.

6.3 Homogeneous Video Transcoding

Homogeneous video transcoding algorithms aim to reduce the bit rate, frame rate and/or resolution of the pre-encoded video stream. The reason they are called homogeneous transcoding methods is that they do not involve any kind of syntax modifications to coded video data. Therefore, the incoming compressed video stream preserves its format and compression characteristics after it has been converted to a lower rate or resolution, as illustrated in Figure 6.3.

By using the incoming video bit stream as input to the video transcoder, it is possible to transmit the transcoded video data onto the communication channels that have different bandwidth requirements, and at various output bit rates. This very important feature gives support for multipoint video conferencing scenarios.

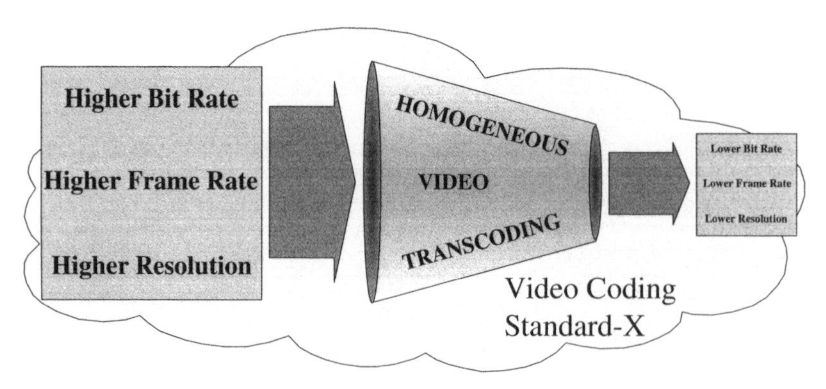

Figure 6.3 Homogeneous video transcoding

There are two methods for combining multiple video streams to achieve successful video conferencing, namely the coded domain combiner and transcoding. The former is rather a simple and a less complex process, whereby the outgoing video stream is obtained by concatenating the incoming multiple video streams. Thus, the combined bit rate is the sum of bit rates of all the incoming video streams. This method distributes the available bandwidth evenly among all the participants of a videoconferencing session. Therefore, the input/output bit rates for each user become highly asymmetric, yet allocating bandwidth to video sources regardless of their activity. On the other hand, the latter method, namely transcoding, partially decodes each of the incoming video streams, combines them in the pixel domain and re-encodes the video data in the form of a single video stream. This method provides every user with full bandwidth and uniform video quality due to the re-encoding of high motion areas of active conference participants with higher bit rates. Obviously, this second method incurs a higher complexity than the simpler combination method (Sun, Wu and Hwang, 1998).

Similarly, Lin, Liou and Chen (2000) present a dynamic rate control method that operates in the video transcoder to enhance the visual quality and allow region of interest (ROI) coding in multipoint video conferencing. This method firstly identifies the active conference participants from the multiple incoming video streams. Then the motion active streams are transcoded with a more optimised bit allocation approach at the expense of relatively reduced qualities provided to inactive users.

Research into homogeneous video transcoding has been boosted by the increasing popularity of VoD applications. Since VoD data is encoded as a high quality, high resolution and high bit rate MPEG-2 stream (i.e. a few Mbit/s), reducing the rate is at times necessary, particularly when an end-user cannot handle the rate of the original video stream. This rate reduction is also necessary in bandwidth-limited networks or even at congested network nodes. Not only the original rate, but also sometimes the original spatial video resolution need to be reduced (such as CIF to QCIF) as end-users are equipped with smaller resolution displays.

6.4 Bit Rate Reduction

Bit rate reduction algorithms have been the most popular research topic among all the different video transcoding schemes available so far, due to considerable interest in VoD applications. The examples of standard rate conversions can easily be found in literature for high bit rate video transmissions, such as conversions from a few Mbit/s down to a few hundred kbit/s. However, due to the deployment of mobile wireless interfaces and satellite links, conversions from high to low rates and from low to very low bit rates (i.e. from a few Mbit/s to a few hundred kbit/s or from a few hundred kbit/s to a few ten kbit/s) have also become increasingly important.

As described in Chapter 3, the incoming bit rates can be down-scaled either by arbitrarily selecting the high-frequency discrete cosine transform (DCT) coefficients first and then simply discarding (truncating) them (Assuncao and Ghanbari, 1997) or by performing a re-quantisation process with a coarser quantisation step-size (Nakajima, Hori and Kanoh, 1995; Sun, Wu and Hwang, 1998; Werner, 1999). Both methods reduce the number of DCT coefficients by causing a number of them to become zero coefficients, thereby reducing the number of non-zero coefficients to be coded. This gives rise to a lower bit rate at the output of the transcoder.

One of the bit rate reduction methods is the re-quantisation of the transform coefficients, as already discussed in Chapter 3. Re-quantisation is achieved by the use of built-in scalar quantisers in MPEG video standards. A second approach has been introduced by Lois and Bozoki (1998). Instead of using the scalar quantisation in the transcoder, a lattice vector quantiser (LVQ) is applied to exceed the MPEG compression capabilities while providing acceptable quality. LVQ is a multidimensional generalisation of uniform step scalar quantisers which produces minimal distortion for a certain input of uniform distribution. The codebook storage is not required and the search complexity is simplified. LVQ allows the quantisation errors to be more uniform in the transcoded pictures, and hence smaller artefacts are visible on the edges. However, the drawback of the algorithm is that LVQ transcoding leads to MPEG-incompatible bit streams. Therefore, a low complexity and low cost user interface is also needed, which involves the LVQ decoder and the MPEG entropy encoding engine. The output can then be directly fed into an MPEG video decoder at the very end of the telecommunication system.

The DCT is a widely used method in most of the current image and video compression standards, such as JPEG, MPEG, H.26X series, etc. Guo, Au and Letaief (2000) present three distribution parameter estimation methods based on the de-quantised values of DCT coefficients used in the transcoding schemes. The methods achieve good transcoding qualities even for fixed rate scenarios.

Bit rate reduction can be accomplished using one of five different schemes. The first one is the conventional cascaded fully decoding/re-encoding scheme. The remaining four schemes consist of low-complexity straightforward transcoding methods. These schemes are used for fixed quality and hence variable bit rate conditions. For fixed rate and hence variable quality applications, the same methods can also be exploited while taking into consideration the changing quality factor in the video transcoder. The target bit rates generated by fixed rate operations can be achieved by using simple mathematical equations given in (Assuncao and Ghanbari, 1997; Fu *et al.*, 1999; Lee, Pattichis and Bovik, 1998).

6.5 Cascaded Fully Decoding / Re-encoding Scheme

The cascaded method of fully decoding and then re-encoding of the incoming compressed video stream is the conventional tandem operation of two video networks, as seen in Figure 6.4. This scheme comprises the full decoding of the input bit stream, and then performs re-sizing and/or re-ordering of the decoded sequence before fully re-encoding it. This scheme involves complex frame re-ordering and full-scale (± 16 pixels) motion re-estimation operations. Therefore, it is the scheme that has the highest complexity, a high processing time and power consumption, causing a significant delay and low-quality pictures due to the motion re-estimation mechanism that is performed by reference to the reduced quality decoded pictures.

In conclusion, this scheme is a sub-optimal scheme with a high level of complexity. It performs two separate operations on the incoming video stream, namely full decoding and re-encoding processes. As a result, the video frame headers and the MB headers are modified by the re-encoding process. Figure 6.4 shows the difference between the cascaded decoding/re-encoding method and the transcoding algorithm where the decoder and the re-encoder blocks are replaced by a lower complexity approach.

6.6 Transcoding with Re-quantisation Scheme

The transcoding method that employs simple or direct re-quantisation is also referred to as the open-loop transcoding algorithm. The reason for such classification is that the scheme depends on a straightforward simple transcoding operation without any feedback loop, as illustrated in Figure 6.5. Using this algorithm, only the DCT coefficients are decoded while other video parameters (such as motion vectors) remain in the VLC domain. Then the decoded transform coefficients are inverse zigzag-scanned and inverse quantised with the quantisation parameter of the video encoder. Preceding the zigzag re-scanning operation, the DCT coefficients are re-quantised with a coarser quantiser in order to reduce the video transmission rate, as stated earlier. Eventually, the re-quantised coefficients need

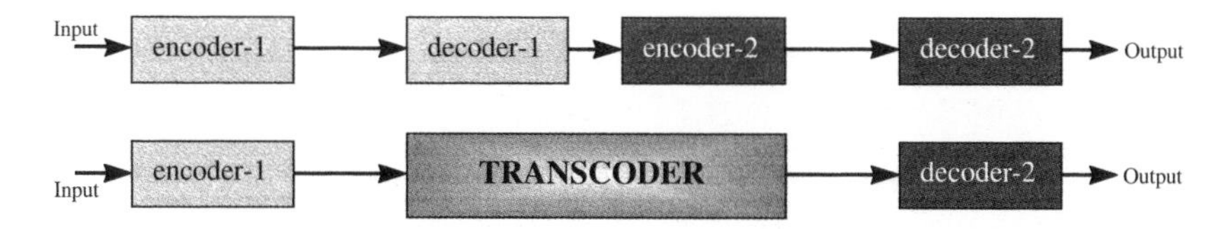

Figure 6.4 Cascaded fully decoding/re-encoding scheme versus transcoding

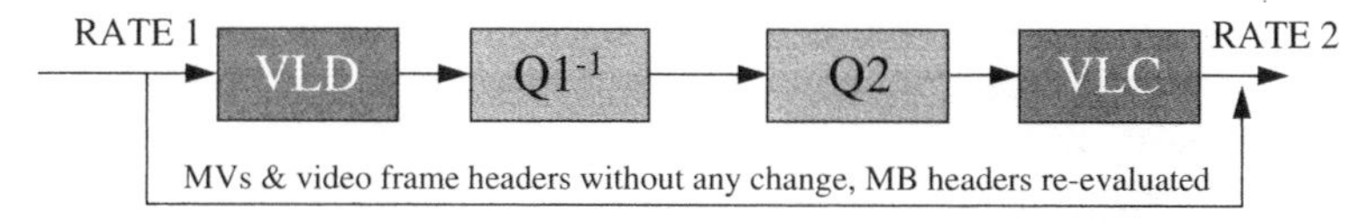

Figure 6.5 Transcoding with re-quantisation scheme

to be Huffman re-encoded. Here, the transcoding operation does not involve complex frame re-ordering, or full-scale (± 16 pixels) motion re-estimation operations. Therefore, the open-loop transcoding comprises the simplest and most straightforward transcoding mechanism with the lowest complexity, plus a very small processing time and little power consumption.

In this method of homogeneous transcoding, original motion vectors (MVs) and video frame headers are preserved and re-used without any modification. On the other hand, macroblock (MB) headers are required to be re-evaluated since an originally encoded MB may turn out to be skipped (uncoded) due to the coarser re-quantisation process. There are a few critical points in selecting the MB types during MB re-evaluation. An originally skipped MB should be transcoded to a skipped MB and an INTRA MB should be transcoded to an INTRA MB. However, an INTER MB can be transcoded to an INTER, INTRA or a skipped MB, depending on the transcoding conditions.

Since the open-loop transcoding is achieved in the coded domain, its implementation is a simple, fast and a low-complexity process. However, the direct re-quantisation algorithm with open-loop transcoding has some drawbacks, such as producing an increasing distortion in the predicted pictures caused by the picture drift phenomenon. Drift occurs due to the mismatch between the locally reconstructed pictures at the encoder and the transcoded pictures in the system containing two different quantisers (Assuncao and Ghanbari, 1997; Sun, Kwok and Zdepski, 1996). This detrimental impact on transcoded video quality has to be minimised for better transcoding performance. The following section analyses the drift problem both conceptually and mathematically, and presents drift-free transcoding algorithms.

6.6.1 Picture drift effect

Picture drift in transcoded video has been addressed in numerous publications (Assuncao and Ghanbari, 1997; Bjork and Christopoulos, 1998; Sun, Kwok and Zdepski, 1996). Drift is an accumulative effect of distortion that occurs due to the mismatch between the reconstructed images of originally encoded and transcoded video frames. This mismatch is an eventual result of the quantisation level differences between the originally encoded and transcoded video frames. As depicted in Figure 6.5, the rate reduction algorithm within the video transcoder starts with the de-quantisation of the DCT coefficients using the original quantiser levels. As explained earlier, these coefficients are re-encoded with a different quantiser for output bit rate reduction. This simply causes distorted reconstruction at the very end decoder. Nevertheless, this quality-destructive effect should not be confused with the quality degradation resulting from the existence of one decoding/re-encoding cycle within the transcoding operation. A single decoding/re-encoding stage between the two end-points introduces some quality loss since the re-encoding operation relies on the already decoded lower quality video data. Since the quantisation of DCT coefficients is a lossy operation, the lower quality achieved by decoding the coefficients prior to re-encoding them is a predicted outcome. Thus, this occasion should clearly be distinguished from the picture drift caused by the mismatch between the encoder and the decoder ends.

However, it is significant that drift occurs only in open-loop transcoding where there is not a feedback loop to compensate for this unwanted picture quality deterioration effect. Moreover, this is a highly prediction-oriented problem which is only caused by the transcoding operation of INTER frames. Therefore, the quality deterioration gradually increases until an INTRA coded frame refreshes the video scene. The transcoding of INTRA frames and bi-directional (B) frames do not contribute to this particular problem, the reason being that I-frames are encoded with reference to themselves, B-frames are not used for predicting forthcoming frames. One very simple way of counteracting the drift effect is the regular and frequent insertion of INTRA frames. However, this is not the optimal solution to the drift problem, as it imposes additional data onto the video stream. This causes an eventual increase in the bit rate which defeats the objective of bit rate reduction and hence, the functionality of the video transcoder.

The other more practical and widely accepted solution is to design a video transcoding algorithm which efficiently resolves the picture drift problem. A description of this kind of transcoder architecture is presented in the next section, following the mathematical analysis of the drift phenomenon.

The analysis of the drift error has been given by Assuncao and Ghanbari (1997). In this analysis, the decoder is assumed to be similar to the local decoder at the encoder. Consequently, in the case of an error-free environment, the reconstructed pictures at the decoder should be the same as the ones at the encoder without any transcoding operation. Thus:

$$RP_n^d = RP_{n}^e, n = 0, 1, \ldots, N - 1 \tag{6.1}$$

where RP^d, RP^e and N represent the reconstructed pictures at the decoder, at the encoder and the number of total video frames, respectively. The reconstruction of a picture can be represented by some prediction error, e_n, together with a motion-compensated prediction $MCpred$ term for an INTER frame:

$$RP_n^d = RP_n^e = e_n + MCpred(RP_{n-1}^d), 1 \le n \le N - 1 \tag{6.2}$$

whereas for an INTRA frame:

$$RP_n^d = RP_{n}^e, n = 0 \tag{6.3}$$

since an I-frame is encoded without the need for any motion compensation or prediction operations.

Rate reduction with an open-loop transcoding algorithm naturally modifies the above equations due to the addition of the transcoding distortion. Therefore, the reconstructed images at the decoder and the encoder can no longer be the same as above. Instead, the following equations can be derived:

$$\begin{aligned}
RP_n^{d_distorted} &\ne RP_n^e \\
RP_0^{d_distorted} &= RP_0^e + t_0^{distort}, 1^{st}\ frame_INTRA \\
RP_1^{d_distorted} &= e_1 + t_1^{distort} + MCpred(RP_0^{d_distorted}), 2^{nd}\ frame_INTER \\
RP_1^{d_distorted} &= e_1 + t_1^{distort} + MCpred(RP_0^e + t_0^{distort}) \\
RP_1^{d_distorted} &= e_1 + t_1^{distort} + MCpred(RP_0^e) + MCpred(t_0^{distort})
\end{aligned} \tag{6.4}$$

where $MCpred$ is assumed to be a linear operation. From the first two lines of Equation 6.4, it is clearly seen that the transcoding distortion $t^{distort}$ is the difference between the current pictures of the decoder and the encoder. The remaining lines of the equation indicate that for the next P-frame, the reconstructed picture at the decoder is not only the motion-compensated previous I-frame together with the prediction error, but also the transcoding distortion of the current frame and the previous motion-compensated frame. The latter distortion term is referred to as the residue of transcoding distortion from the previous frame and is represented as:

$$\Delta_1 = MCpred(t_0^{distort}) \tag{6.5}$$

where Δ is referred to as the drift error in the picture. Similarly, the drift error for the 3rd frame (2nd P-frame) can be written as:

$$\begin{aligned} RP_2^{d\_distorted} &= e_2 + t_2^{distort} + MCpred(RP_1^{d\_distort}),\ 3^{rd}\ frame\_INTER \\ \Delta_2 &= MCpred[t_1^{distort} + MCpred(t_0^{distort})] \end{aligned} \tag{6.6}$$

Thus, as also observed in Equation 6.6, the drift error presents an accumulative behaviour throughout a predictive video sequence and it can be given for any picture by:

$$\Delta_n = MCpred\{t_{n-1}^{distort} + MCpred[t_{n-2}^{distort} + \ldots + MCpred(t_0^{distort})]\} \tag{6.7}$$

6.6.2 Drift-free transcoder

Having identified the problem, the design of a drift-free video transcoding algorithm is quite a straightforward technique. As analysed by Assuncao and Ghanbari (1997), the drift error can be corrected with the use of a drift error correction loop, as depicted in Figure 6.6. This particular figure shows a very primitive configuration of a drift-free video transcoder. The basic structure simply includes two major components, namely a decoding block and a re-encoding block. Thus, a homogeneous video transcoder comprises a decoder end as an input and an encoder end as an output. However, these blocks are not proper decoder and encoder blocks as configured in the cascaded fully decoding/re-encoding scheme (refer to Figure 6.4), but indeed they form a partial decoding and encoding structure. The drift-free operation is achieved by the use of a feedback loop (within the re-encoding block) that compensates for this error. However, although this implementation provides drift-free transcoding with the use of the feedback loop, it also incurs extra complexity due to the need for DCT/IDCT operations and a frame buffer that is used to store the locally reconstructed frames. Since the picture reconstruction is carried out in the pixel domain, the DCT/IDCT operations are inevitable. Nevertheless, a few proposals of DCT domain drift-free video transcoding algorithms (Acharya and Smith, 1998; Assuncao and Ghanbari, 1997, 1998) have also been presented. These schemes, however, do not employ less complex techniques (Bjork and Christopoulos, 1998; Senda and Harasaki, 1999).

Referring to Figure 6.6, the input rate R_1 is decoded in the first loop and then re-encoded with a coarser quantisation Q_2 for a reduced output rate R_2. Therefore:

$$Q_2 > Q_1 \Rightarrow R_2 < R_1 \tag{6.8}$$

Moreover, two new equivalent rates R'_1 and R'_2 can be defined after the inverse quantisation points 1 and 2. Hence:

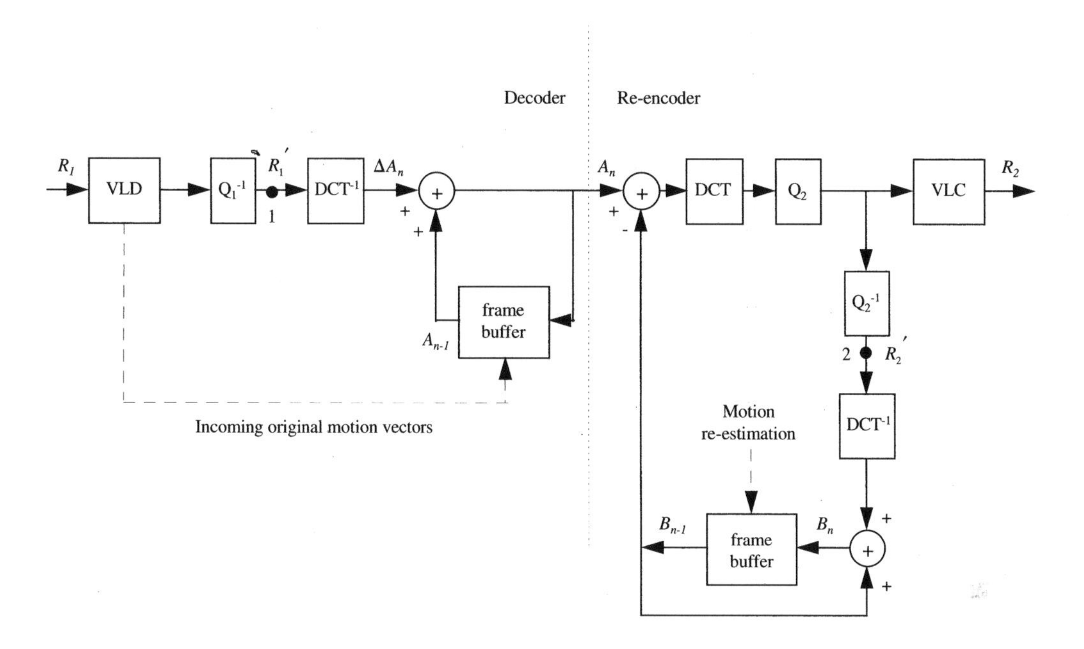

Figure 6.6 Block diagram of a drift-free homogeneous video transcoder

$$\begin{aligned} \Delta A_n &= A_n - A_{n-1} \\ R'_1 &= DCT(\Delta A_n) \end{aligned} \tag{6.9}$$

where ΔA_n and DCT represent the reconstruction error from the incoming bit-stream and the transform operator, respectively. Meanwhile, A_n, A_{n-1}, B_n and B_{n-1} stand for the locally decoded current and previous pictures of the input and output streams, respectively. Similar relations can also be obtained for R'_2 as:

$$\begin{aligned} R'_2 &= DCT(A_n - B_{n-1}) \\ R'_2 &= DCT(\Delta A_n + A_{n-1} - B_{n-1}) \\ R'_2 &= DCT(\Delta A_n) + DCT(A_{n-1} - B_{n-1}) \\ R'_2 &= R'_1 + DCT(A_{n-1} - B_{n-1}) \end{aligned} \tag{6.10}$$

where the transform operation is considered to be a linear operation. The last line of Equation 6.10 is particularly significant as it hints at the direct use of the incoming original rate without the need for fully decoding it. This feature intensely simplifies the drift-free video transcoder structure in a way that the fully decoding and re-encoding operations in cascade are not required at all. Thus, this structure also reduces the complexity notably. The simplified structure for this kind of

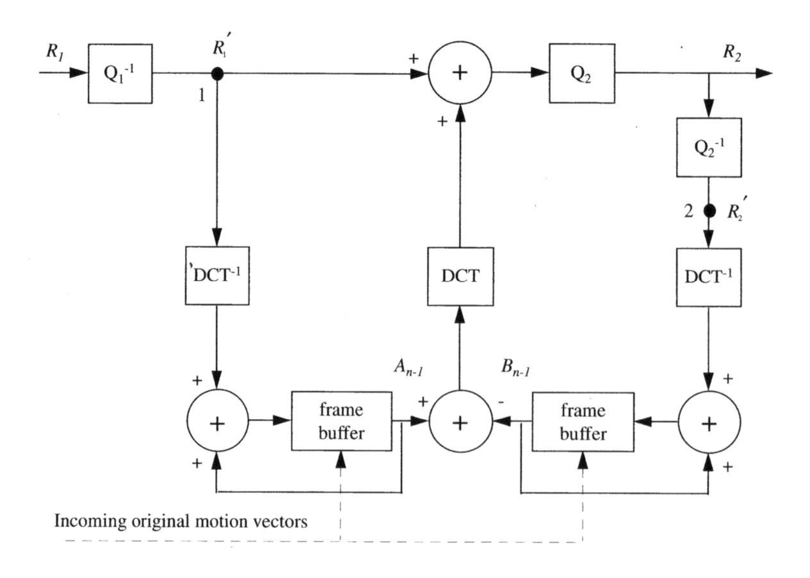

Figure 6.7 Simplified drift-free video transcoder block diagram

transcoder can be observed in Figure 6.7. Since the two loops are symmetrical, they can also be combined for further simplification. The right-hand side loop, shown in Figure 6.7, is the feedback loop which gives the closed-loop transcoder its name. This loop accumulates the errors introduced by the different quantisation factors and adds them back into the next frames following the motion-compensation process. The feedback of these errors compensates for the picture drift and stops its accumulative effects throughout the video sequence.

6.7 Transcoding with Motion Data Re-use Scheme

The motion data re-use scheme comprises the simplest algorithm of all the drift-free video transcoding methods. This is due to the fact that it does not include any kind of motion re-evaluation process. In this scheme, the video data is partially decoded, as mentioned earlier in Section 6.6. Thus, the incoming compressed video stream is partially decoded (only DCT coefficients) using the original quantiser level. Following this process, the decoded coefficients are re-quantised with a different quantisation level in order to yield a certain target output bit rate. Naturally, these re-quantised coefficients need to be re-encoded using Huffman coder.

Up to this point, the transcoder operation presents a close resemblance to the previously discussed open-loop transcoding method. However, the discrepancy arises in the presence of an additional feedback loop as part of the overall

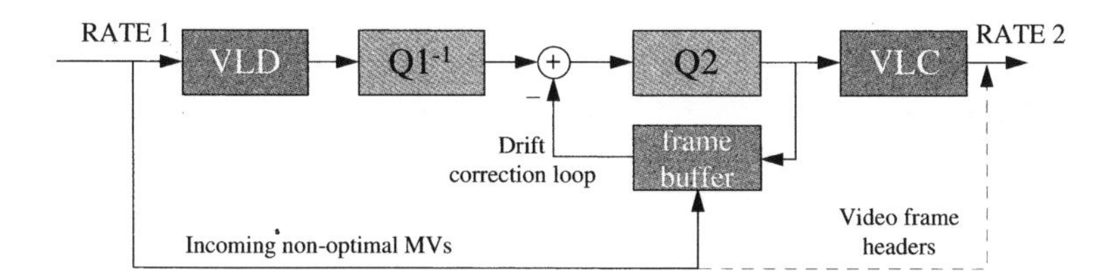

Figure 6.8 Transcoding with motion data re-use scheme

transcoding system, as depicted in Figure 6.8. The feedback loop corrects the drift error caused by the use of different quantiser levels. The drift error correction is simply carried out by the method discussed in Section 6.6.2. Thus, the feedback loop corrects the accumulated mismatch errors between the reconstructed images of the source coder and those of the transcoder. In Figure 6.8, the loop is depicted with the frame buffer block containing the previously reconstructed video frames. These video frames are reconstructed with the new quantiser levels that are set by the video transcoder itself to achieve the necessary amount of bit rate reduction. Therefore, the scheme requires a further step to the open-loop transcoding as it also needs to store the previously reconstructed frames in the pixel domain. This simply means that closed-loop algorithms comprise both pixel domain and DCT-domain operations. Even though the transcoding with motion data re-use scheme does not involve complex frame re-ordering or full-scale (± 16 pixels) motion re-estimation operations, it incurs higher complexity than the simple re-quantisation method, but much lower complexity than the cascaded fully decoding/re-encoding, and a low processing time and power consumption in addition to a small amount of delay.

The reason why the scheme exhibits a less complex behaviour than the closed-loop drift-free transcoder schemes with motion data re-evaluation methods is that the incoming original MVs are re-used without any modification. However, this implies a sub-optimal motion prediction that leads to some quality degradation in the transcoded video despite the existence of the drift correction loop. Input MVs are not optimal because the differential reconstruction errors cause the incoming MVs to deviate from their optimal values (Bjork and Christopoulos, 1998; Youn, Sun and Lin, 1998). In simple terms, the original MVs may not sometimes be the most suitable MVs for the new set of quantiser levels and they may point to the wrong blocks within a video frame. The quality of transcoding can be further improved by taking this fact into consideration. The originally received MVs can be refined, as will be discussed in Section 6.9.1. Moreover, MB headers should also be re-evaluated to optimise their values in accordance with the new quantiser levels used.

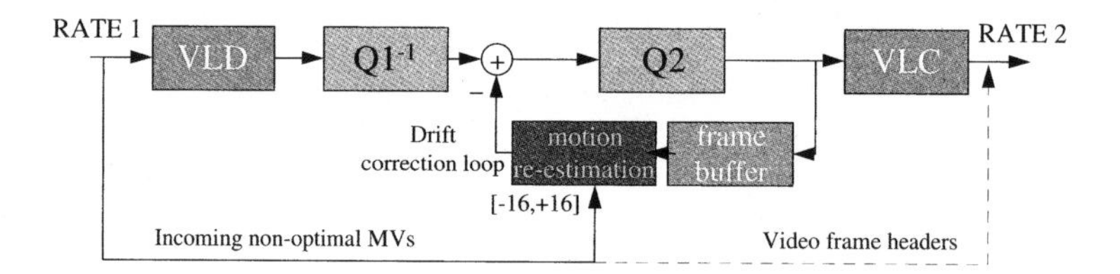

Figure 6.9 Transcoding with motion data re-estimation scheme

6.8 Transcoding with Motion Data Re-estimation Scheme

This scheme demonstrates similar characteristics to the previous motion data re-use scheme as it also includes a feedback loop for drift error correction. However, the motion data re-estimation scheme, as the name implies, comprises a full-scale re-estimation of the new MVs. Thus, the received video motion data is not used, and new MVs are estimated during the transcoding process. The new motion estimation is carried out for a full size MV search window, which is ± 16 pixels around the candidate block for which motion is being estimated. Therefore, the scheme does not involve complex frame re-ordering. However, it accomplishes full-scale (± 16 pixels) motion re-estimation operation.

Eventually, this scheme incurs a much higher complexity than the simple re-quantisation and the motion data re-use methods, but lower complexity than the cascaded fully decoding/re-encoding scheme, plus a considerable amount of processing time and power consumption with a substantial amount of delay.

Moreover, as illustrated in Figure 6.9, due to the existence of the motion re-estimation block within the drift correction loop, it is possible to reduce the effects of non-optimal MVs on the transcoding quality. It is for this reason that motion data re-estimation allows estimating and hence selecting the most suited MVs for the modified quantiser levels. MB headers are also required to be re-evaluated.

6.9 Transcoding with Motion Refinement Scheme

So far, it has been shown that it is possible to resolve the drift problem created by the mismatch errors using a feedback loop. Two closed-loop schemes have been presented in the preceding two sections to reduce the drift effects on the transcoded video quality. The first scheme produces non-optimal MVs that further reduce the service quality. On the other hand, the second scheme improves the transcoding quality with the full-scale re-estimation of motion data at the expense of added complexity. Therefore, it is generally accepted that often a correction loop alone is not sufficient for optimal QoS, and a further MV refinement process needs also to be integrated into the system. MV refinement can only be accomplished in the pixel domain with the use of a locally reconstructed video frame, and therefore DCT-domain transcoding algorithms fail to provide an acceptable motion data refinement (Senda and Harasaki, 1999).

The motion data refinement block accomplishes this operation, as depicted in Figure 6.10. Unlike the previous scheme where the original MVs were discarded, in this method the received motion data is used in the refinement process. Since the direct re-use of these non-optimal vectors has a negative impact on the video quality, these vectors have to be refined first. Refinement is carried out around the non-optimal MVs to yield more accurate values of predicted motion. Thus, transcoding with motion refinement scheme does not involve complex frame re-ordering or full-scale (± 16 pixels) motion re-estimation operations. However, it performs small-scale ($\pm 1, 2, 3, 4$ pixels) motion refinement.

The scheme incorporates higher complexity than the simple re-quantisation and the motion data re-use methods, but lower complexity than both the transcoding with full motion data re-estimation and the cascaded fully decoding/re-encoding schemes. It also has a moderate processing time and power consumption with some amount of delay.

The complexity, and hence the processing delay, increase when the motion refinement window size increases. It is also important to note that the MB headers need to be re-evaluated as in the previous schemes in order to ensure the appropriate MB types given the outgoing bit rate requirements.

6.9.1 *MV refinement algorithm*

This algorithm is based on fine-tuning the incoming MV values. The original MVs are referred to as the non-optimal MVs. MV refinement is a process whereby the non-optimal MVs are refined within a small search window around the blocks to which these initial vectors point. This is required due to the fact that the re-quantisation results in non-optimal MVs that were originally estimated by the encoder that used a different quantiser level. It may therefore be possible that the initially estimated vectors may not be able to point to the right blocks or MBs within a video frame due to the content changes caused by the differences in the quantisation levels.

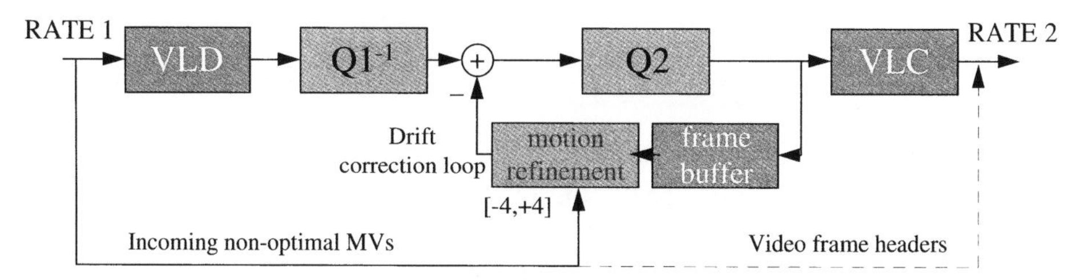

Figure 6.10 Transcoding with motion refinement scheme

The proposed solution for this problem is to refine the non-optimal MVs using their original MVs. In this way, the complexity of re-estimating the motion data is avoided. It is generally accepted that for the refinement procedure, a small MV search window gives the necessary quality improvement with substantially reduced complexity.

The need for the MV refinement procedure was presented by Youn, Sun and Lin (1998), who also elaborated on the mathematical aspects of the problem. In general, the MV set (I_x, I_y) for the encoder is obtained by:

$$
\begin{aligned}
(I_x, I_y) &= \arg\min_{(a,b)\in W} SAD_e(a,b) \\
SAD_e(a,b) &= \sum_h \sum_v |P_e^c(h,v) - R_e^p(h+a, v+b)|
\end{aligned}
\tag{6.11}
$$

where h and v are the horizontal and vertical variables in the motion estimation process. $P_e^c(h,v)$ and $R_e^p(h+a, v+b)$ represent a pixel in the current frame and a displaced pixel by (a,b) in the previously reconstructed reference frame, respectively. Here, the superscripts c and p represent the current and the previous frames, respectively. Finally, the subscript e indicates the encoder block, and W shows the fixed search window range.

Similar equations could also be derived for the transcoded MV set (T_x, T_y) by only replacing the subscript e by t, indicating the transcoder block:

$$
\begin{aligned}
(T_x, T_y) &= \arg\min_{(a,b)\in W} SAD_t(a,b) \\
SAD_t(a,b) &= \sum_h \sum_v |P_t^c(h,v) - R_t^p(h+a, v+b)|
\end{aligned}
\tag{6.12}
$$

As observed in Figure 6.11, the reconstructed picture within the transcoder R_e is also fed into the re-encoding part of the transcoder block, and thus it is similar to the current picture of the transcoder P_t. Therefore

$$
\begin{aligned}
SAD_t(a,b) &= \sum_h \sum_v |P_t^c(h,v) - R_t^p(h+a, v+b)| + SAD_e(a,b) - SAD_e(a,b) \\
SAD_t(a,b) &= \sum_h \sum_v \left| \begin{array}{l} P_t^c(h,v) - R_t^p(h+a, v+b) \\ +[P_e^c(h,v) - R_e^p(h+a, v+b)] \\ -[P_e^c(h,v) - R_e^p(h+a, v+b)] \end{array} \right| \\
SAD_t(a,b) &= \sum_h \sum_v |P_e^c(h,v) - R_e^p(h+a, v+b) + \Delta_e^c(h,v) - \Delta_t^p(h+a, v+b)|
\end{aligned}
\tag{6.13}
$$

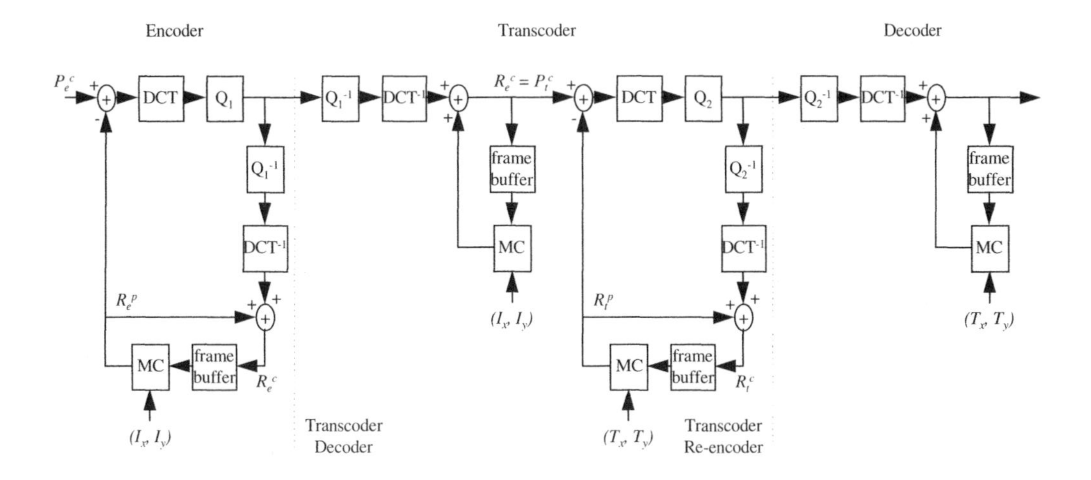

Figure 6.11 A complete video transcoding system

where, bearing in mind that $R_e = P_t$:

$$\begin{aligned} \Delta_e^c(h, v) &= R_e^c(h, v) - P_e^c(h, v) \\ \Delta_t^p(h, v) &= R_t^p(h, v) - P_t^p(h, v) \end{aligned} \tag{6.14}$$

In the above equation, $\Delta_e^c(h, v)$ represents the quantisation error of the current frame in the encoder, and $\Delta_t^p(h, v)$ represents the quantisation error of the previous frame in the transcoder. Thus, the optimal MVs in the transcoder depend on the incoming non-optimal MVs and the quantisation errors in both the encoder and transcoder.

From the observations above, the MVs produced by the transcoding operation can be realised as follows:

$$\begin{aligned} SAD_t(T_x, T_y) &= SAD_e(I_x, I_y) + \sum_h \sum_v |\Delta_e^c(h, v) - \Delta_t^p(h + I_x, v + I_y)| \\ SAD_t(T_x, T_y) &= SAD_e(I_x, I_y) + SDQE \end{aligned} \tag{6.15}$$

where *SDQE* represents the sum of the differential quantisation error, previously referred to as Δ functions in Equation 6.14. Equation 6.15 shows that the direct re-use of the incoming original MVs (I_x, I_y) results in non-optimal outgoing MVs (T_x, T_y) due to the existence of differential quantisation errors.

A small reduction in the bit rate results in slight quality degradation due to the fact that the differential quantisation error would also be small. Conversely, the deterioration in performance becomes significant for a large reduction in bit rate due to high differential quantisation errors. Thus, the quality degradation is proportional to the differential quantisation error.

Figure 6.12 illustrates a possible scenario where refinement is needed. The dashed vector represents the non-optimal MV originally received by the transcoder. Refinement is accomplished by a small search window of ± 3 pixels around the pixel A pointed to by the initial MV. The conventional motion estimation SAD calculations are performed within the refinement window and a best match is sought. If a better matching block is found, resulting in a smaller SAD, then a new pixel (B) is used as a reference for the position of the newly estimated block. Thus, a new MV set is refined, as shown in Figure 6.12. The difference vector between the non-optimal MV and the refined one gives the refinement MV, $\vec{AB}$, referred to as the refining MV. The following basic vector operation can be used to obtain the refining MV (Youn, Sun and Lin, 1999):

$$\vec{OB} = \vec{OA} + \vec{AB} \tag{6.16}$$

where $\vec{AB}$ represents the refining MV for the corresponding video block.

6.9.2 *Effects of refinement window size on transcoding quality*

The MV refinement window has conventionally been chosen to be small to avoid the high complexity operations of the motion re-estimation process in the video transcoder. Indeed, small window sizes give reasonable quality improvement and achieve a good trade-off between complexity and transcoding quality.

If the refinement window size is increased, only a small degree of quality improvement can be achieved. This is expected, since a larger window size allows for more accurate motion estimation. However, while a larger search range achieves better performance, it also leads to an increase in bit rate due to the larger

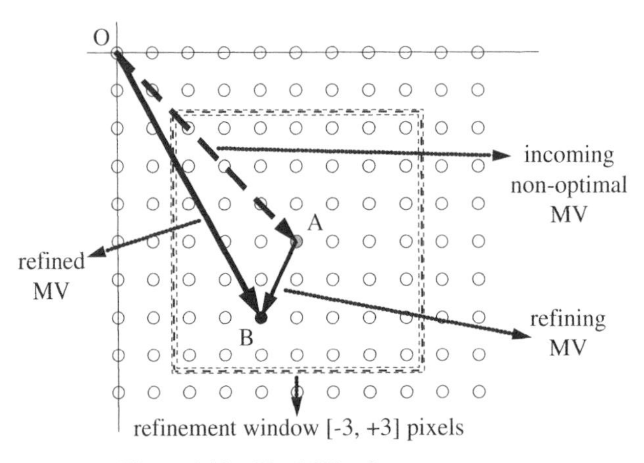

Figure 6.12 The MV refinement process

motion vectors obtained. Therefore, a trade-off is required between bit rate and complexity for optimal transcoding performance. Figure 6.13 demonstrates the average PSNR levels obtained at different refinement window sizes, while Figure 6.14 shows the effect of the window size on the average bit rate of the transcoder.

Figure 6.13 shows that an increase in the refinement window size results in a slight increment in PSNR values at the cost of higher average bit rate, as observed in Figure 6.14. However, the increase in PSNR values is only noticeable when the transcoder is required to achieve a considerable reduction in bit rate, such as from 116 kbit/s to 47–43 kbit/s. Conversely, neither the PSNR nor the bit rate graphs

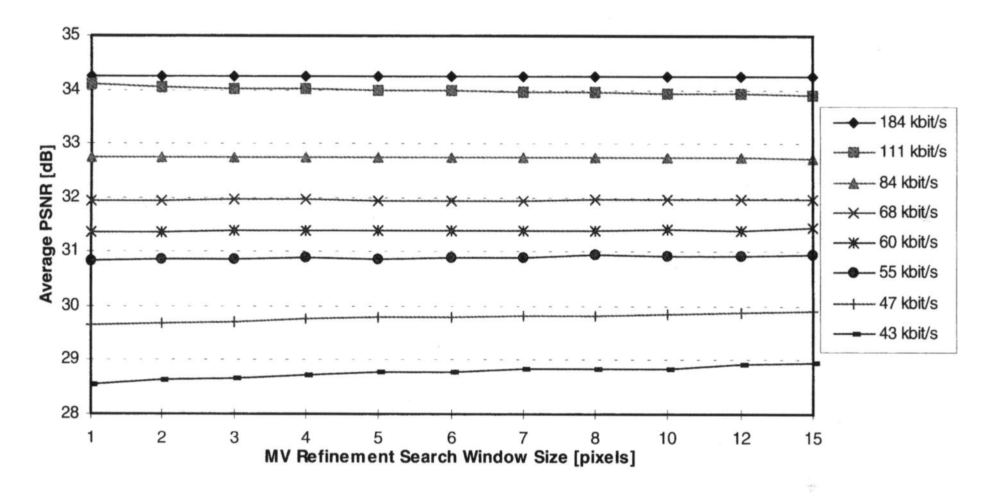

Figure 6.13 Average PSNR values of the transcoded Foreman, originally encoded at an average of 116 kbit/s and 25 f/s, versus various refinement window sizes

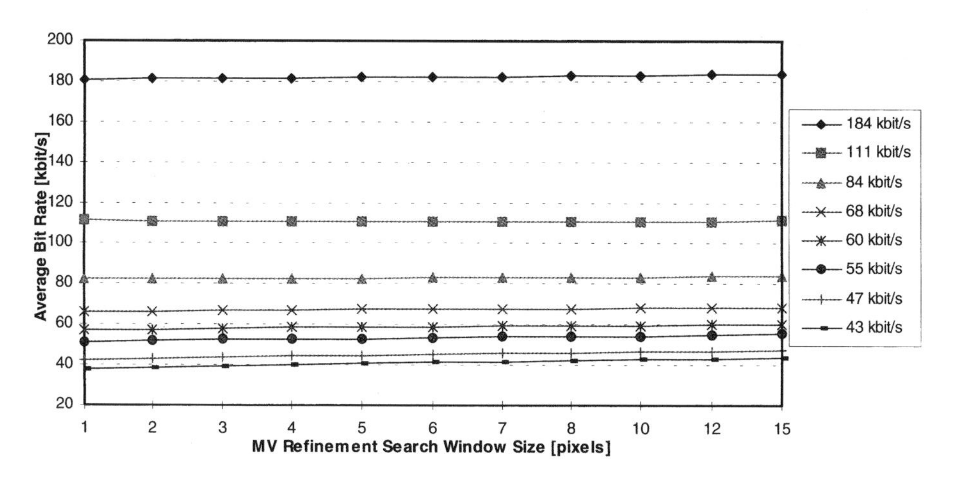

Figure 6.14 Average bit rates of the transcoded Foreman sequence of Figure 6.13 versus various refinement window sizes

show significant changes with the varying refinement window size when the transcoder is required to produce a small reduction in transcoded bit rate.

6.10 Performance Evaluation of Rate Reduction Transcoding Algorithms

The performance of real-time homogeneous video transcoders has been investigated (Gopala Krishnan *et al.*, 1999; Tudor *et al.*, 1997; Yeadon *et al.*, 1998) by producing and analysing experimental results. In this section, the performance of the presented homogeneous transcoding algorithms is evaluated. This evaluation is performed using the 150-frame Suzie sequence originally encoded at 64 kbit/s and a frame rate of 25 f/s. The 150-frame sequence is encoded as an I-P-P-P-... sequence, with QCIF (176 × 144 pixels) format using the MPEG-4 standard. The video transcoder employed also produces MPEG-4 compliant output streams. The bit rate reduction is achieved by transcoding the 64 kbit/s stream down to 32 kbit/s.

Figure 6.15 shows the objective results and Figure 6.16 presents some frame extracts from the transcoded sequences. In both figures, the effect of the picture drift is clearly represented by the significant reduction in video quality. This quality degradation reaches 7 dB on average compared to the directly encoded and decoded sequence at 32 kbit/s. Figure 6.16 contains two reference pictures taken from the directly encoded/decoded sequences at 64 kbit/s and 32 kbit/s, respectively. Figures 6.16(a) and (b) show the reference pictures for comparison with the transcoded pictures.

Table 6.1 lists the average bit rates and PSNR values depicted in Figures 6.15 and 6.16. It can be noted that, apart from the drift-prone open-loop transcoder, all the other transcoding schemes have almost a similar performance. However, despite their identical quality and bit rate achievements, the transcoding algorithms differ considerably in their complexity. Therefore, the choice of a particular transcoding scheme is a compromise between complexity of the algorithm and its performance.

6.11 Frame Rate Reduction

Conventionally, frame rate reduction is mostly used to enhance the spatial quality of individual video frames in bandwidth-limited channels. Similarly, a video transcoder can also be used to reduce the frame rate of the video stream before it reaches a network of lower supportable rates. In certain situations, reducing the bit rate may not be enough and the video transcoder has to reduce the frame rate of the incoming bit stream for useful transcoding results. Frame reduction enables

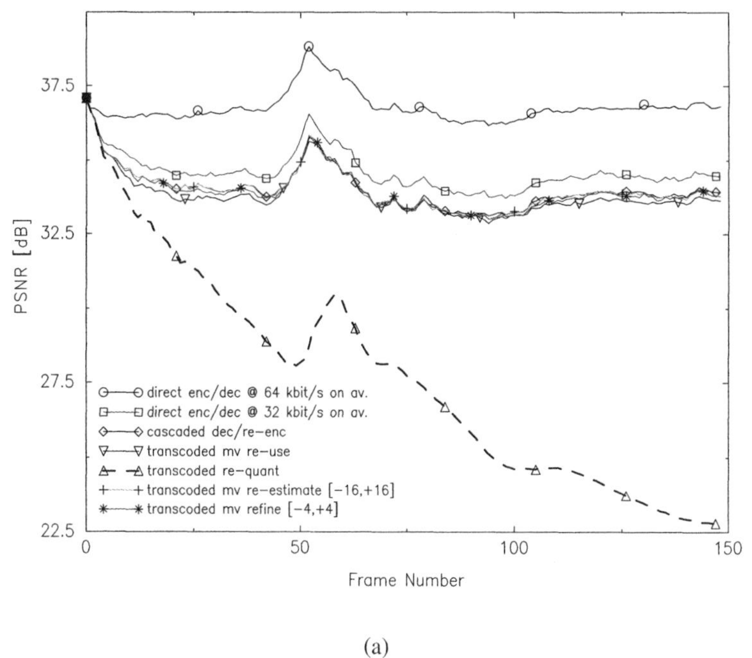

(a)

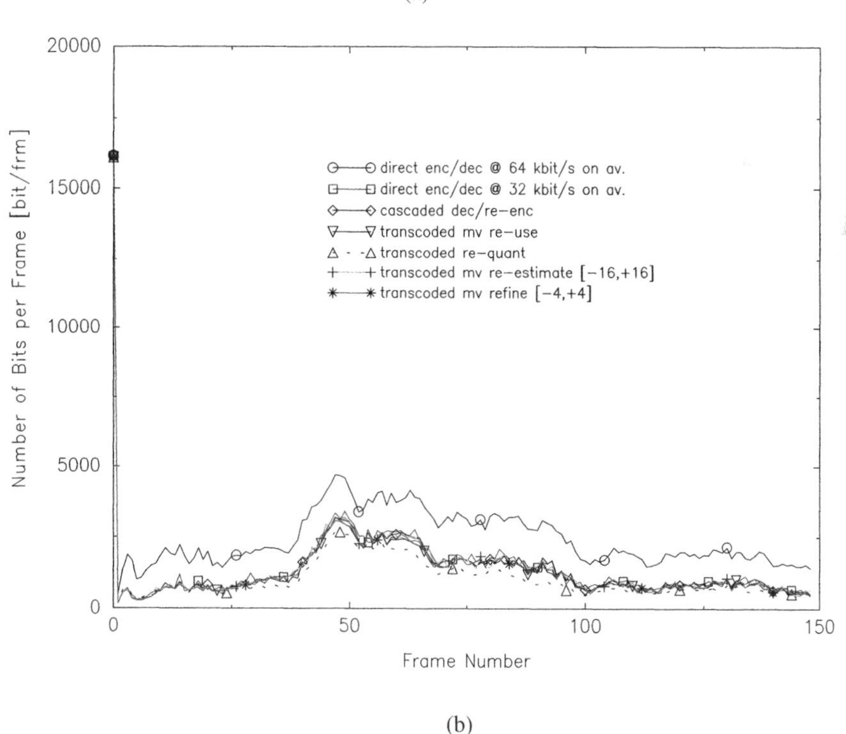

(b)

Figure 6.15 Performance evaluation of the homogeneous video transcoding algorithm using the Suzie sequence: (a) PSNR, (b) number of output bits per frame

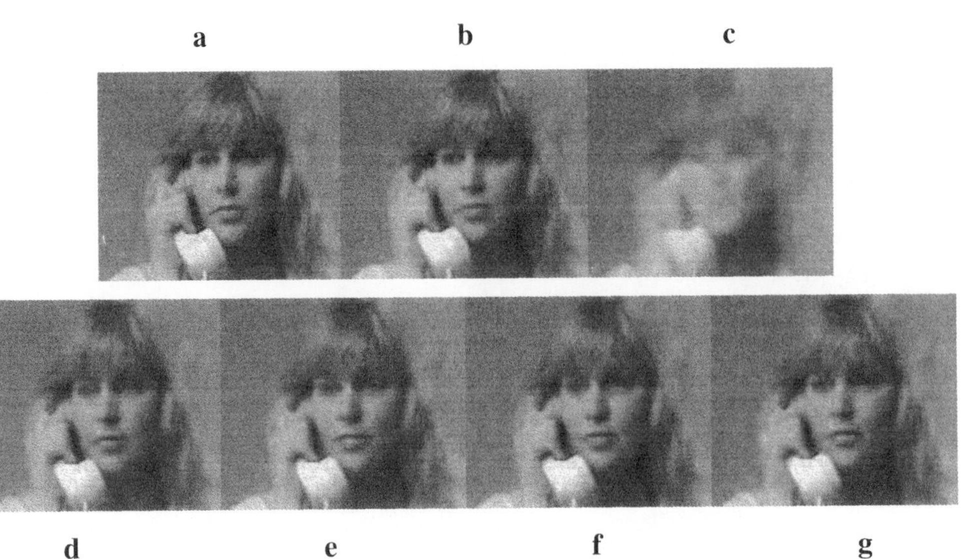

Figure 6.16 Subjective results of the 150th frame of the Suzie sequence: (a) direct enc/dec (64 kbit/s), (b) direct enc/dec (32 kbit/s), (c) re-quantisation scheme (64 → 32 kbit/s), (d) cascaded dec/re-enc scheme (64 → 32 kbit/s), (e) MV re-use scheme (64 → 32 kbit/s), (f) MV re-estimation scheme (64 → 32 kbit/s), (g) MV refinement scheme (64 → 32 kbit/s)

Table 6.1 Average bit rate and PSNR values of Figures 6.15 and 6.16

Type of scheme	Av. bit rate (kbit/s)	Av. PSNR (dB)
direct enc/dec @ QP = 8	62.92	36.803
direct enc/dec @ QP = 14	35.32	34.619
cascaded dec/re-enc	34.02	34.038
transcoded MV re-use	32.68	33.850
transcoded re-quant (open-loop)	27.55	27.702
transcoded MV re-estimate ±16p	33.76	34.019
transcoded MV refinement ±4p	32.74	33.978

the transcoder to allocate more bits for the remaining video frames in the sequence, particularly those with high motion activity. Frame rate reduction can simply be accomplished by arbitrary frame dropping. For instance, dropping every other frame in a sequential order leads to a half rate reduction in the transcoded sequence.

Although reducing the frame rate is a straightforward process, its side-effects can be disturbing to decoded video quality. This is because dropping some frames leads to a mismatch between the decoded sequence and the locally reconstructed one at the encoder. This mismatch has a damaging accumulative effect on the decoded video quality. The damage mainly occurs due to the loss of frames that are required for the successful decoding of the sequence. Although these frames are intentionally dropped by the transcoder for frame rate reduction, the end decoder still expects to receive them. The predictive encoding increases the importance of these dropped frames for the accurate reconstruction of the sequence. To mitigate the damaging effects of frame dropping, the transcoder has to re-estimate the motion of the transcoded video in reference to preserved frames. The transcoder makes use of the incoming MVs in order to estimate their new values after frame reduction (Youn, Sun and Lin, 1998, 1999). A frame is transcoded by taking the last preserved (undropped) frame as a reference. The new MV value is the sum of the corresponding MVs in the skipped and the reference frames. Furthermore, the resulting MV value is also refined using the MV refinement algorithm described in Section 6.9.1. Moreover, (Hwang and Wu, 1998) presents a bilinear interpolation method, which re-estimates the MVs of the currently transcoded frame by interpolating their values using MVs in previous frames down till the previous non-skipped frames. Using this method, the interpolated MV serves as the new search centre, thereby reducing the MV refinement search range significantly. The size of the refinement window is determined by the number of skipped frames and the magnitudes of their MVs (Hwang and Wu, 1998).

Table 6.2 shows the bit rate and frame rate reductions obtained by using different homogeneous transcoding techniques. The transcoding operation is performed by reducing the original frame rate (25 f/s) down to 12.5 f/s. Evidently, using the same quantiser levels, the transcoded sequence has a lower bit rate of 54 kbit/s. It can be noted that the frame reduction method produces a slightly better objective quality than the bit rate reduction method at the same bit rate. This is due to the fact that the frame rate reduction allows the transcoder to allocate more bits to the preserved frames. However, these extra bits could only contribute to quality improvement when there is a relatively high amount of motion in the video scene. Moreover, the results in Table 6.2 imply that the combination of bit and frame rate transcoding schemes could provide very low output rates by introducing a graceful degradation to perceptual quality. Transcoding without the use of previous MV data achieves a slightly worse quality than its counterpart with MV re-use. This is mainly due to the higher accuracy in estimating the MVs of transcoded frames.

Table 6.2 Frame rate transcoding of the 200-frame Foreman sequence, MV refinement window = ± 3 pixels

Type of the scheme	Av. bit rate (kbit/s)	Frame rate (f/s)	Quantisation parameter	Av. PSNR (dB)
direct encoding/decoding	87.403	25	10	33.582
frame rate transcoding only (without MV addition)	52.422	$25 \rightarrow 12.5$	$10 \rightarrow 10$	31.129
frame rate transcoding only (with MV addition)	54.028	$25 \rightarrow 12.5$	$10 \rightarrow 10$	31.268
bit rate transcoding only (drift-free MV ref. scheme)	55.655	$25 \rightarrow 25$	$10 \rightarrow 15$	31.135

6.12 Resolution Reduction

A video transcoder can also be used to achieve spatial scaleability due to its resolution reduction feature. For this purpose, a down-sampling filter within the video transcoder is required. The down-sampling filter is situated between the decoder and the re-encoder stages of the transcoder (Bjork and Christopoulos, 1998). Its objective is to clearly downsample the incoming video to reduce its bit rate. This is necessary when large resolution video is to be delivered to end-users who have limited display capabilities. In this case, reducing the resolution of the video frame size allows for the successful delivery and display of the requested video material. When the picture resolution is reduced by the transcoder, some quality impairment may be noticed as a result (Mokry and Anastassiou, 1994; Vetro and Sun, 1998). This quality degradation is accumulative in a similar way to the drift error effects explained in Section 6.6.1. The main difference between this kind of artefact and the drift effect is that the former results from the downsampling inaccuracies, whereas the latter is a consequence of quantiser mismatches in the rate reduction process. For instance, a CIF to QCIF conversion process requires a resolution reduction by a factor of two in each spatial dimension of the video frame. Therefore, the number of MBs in the resulting QCIF frame is quartered. Consequently, the selection of the new MB types and MV values is a critical factor that dictates the efficiency of the transcoding algorithm.

Figure 6.17(a) illustrates the first scenario where four MBs are to be replaced by a single MB in the transcoded QCIF frame. This requires the selection of the most suitable MV out of four existing MVs in the CIF frame. Either one of the following three solutions could be adopted:

- Averaging the four MVs and scaling the average (i.e. dividing by two in each dimension).
- Taking the median of any three MVs and scaling the median (i.e. dividing by two in each dimension).
- Randomly picking one MV out of four and scaling it (i.e. dividing by two in eachdimension).

More advanced and accurate MV selection can also be accomplished at the expense of additional complexity (Hashemi *et al.*, 1999; Hwang and Wu, 1998; Senda and Horasaki, 1999; Shen *et al.*, 1999). However, the MV refinement process is still required in this scheme as the selected MVs are not optimally estimated.

A similar scenario occurs when only one MB type has to replace four different MB types, as illustrated in Figure 6.17(b). A possible solution to this problem consists of two steps (Bjork and Christopoulos, 1998):

1. If at least one INTRA type exists among the four MBs, select the I type. If there

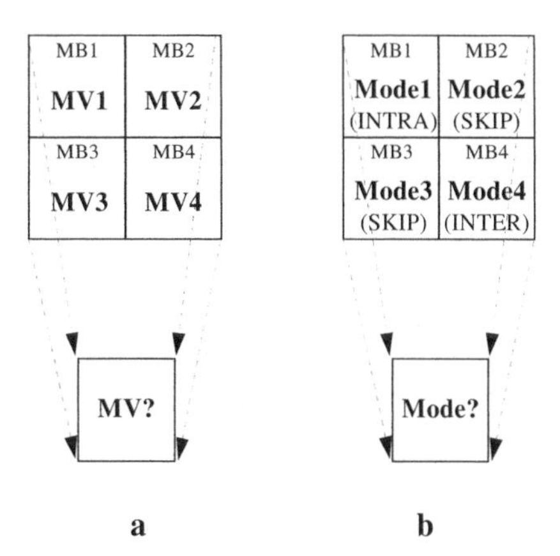

Figure 6.17 Downsampling by a factor of two in each dimension: (a) MV selection problem, (b) MB type (mode) selection problem

is no I-MB and at least one INTER type MB exists, select the P type. If all four MBs were originally skipped, select the skipped type.

2. The MB types are re-evaluated in the re-encoder following step 1.

Downsampling could also be achieved in the coded (DCT) domain, resulting in a significant reduction in complexity (Zhu, Yang and Beacken, 1998). In the DCT domain downsampling process, the motion compensation is performed in the DCT domain and the down-conversion is applied on an MB-by-MB basis. Thus, all four luminance (Y) blocks are reduced to one Y block, and the chrominance blocks are left unchanged. Once the conversion is complete for four neighbouring MBs, then the corresponding four chrominance blocks are also reduced to one chrominance block (one individual block for Cb and one for Cr). This is a low-complexity method for resolution reduction, as it does not require any motion estimation and DCT/IDCT operations. DCT motion compensation is one of the most important features of this down-conversion algorithm whereby a MV is selected for each INTER MB in the downsampled frame (Zhu, Yang and Beacken, 1998; Assuncao and Ghanbari, 1997). Furthermore, a fast DCT-domain algorithm for down-scaling an image by a factor of two has been proposed by Natarajan and Vasuder (1995). This algorithm makes use of pre-defined matrices to do the down-sampling in the DCT domain at fairly good quality and low complexity. Figure 6.18 shows two pictures that are downsampled using the DCT-domain downsampling algorithm. The original movie sequence is captured at 4CIF (704 × 576) spatial resolution. Figure 6.18(a) illustrates a frame from the sequence downsampled by a factor of two to give the CIF (352 × 288) frame resolution. The resulting CIF sequence is further down-scaled by a factor of two to produce the QCIF (176 × 144) frame resolution illustrated in Figure 6.18(b).

(a)

(b)

Figure 6.18 DCT-domain down-scaling of the Harry sequence from 4CIF (704 × 576 pixels) to: (a) CIF (352 × 288 pixels), (b) QCIF (176 × 144 pixels)

The resolution reduction transcoding would be ideal for point-to-multipoint video conferencing scenarios. In this kind of communication scenario, the video transcoder receives the high-resolution video stream from the source and generates a number of lower-resolution transcoded streams to videoconferencing participants, for instance, in accordance with their bandwidth requirements and display capabilities.

6.13 Heterogeneous Video Transcoding

The seamless interconnection of various communication networks has become a challenging issue in both the research and development arenas. Similarly, video transcoding has also received its share of attention for the provision of video communication services across asymmetric networks. The heterogeneous video transcoding algorithms provide solutions for the incompatibility problem caused by the use of different video coding standards across different networking platforms.

Therefore, the heterogeneous video transcoding involves video coding standard conversions for inter-network communications. As illustrated in Figure 6.19, a video gateway embedding the heterogeneous video transcoder is located at the interconnection point between different networks. The operating video coding standards within these networks can be different from each other. In such a case, the video proxy performs the necessary syntax translations between the different standards in order to achieve the required interoperability.

The problem of standard incompatibility has conventionally been overcome by employing a full decoder/re-encoder pair. However, as stated earlier, this kind of cascade introduces a quality loss in video communications. On top of quality loss, decode/re-encode stage introduces a considerable amount of additional complexity resulting from the DCT/IDCT, motion compensation and re-estimation processes. On the contrary, heterogeneous video transcoding is a straightforward algorithm which merely comprises video syntax conversions in the compressed domain. Therefore, the conversion algorithm consists of the following steps, illustrated in Figure 6.20:

- video frame header adjustment
- video data translation from one syntax to another

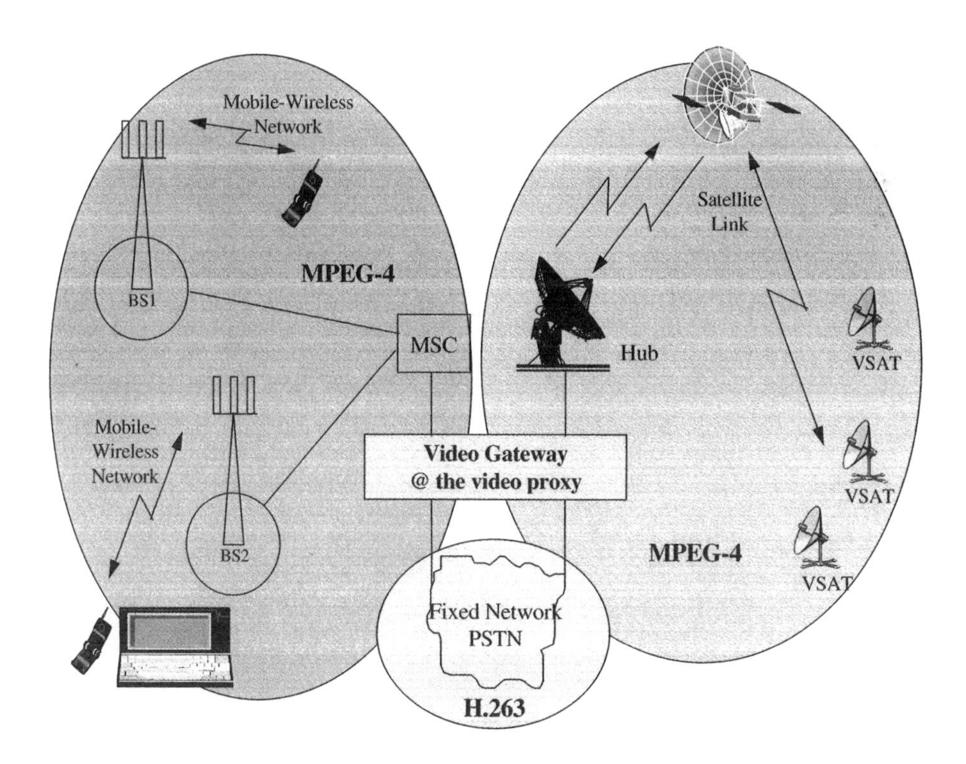

Figure 6.19 Heterogeneous multimedia networks scenario

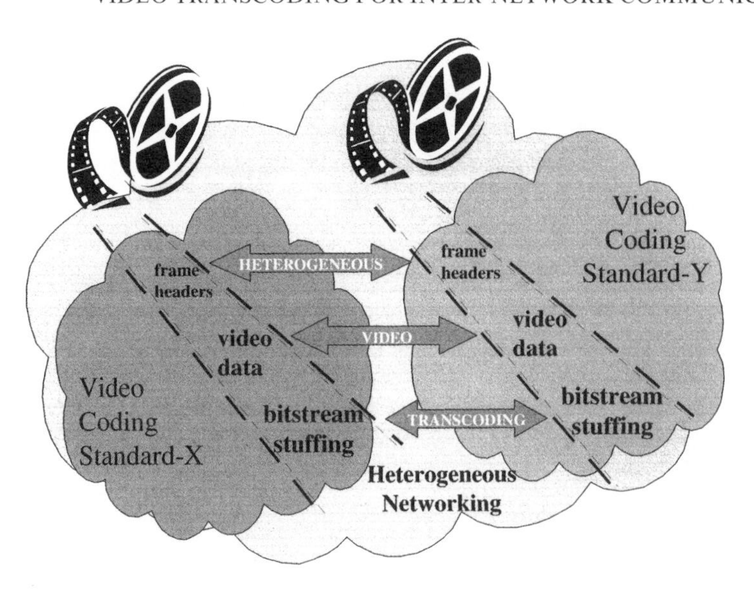

Figure 6.20 Inter-network heterogeneous video transcoding

- necessary bit stream stuffing for different synchronisation requirements of different standards

Video data translation is the major process of the entire transcoding scheme. This process consists of enhanced mapping operations that involve transforming video parameters from one syntax to another (Dogan, Sadka and Kondoz, 1999). However, this mapping process is still a much less complex and hence less time- and power-consuming scheme than the full decode/re-encode technique. This is due to the fact that transcoding does not involve any computationally intensive transformations between the pixel and frequency domains, or any motion re-estimation and compensation processes.

Moreover, syntax conversion does not require the inverse quantisation and re-quantisation of transform coefficients, except when bit rate reduction is also required. Consequently, when bit rate reduction is not an objective in the transcoding operation, the picture drift is avoided. However, when heterogeneous transcoding is combined with the homogeneous transcoding operation, the syntax conversion algorithm must be accompanied by one of the drift-free bit rate reduction schemes discussed in earlier sections of this chapter.

Compatibility between different video standards can simply be achieved when the coders operate in their baseline mode (Dogan, Sadka and Kondoz, 1999; Schafer, 1998). In this way, the difficulty of mapping some features of one particular standard, which are not supported by the other, is considerably simplified. Dogan, Sadka and Kondoz (1999) present a bi-directional transcoder between the

H.263 and MPEG-4 video standards at low bit rates. For the high to low bit rate standard conversion mechanism between MPEG-1,2 and H.261/H.263, see Shanableh and Ghanbari (1999, 2000). Feamster and Wee (1999) take the MPEG-2 to H.263 video transcoder one step further by accomplishing the transcoding of an interlaced MPEG-2 bit stream to a lower bit rate progressive H.263 bit stream. This algorithm presents a field to frame conversion with the associated rate reduction and spatial down-sampling operations. Furthermore, Wu *et al.* (1996) introduce a method for transcoding JPEG pictures into MPEG-1 video. A fast compressed-domain MPEG-1 video to M-JPEG (Lei and Ouhyoung 1994) transcoding algorithm without fully decompressing the MPEG-1 source is presented by Acharya and Smith (1998). Current research in this area is focusing on MPEG-2 to MPEG-4 transcoders with bit rate management capabilities.

Figures 6.21 and 6.22 show the objective and subjective quality obtained by using a bi-directional heterogeneous video transcoding algorithm between the H.263 and MPEG-4. 150 frames of the Suzie sequence are initially encoded at an average bit rate 56 kbit/s and a frame rate of 25 f/s. It can be clearly seen that the transcoder performance is almost as good as the direct encoding/decoding scheme. Conversely, the cascaded decoding/re-encoding PSNR values are on average 1–1.5 dB less than those of the transcoder. This is because transcoding uses the DCT coefficients and the MVs of the incoming bit stream (without fully decoding them), while the cascaded decoding and re-encoding scheme re-estimates the MVs and re-calculates the DCT coefficients based on the lossy reconstructed pictures (not the originals). A 0.5 dB improvement is noticed in the MPEG-4 to H.263 conversion due to the use of the advanced prediction mode that helps improve the picture quality, particularly in high-motion areas.

6.14 Video Transcoding for Error-resilience Purposes

Video transcoders can also be used to enhance the resilience of compressed video streams to transmission errors (de los Reyes, 1998; Dogan, Sadka and Kondoz, 2000; Talluri, 1998). The error-resilient operation of a video transcoder is required in a typical scenario shown in Figure 6.23. The proxy is a video gateway between low bit-error-ratio (BER) and high bandwidth network, i.e. PSTN (on the left-hand side) and a relatively high BER and low bandwidth network, i.e. mobile radio network (on the right-hand side). Furthermore, the video proxy can also apply error concealment on the incoming video to alleviate the effects of errors encountered in the originating network before the transcoding operation is started, as shown in Figure 6.24. Error concealment can be applied on the partially decoded video data using one of the techniques discussed in Chapter 4. The output rate can be controlled by means of a frame buffer placed after the video transcoder. Depending on the status of the output buffer, the output rate can adaptively be

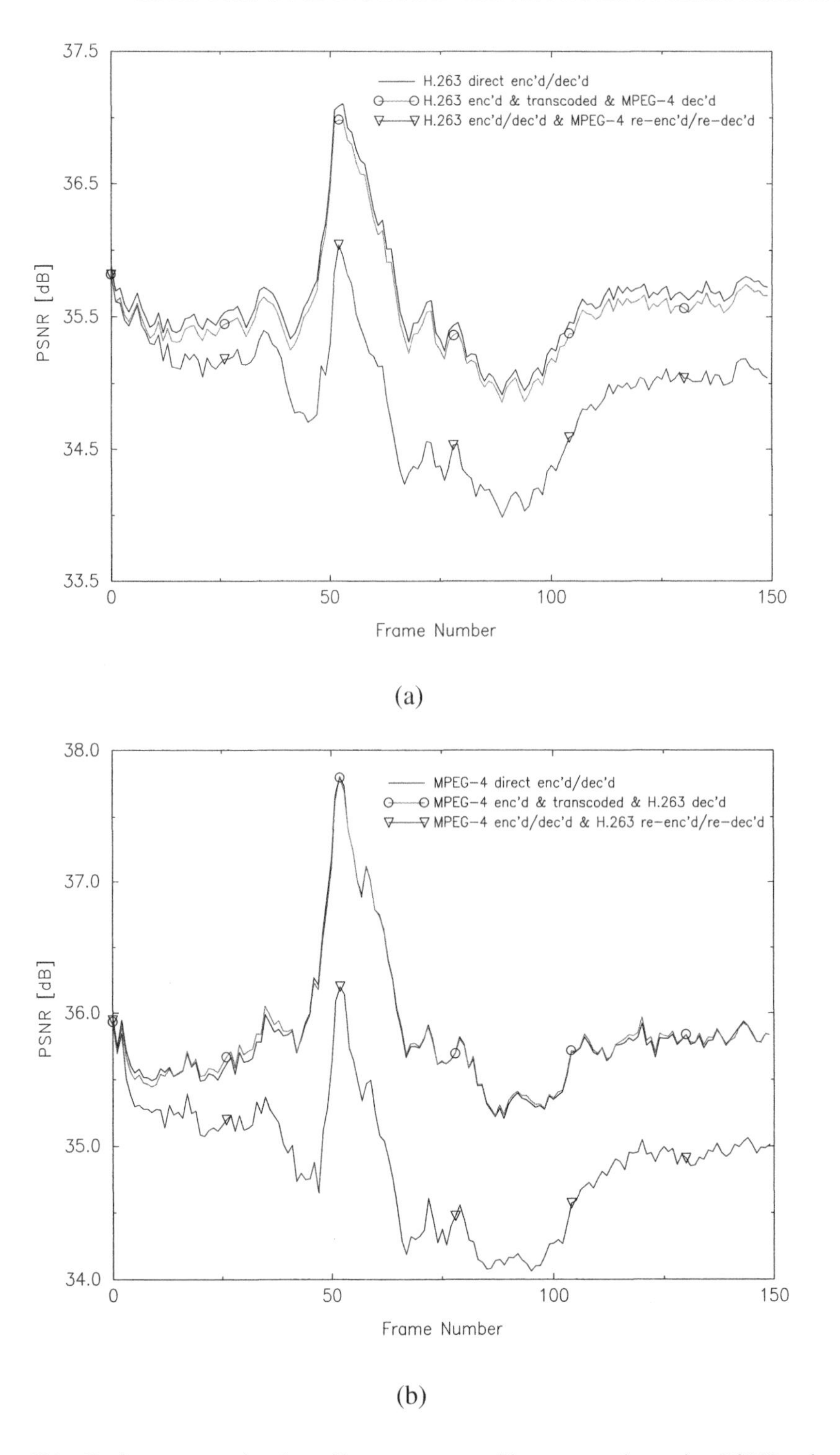

Figure 6.21 Performance evaluation of heterogeneous video transcoders using PSNR values: (a) H.263 to MPEG-4, (b) MPEG-4 to H.263

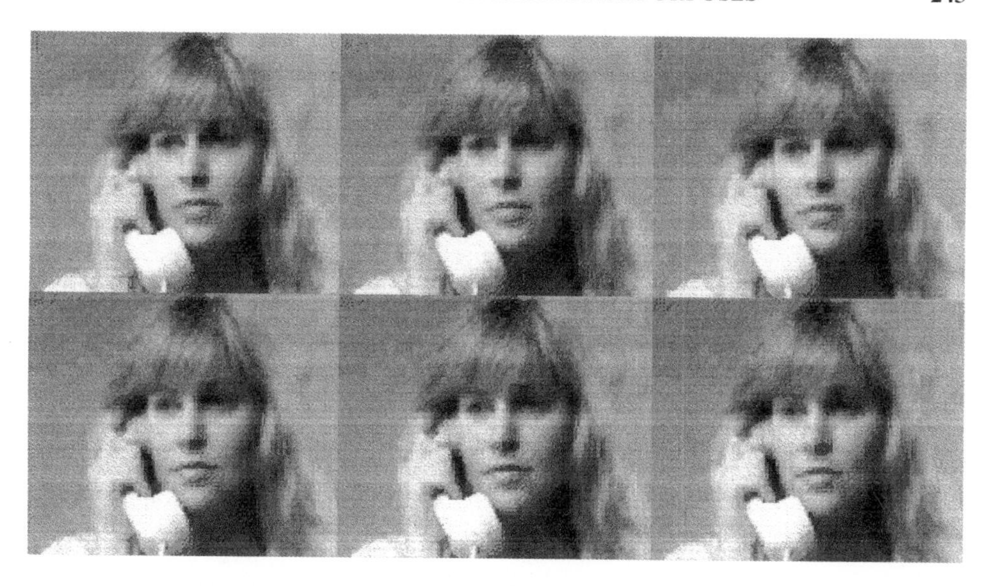

Figure 6.22 Subjective performance evaluation of heterogeneous video transcoding using 150 frames of the Suzie sequence at 56 kbit/s: (top row) MPEG-4 to H.263, (bottom row) H.263 to MPEG-4, (first column) direct encoding/decoding, (second column) transcoding, (third column) cascaded decoding/re-encoding

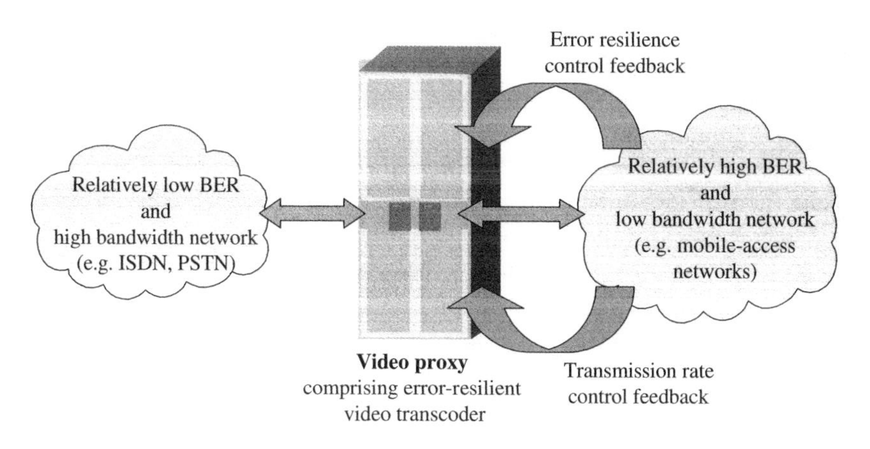

Figure 6.23 Error-resilient video transcoding scenario

adjusted in compliance with the varying bandwidth conditions and channel error characteristics. Similarly, the amount of error protection to be added to coded video data can also be controlled by monitoring the output rate of the transcoder and the changing error conditions using feedback messages from the network, as shown in Figures 6.23 and 6.24.

For error resilience purposes, the video transcoder can apply data partitioning

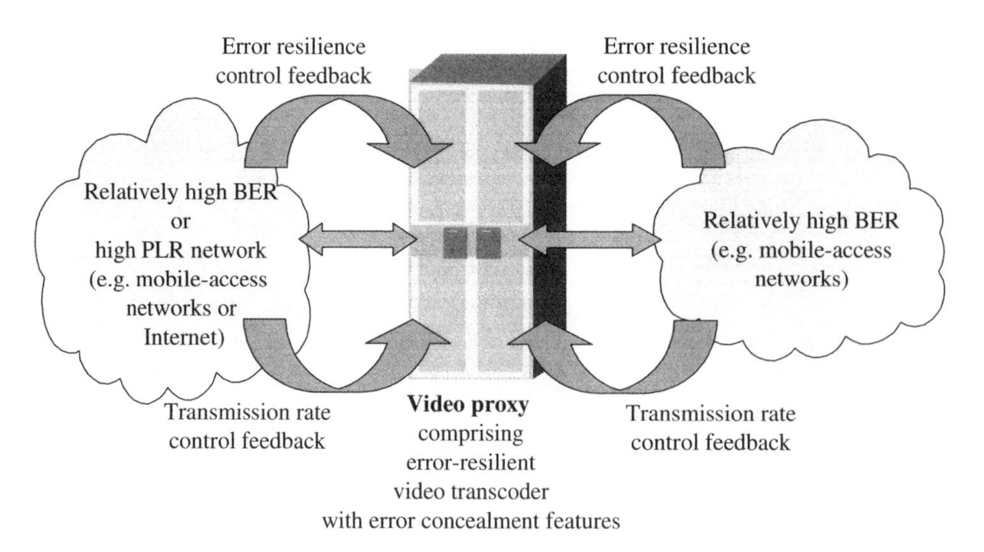

Figure 6.24 Error-resilient video transcoding with error concealment capabilities

and insertion of re-synchronisation markers into the incoming bit stream. Furthermore, the transcoder can apply unequal error protection to various segments of video data. Thus, the header data (followed by MVs) is assigned the highest protection, whereas the DCTs are transcoded with a negligible level of protection. Moreover, the video proxy can forward the transcoded data in the form of separate sub-streams sent over different bearers in accordance with their sensitivity to errors and contribution to overall quality (as described in Chapter 5). This allows for the transmission of high-priority error sensitive video data over high-quality secure bearers for optimal video quality.

The adaptive intra refresh (AIR) algorithm can also be applied to the transcoded video frames to stop the temporal propagation of errors throughout the video sequence (Dogan *et al.*, 2001; de los Reyes *et al.*, 2000). The increase in bit rate caused by AIR is controlled by the rate management feature of the transcoding algorithm. Consequently, the video proxy produces an error-resilient video stream with a controlled output rate. Moreover, de los Reyes *et al.* also proposed a spatial error protection for the transcoded data using data partitioning. On the other hand, when the transcoded frames are subject to loss, the video proxy makes use of feedback messages sent from the destination networks to adapt the transcoding process and parameters accordingly. Adaptive error-resilient transcoding is performed by tracing the lost frames and then avoid using them in the transcoding process. This kind of transcoder operation requires the storage of a number of transcoded frames to guarantee the presence of at least one reference frame in case of huge frame losses. Moreover, the error-resilient entropy coding (EREC) algorithm (Redmill and Kingsbury, 1966) has also been exploited in video transcoding schemes for error resilience purposes (Swann and Kingsbury, 1996). The error resilience tools applied to the video transcoder are all compliant to video standards so that the transcoded streams could be decoded without facing any incompatibility problems with standard-compliant decoders.

Figure 6.25 shows the PSNR values achieved by transcoding 200 frames of the Mother and Daughter sequence encoded at 70.5 kbit/s and a frame rate of 25 f/s. The PSNR values reflect the performance of the transcoder for different BER values. The sequence is transcoded down to 27 kbit/s. The error resilience is achieved by employing AIR during the transcoding operation. As observed in Figure 6.25, the error-resilient transcoder provides 4 dB improvement on average when compared to the non-resilient transcoder. For fair comparison, both versions of the transcoder (resilient and non-resilient) shown in Figure 6.25 are made to generate the same output rate. Figure 6.26 depicts some frames taken from the Suzie sequence transcoded from 94.5 kbit/s down to 38 kbit/s. The subjective quality improvement of Figure 6.26(c) compared to (b) verifies the efficiency of the error-resilient transcoder.

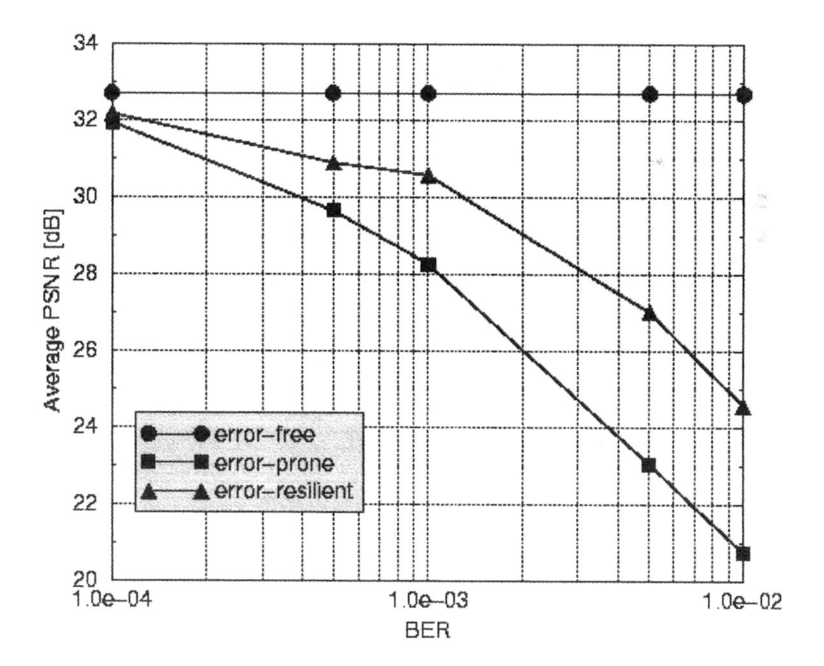

Figure 6.25 PSNR versus BER for the 200-frame Mother and Daughter sequence transcoded down to 27 kbit/s

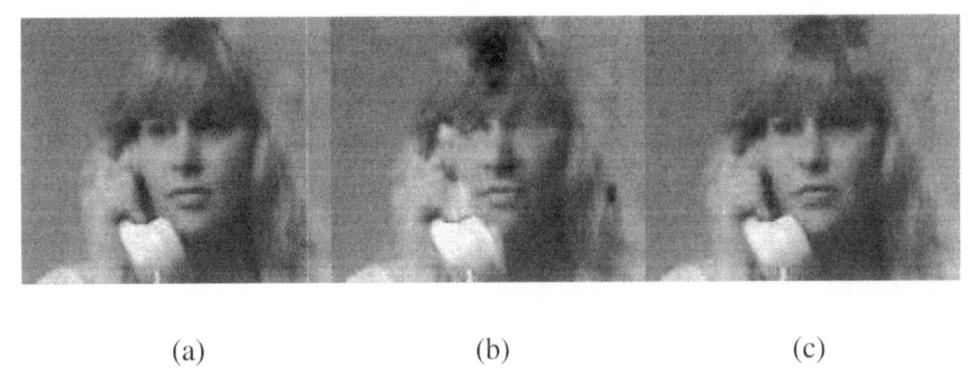

(a) (b) (c)

Figure 6.26 Frame 150 of the Suzie sequence transcoded from 94.5 kbit/s to 38 kbit/s with BER = 10^{-3}: (a) error-free, (b) non-resilient error-prone, (c) error-resilient sequences

6.15 Video Transcoding for Multimedia Traffic Planning

Multimedia traffic planning has recently become a popular topic of research due to the introduction and deployment of numerous diverse multimedia networks. Naturally, these many different networks differ in their features and inherent characteristics. One of the most significant characteristics of a network is its bandwidth, which clearly controls the traffic load and determines the congestion status at any time. Congestion is a bottleneck caused by several users sending their traffic over a limited-bandwidth channel (Halsall, 1996). This bottleneck often forces the running applications to reduce their transmission rates over a congested network. This enforcement mostly results in reduced QoS levels for each of the network clients, and hence causes them to suffer from the congestion individually (Sisodia *et al.*, 2000).

Among all the multimedia traffic types, coded video normally experiences the worst effects of congestion within a network (ElGebaly, 1999). Congestion causes the decoded video to freeze for some time until the congestion is resolved. Once congestion has been cured and the streaming of video is resumed, the video encoder eventually skips all the missing video frames discarded by the network. This results in a leap in the video sequence, with an annoying jittering effect. The depth of this sudden leap is directly related to the duration of congestion and hence the number of dropped video frames. Due to predictive coding, dropping some frames due to congestion results in the incorrect reconstruction of all the subsequent INTER frames, thereby leading to temporal propagation of errors. However, this quality degradation stops when the scene is refreshed with an INTRA coded frame.

The impacts of congestion on two-way low bit rate communications, on the other hand, are much more severe, as I-frames cannot be frequently accommodated by the network. Frame losses due to congestion result in accumulative prediction errors throughout the entire video session. A multiparty video telephony/conferencing scenario, as seen in Figure 6.27, provides a very typical example for such kind of low bit rate communications.

Assuming asymmetric network characteristics with different transmission rates, multiparty video communications can only be achieved with the use of a video transcoder situated at the multipoint control unit (MCU), as depicted in Figure 6.27. In case the incoming video bit stream has a higher rate than the destination networks can handle, the video transcoder has to perform rate reduction, as discussed earlier.

This process is referred to as multimedia traffic planning and is achieved by exploiting the useful features of video transcoders (Assuncao and Ghanbari, 1997; Yeadon *et al.*, 1998).

Clearly, a layered structure of video transcoders can be designed to meet the requirements of various links with different bandwidth characteristics, as illustrated in Figure 6.28. Thus, for one input stream, a stack of video transcoders produces several outputs at varying rates. With the use of the techniques described by Iannaccone (1999), the replication of the similar video data in each of the output streams can also be optimised for multi-rate video transmission. Moreover, with the use of a feedback channel to report the changing congestion conditions of a particular network, the video transcoder is able to dynamically adapt its output rate to the reported channel conditions. Thus, the transcoder provides a varying target output rate, and hence a variable video quality depending on the congestion feedback report. This kind of system acts as a transmission rate regulator rather

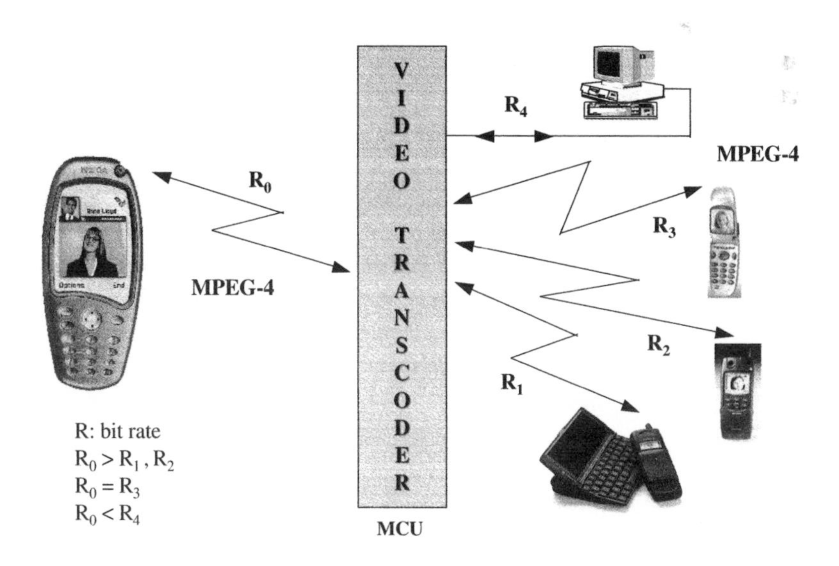

Figure 6.27 MPEG-4 multiparty video-telephony scenario, using a video transcoder at the multipoint control unit (MCU) for multimedia traffic planning

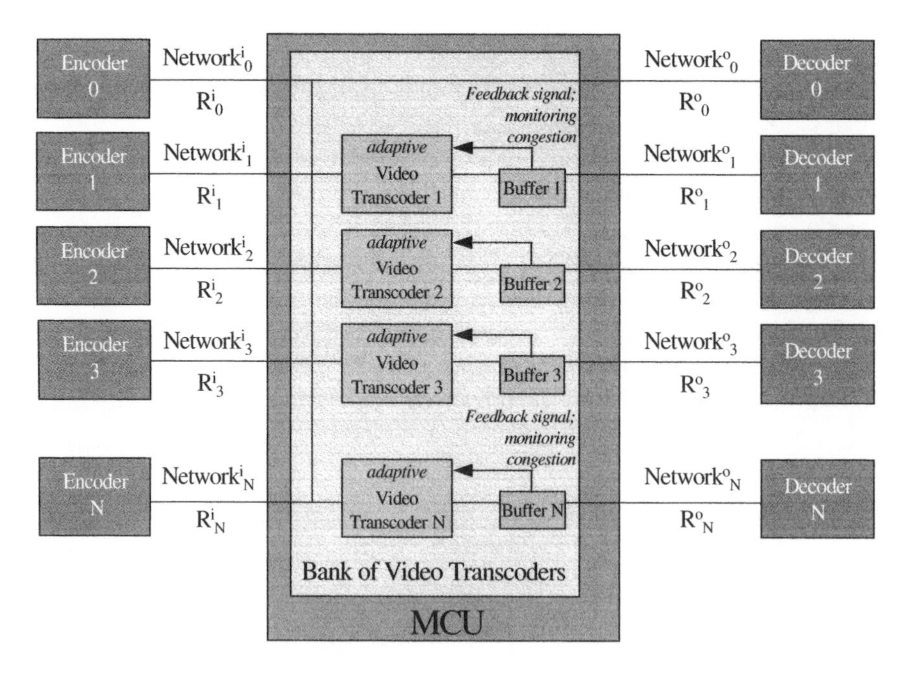

Figure 6.28 System architecture; the stack of video transcoders (superscripts *i* and *o* representing inputs and outputs, respectively, and subscripts representing network types)

than a conventional rate reduction transcoder, since the varying rate can also be in some cases increased for better QoS. The proposal of such a transcoding architecture that carries out the bit rate management for multimedia traffic planning is presented by Dogan, Sadka and Kondoz (2001).

Figure 6.29 shows the PSNR and bit rate variations achieved by the adaptive congestion control scheme for 200 frames of the Foreman sequence originally coded at 116 kbit/s with a frame rate of 25 f/s. The video transcoder stack produces a reduced rate $R^o = 82$ kbit/s from the input rate $R^i = 116$ kbit/s. However, due to exacerbating congestion, the supported rate is further reduced to an average of 66 and 53 kbit/s, respectively. This is evident in Figure 6.29 between frames 50–100 and 150–175. In these two particular cases, the video transcoder produces reduced rates at the cost of some quality degradation. On the other hand, between frames 100–125 where congestion is reduced, the adaptive transcoder bank increases the bit rate, providing the best achievable quality during this short period of the video sequence. It is clear that the adaptive transcoder follows well the variations in the congestion conditions adaptively with the use of feedback network reports.

The instantaneous bit rate and PSNR variations are also shown in Figure 6.29 with the $-$ sign representing the decrease, and the $+$ sign showing the increase with respect to the fixed bit rate levels. For comparison purposes, the PSNR figure also shows the values for certain fixed rate outputs of the transcoder. The adaptive

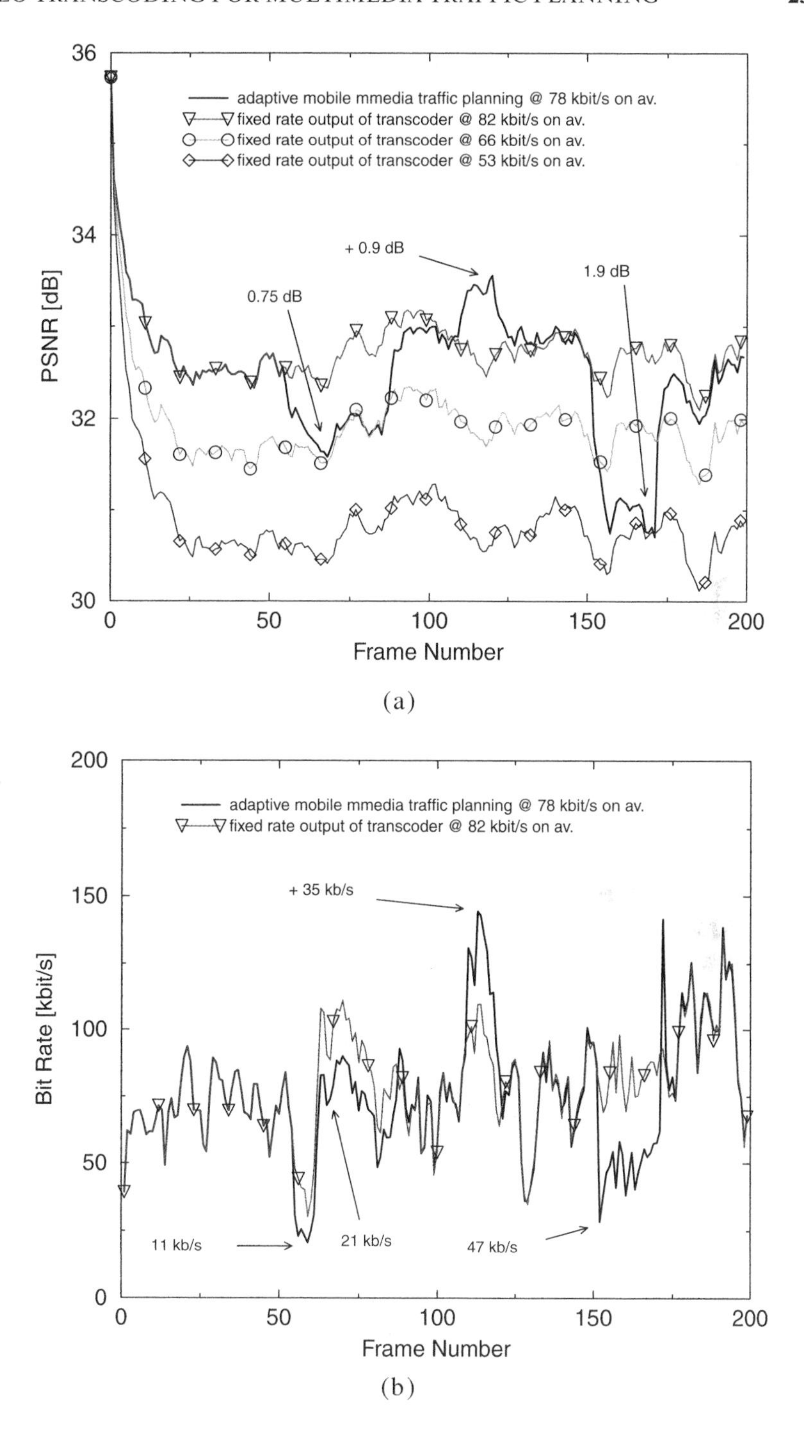

Figure 6.29 Objective results of the Foreman sequence for the adaptive multimedia traffic planning: (a) PSNR, (b) bit rate

scheme demonstrates an improved overall PSNR (32.483 dB at an average of 78 kbit/s), verifying that with this technique, the video quality degrades only during the period of congestion. Conversely, fixed rate transcoding presents a constantly lower quality performance.

6.16 Conclusions

Video transcoding at its best is yet to come, as the number of diverse multimedia networks rapidly increases. It has been a hot topic of research for some time and will remain a challenging and attractive issue for years to come, since interoperability of heterogeneous networks is of concern in the third-generation mobile multimedia communications and beyond. From the end-user's point of view, the underlying architecture, on which a particular application runs, is not regarded as a significant factor. The most important aspect is that the service quality reaches a level that satisfies the user requirements. Therefore, video transcoding should be made capable of operating transparently to the network clients and provide standards-compliant services.

The user transparent operation of a video transcoder has been described in this chapter. Furthermore, the standards-compliant services offered by the video transcoding algorithms have also been discussed. These services include rate reduction, resolution reduction and syntax conversions. In addition to error-free transmissions, error-resilient handling of the video data has also been examined at the transcoder for robust video transmissions. Since the communications trend has already converged onto the mobile wireless scenarios, the error-resilient video transcoding has become a requirement rather than an option due to the high BER encountered in mobile environments. Finally, the valuable role of a video transcoder for multimedia traffic planning has been demonstrated. It has been shown that congestion control can be carried out within the network at the video proxy. Thus, the source encoder and end-user terminals can be kept free from the networking issues which are resolved at the central proxy unit in the network. This results in supporting compact and simple network interfaces while providing dynamic and rapid solutions to congestion problems and transmission errors.

6.17 References

Acharya, S., and Smith, B., Compressed domain transcoding of MPEG, *Proceedings of the IEEE International Conference on Multimedia Computing and Systems*, 295–304, Austin, USA, June 1998.

Assuncao, P. A. A., and Ghanbari, M., Post-processing of MPEG-2 coded video for transmission at lower bit rates, *Proceedings of the IEEE International Conference on Acoustics, Speech, and Signal Processing, ICASSP '96*, **4**, 1998–2001, Atlanta, USA, May 1996.

Assuncao, P. A. A., and Ghanbari, M., Transcoding of single-layer MPEG video into lower rates, *IEE Proceedings – Vision, Image and Signal Processing*, **144**, No. 6, 377–383, Dec. 1997.

Assuncao, P. A. A., and Ghanbari, M., Transcoding of MPEG-2 video in the frequency domain, *Proceedings of the IEEE International Conference on Acoustics, Speech, and Signal Processing, ICASSP'97*, **4**, 2633–2636, Munich, Germany, April 1997.

Assuncao, P. A. A., and Ghanbari, M., A frequency-domain video transcoder for dynamic bit-rate reduction of MPEG-2 bit streams, *IEEE Transactions on Circuits and Systems for Video Technology*, **8**, No. 8, 953–967, Dec. 1998.

Assuncao, P. A. A., and Ghanbari, M., Fast computation of MC-DCT for video transcoding, *IEE Electronics Letters*, **33**, No. 4, 284–286, Feb. 1997.

Assuncao, P. A. A., and Ghanbari, M., Congestion control of video traffic with transcoders, *Proceedings of the IEEE International Conference on Communications, ICC'97*, **1**, 523–527, Montreal, Canada, June 1997.

Bjork, N., and Christopoulos, C., Transcoder architectures for video coding, *IEEE Transactions on Consumer Electronics*, **44**, No. 1, 88–98, Feb. 1998.

CCITT Recommendation H.261: Video codec for audio visual services at p × 64 kbits/s, COM XV-R 37-E, 1990.

de los Reyes, G., Reibman, A. R., Chuang, J. C., and Chang, S. F., Video transcoding for resilience in wireless channels, *Proceedings of the IEEE International Conference on Image Processing, ICIP'98*, **1**, 338–342, Chicago, USA, Oct. 1998.

de los Reyes, G., Reibman, A. R., Chang, S. F., and Chuang, J. C. I. Error-resilient transcoding for video over wireless channels, *IEEE Journal on Selected Areas in Communications*, **18**, No. 6, 1063–1073, June 2000.

Dogan, S., Sadka, A. H., and Kondoz, A. M., Tandeming/transcoding issues between MPEG-4 and H.263, Mobile and Personal Satellite Communications 3, *Proceedings of the Third European Workshop on Mobile/Personal Satcoms, EMPS'98*, Edited by M. Ruggieri, 339–346, Springer-Verlag, UK, 1999.

Dogan, S., Sadka, A. H., and Kondoz, A. M., Efficient MPEG-4/H.263 video transcoder for interoperability between heterogeneous multimedia networks, *IEE Electronics Letters*, **35**, No. 11, 863–864, May 1999.

Dogan, S., Sadka, A. H., and Kondoz, A. M., Video transmission over mobile satellite systems, *International Journal of Satellite Communications*, **18**, No. 3, 185–205, May 2000.

Dogan, S., Cellatoglu, A., Sadka, A. H., and Kondoz, A. M., Error-resilient MPEG-4 video transcoder for bit rate regulation, *Proceedings of the Fifth World Multi-conference on Systemics, Cybernetics and Informatics, SCI'2001*, vol. XII, parat 2, pp 312–317 Orlando, USA, July 2001.

Dogan, S., Sadka, A. H., and Kondoz, A. M., MPEG-4 video transcoder for mobile multimedia traffic planning, *Proceedings of the IEE Second International Conference on 3G Mobile Communication Technologies, 3G'2001*, IEE Conference Publication No. 477, 109–113, London, UK, Mar. 2001.

ElGebaly, H., Reactive mechanisms for recovering audio performance in multimedia conferencing over packet switched networks, *Intel Technology Journal*, **Q3**, 1–10, 1999.

Feamster, N., and Wee, S., An MPEG-2 to H.263 transcoder, *Proceedings of the Second Multimedia Systems and Applications, SPIE-International Society for Optical Engineering*, **3845**, 164–175, Boston, USA, Sept. 1999.

Fu, H. C., Chen, Z. H., Xu, Y. Y., and Wang, C. H., A neural network based transcoder for MPEG2 video compression, *Proceedings of the IEEE International Conference on Acoustics, Speech and Signal Processing, ICASSP'99*, **2**, 1125–1128, Phoenix, USA, Mar. 1999.

Ghanbari, M., Two-layer coding of video signals for VBR networks, *IEEE Journal on Selected Areas in Communications*, **7**, No. 5, 771–781, June 1989.

Gopalakrishnan, S., Reininger, D., and Ott, M., Realtime MPEG system stream transcoder for heterogeneous networks, *Proceedings of the Packet Video'99 Workshop*, NYC, USA, April 1999.

Guo, Z., Au, O. C., and Letaief, K. B., Parameter estimation for image/video transcoding, *Proceedings of the IEEE International Symposium on Circuits and Systems, ISCAS'2000, Emerging Technologies for the 21st Century*, **2**, 269–272, Geneva, Switzerland, May 2000.

Halsall, F., *Data Communications, Computer Networks and Open Systems*, Addison-Wesley, USA, 1996, 4th edition.

Hashemi, M. R., Winger, L., and Panchanathan, S., Compressed domain motion vector resampling for down-scaling of MPEG video, *Proceedings of the IEEE International Conference on Image Processing, ICIP'99*, **4**, 276–279, Kobe, Japan, Oct. 1999.

Hwang, J.-N., and Wu, T.-D., Motion vector re-estimation and dynamic frame-skipping for video transcoding, *Proceedings of the 32nd Asilomar Conference on Signals, Systems and Computers*, **2**, 1606–1610, Pacific Grove, USA, Nov. 1998.

Iannaccone, G., Reducing replication of data in a layered video transcoder, *Proceedings of the First International Networked Group Communication COST264 Workshop, NGC'99*, 144–151, Pisa, Italy, Nov. 1999.

ISO/IEC JTC1 10918-1, ITU-T Recommendation T.81: Information Technology – Digital compression and coding of continuous-tone still images: requirements and guidelines, 1994.

ISO/IEC CD 11172: Coding of moving pictures and associated audio for digital storage media at up to 1.5 Mbits/sec, Dec. 1991.

ISO/IEC CD 13818-2: Generic coding of moving pictures and associated audio, Nov. 1993.

ISO/IEC JTC1/SC29/WG11 N2802: Information technology – Generic coding of audio-visual objects – Part 2: Visual, Vancouver meeting, July 1999.

ISO/IEC JTC/SC29/WG11 MPEG97/N1646: Description of error resilient core experiments, Bristol meeting, April 1997.

ITU-T Recommendation H.263 (draft): Video coding for low bitrate communication, May 1996.

ITU-T Recommendation H.263 (draft): Video coding for low bitrate communication, Jan. 1998.

ITU-T Recommendation H.263 (draft): Video coding for low bitrate communication, Nov. 1998.

Kan, K.-S., and Fan, K.-C., Video transcoding architecture with minimum buffer requirement for compressed MPEG-2 bit stream, *Signal Processing*, **67**, No. 2, 223–235, June 1998.

Keesman, G., Hellinghuizen, R., Hoeksema, F., and Heideman, G., Transcoding of MPEG bit streams, *Signal Processing: Image Communication*, **8**, No. 6, 481–500, Sept. 1996.

Lee, S., Pattichis, M. S., and Bovik, A. C., Rate control for foveated MPEG/H.263 video, *Proceedings of the IEEE International Conference on Image Processing, ICIP'98*, **2**, 341–345, Chicago, USA, Oct. 1998.

Lei, Y.-W., and Ouhyoung, M., Software-based motion JPEG with progressive refinement for computer animation, *IEEE Transactions on Consumer Electronics*, **40**, No. 3, 557–562, Aug. 1994.

Lin, C.-W., Liou, T.-J., and Chen, Y.-C., Dynamic rate control in multipoint video transcoding, *Proceedings of the IEEE International Symposium on Circuits and Systems, ISCAS'2000, Emerging Technologies for the 21st Century*, **2**, 17–20, Geneva, Switzerland, May 2000.

Lois, L., and Bozoki, S., Transcoding of MPEG video using lattice vector quantisation, *Proceedings of the IEEE International Conference on Image Processing, ICIP'98*, **2**, 365–369, Chicago, USA, Oct. 1998.

Mokry, R., and Anastassiou, D., Minimal error drift in frequency scalability for motion compensation DCT coding, *IEEE Transactions on Circuits and Systems for Video Technology*, **4**, No. 4, 392–406, Aug. 1994.

Nakajima, Y., Hori, H., and Kanoh, T., Rate conversion of MPEG coded video by re-quantisation process, *Proceedings of the IEEE International Conference on Image Processing,*

ICIP'95, **3**, 408–411, Washington, DC, USA, Oct. 1995.

Natarajan, B. K., and Vasudev, B., A fast approximate algorithm for scaling down digital images in the DCT domain, *Proceedings of the IEEE International Conference on Image Processing, ICIP'95*, **2**, 241–243, Washington, DC, USA, Oct. 1995.

Radha, H., and Chen, Y., Fine-granular-scalable video for packet networks, *Proceedings of the Packet Video'99 Workshop*, NYC, USA, April 1999.

Redmill, D. W., and Kingsbury, N. G., The EREC: an error resilient technique for coding variable-length blocks of data, *IEEE Transactions on Image Processing*, **5**, No. 4, 565–574, April 1996.

Schafer, R., MPEG-4: a multimedia compression standard for interactive applications and services, *IEE Electronics and Communication Engineering Journal*, **10**, No. 6, 253–262, Dec. 1998.

Senda, Y., and Harasaki, H., A realtime software MPEG transcoder using a novel motion vector reuse and a SIMD optimization techniques, *Proceedings of the IEEE International Conference on Acoustics, Speech, and Signal Processing, ICASSP'99*, **4**, 2359–2362, Phoenix, USA, Mar. 1999.

Shanableh, T., and Ghanbari, M., Heterogeneous video transcoding MPEG:1,2 to H.263, *Proceedings of the Packet Video'99 Workshop*, NYC, USA, April 1999.

Shanableh, T., and Ghanbari, M., Transcoding of video into different encoding formats, *Proceedings of the IEEE International Conference on Acoustics, Speech, and Signal Processing, ICASSP'2000*, **4**, 1927–1930, Istanbul, Turkey, June 2000.

Shen, B., Sethi, I. K., and Vasudev, B., Adaptive motion-vector resampling for compressed video down-scaling, *IEEE Transactions on Circuits and Systems for Video Technology*, **9**, No. 6, 929–936, Sept. 1999.

Sisodia, G., Ling, G., De, S., and Dowlatshahi, M., Congestion control of compressed video traffic over ATM, *Proceedings of the IEEE Ninth International Conference on Computer Communications and Networks*, 476–479, Las Vegas, USA, Oct. 2000.

Sun, H., Kwok, W., and Zdepski, J. W., Architecture for MPEG compressed bit stream scaling, *IEEE Transactions on Circuits and Systems for Video Technology*, **6**, No. 2, 191–199, Apr. 1996.

Sun, M.-T., Wu, T.-D., and Hwang, J.-N., Dynamic bit allocation in video combining for multipoint conferencing, *IEEE Transactions on Circuits and Systems – II: Analog and Digital Signal Processing*, **45**, No. 5, 644–648, May 1998.

Swann, R., and Kingsbury, N. G., Transcoding of MPEG-II for enhanced resilience to transmission errors, *Proceedings of the IEEE International Conference on Image Processing, ICIP'96*, **2**, 813–816, Lausanne, Switzerland, Sept. 1996.

Talluri, R., Error-resilient video coding in the ISO MPEG-4 standard, *IEEE Communications Magazine*, **36**, No. 6, 112–119, June 1998.

Tudor, P. N., and Werner, O. H., Real-time transcoding of MPEG-2 video bit streams, *Proceedings of the International Broadcasting Convention, IBC 97*, IEE Conference Publication No. 447, 296–301, Amsterdam, Netherlands, Sept. 1997.

Vetro, A., and Sun, H., Generalised motion compensation for drift reduction, *Proceedings of the Visual Communication and Image Processing Annual Meeting, VCIP'98*, **3309**, 484–495, San Jose, USA, Jan. 1998.

Wang, Y., and Zhu, Q.-F., Error control and concealment for video communication: a review, *Proceedings of the IEEE*, **86**, No. 5, 974–997, May 1998.

Warabino, T., Ota, S., Morikawa, D., Ohashi, M., Nakamura, H., Iwashita, H., and Watanabe, F., Video transcoding proxy for 3Gwireless mobile internet access, *IEEE Communications Magazine*, **38**, No. 10, 66–71, Oct. 2000.

Werner, O., Requantisation for transcoding of MPEG-2 intraframes, *IEEE Transactions on*

Image Processing, **8**, No. 2, 179–191, Feb. 1999.

Wu, J.-L., Huang, S.-J., Huang, Y.-M., Hsu, C.-T., and Shiu, J., An efficient JPEG to MPEG-1 transcoding algorithm, *IEEE Transactions on Consumer Electronics*, **42**, No. 3, 447–457, Aug. 1996.

Yeadon, N., Davies, N., Friday, A., and Blair, G., Supporting video in heterogeneous mobile environments, *Proceedings of the ACM Symposium on Applied Computing, SAC '98*, Marriott Marquis, Atlanta, USA, Feb. 1998.

Youn, J., Sun, M.-T., and Xin, J., Video transcoder architectures for bit rate scaling of H.263 bit streams, *Proceedings of the ACM Multimedia '99*, 243–250, Orlando, USA, Oct. 1999.

Youn, J., and Sun, M.-T., Video transcoding with H.263 bit-streams, *Journal of Visual Communication and Image Representation*, **11**, No. 4, 385–403, Dec. 2000.

Youn, J., Sun, M.-T., and Lin, C.-W., Motion estimation for high performance transcoding, *IEEE Transactions on Consumer Electronics*, **44**, No. 3, 649–657, Aug. 1998.

Youn, J., Sun, M.-T., and Lin, C.-W., Motion vector refinement for high-performance transcoding, *IEEE Transactions on Multimedia*, **1**, No. 1, 30–40, Mar. 1999.

Zhu, W., Yang, K. H., and Beacken, M. J., CIF-to-QCIF video bit stream down-conversion in the DCT domain, *Bell Labs Technical Journal*, **3**, No. 3, 21–29, July 1998.

Appendix A

Layering syntax of ITU-T H.263 video coding standard

The syntax and layering structure of ITU-T H.263 video source coder are depicted in Figure A.1. Each of the acronyms that represents a video parameter used in the syntax is then defined under the layer to which it belongs.

A.1 Picture Layer

Each picture consists of a picture header followed by data for Group of Blocks (GOBs), eventually followed by an end-of-sequence (EOS) code and stuffing bits so that the headers are always byte-aligned. For each codeword, the most significant bit is transmitted first, followed by the remaining bits of a codeword.

A.1.1 Picture start code (PSC) (22 bits +0–7 stuffing bits)

PSC is a word of 22 bits. Its value is 0000 0000 0000 0000 1 00000. All picture start codes must be byte-aligned. This can be achieved by inserting less than eight zero-bits at the end of the previous frame, such that the first bit of the start code is the first bit of a byte which represents the start of the current frame. The first 17 bits represent a specific bit pattern that notifies the video decoder of the beginning of a new frame, and the last 5 bits denote the number of the first GOB of the frame.

A.1.2 Temporal reference (TR) (8 bits)

An eight-bit number which can be assigned any of 256 possible values. It is formed by incrementing its value in the previously transmitted picture header by one, plus the number of non-transmitted pictures (at 29.97 Hz) since the previously transmitted one. When the PB-frame option is on, TR addresses only P-frames.

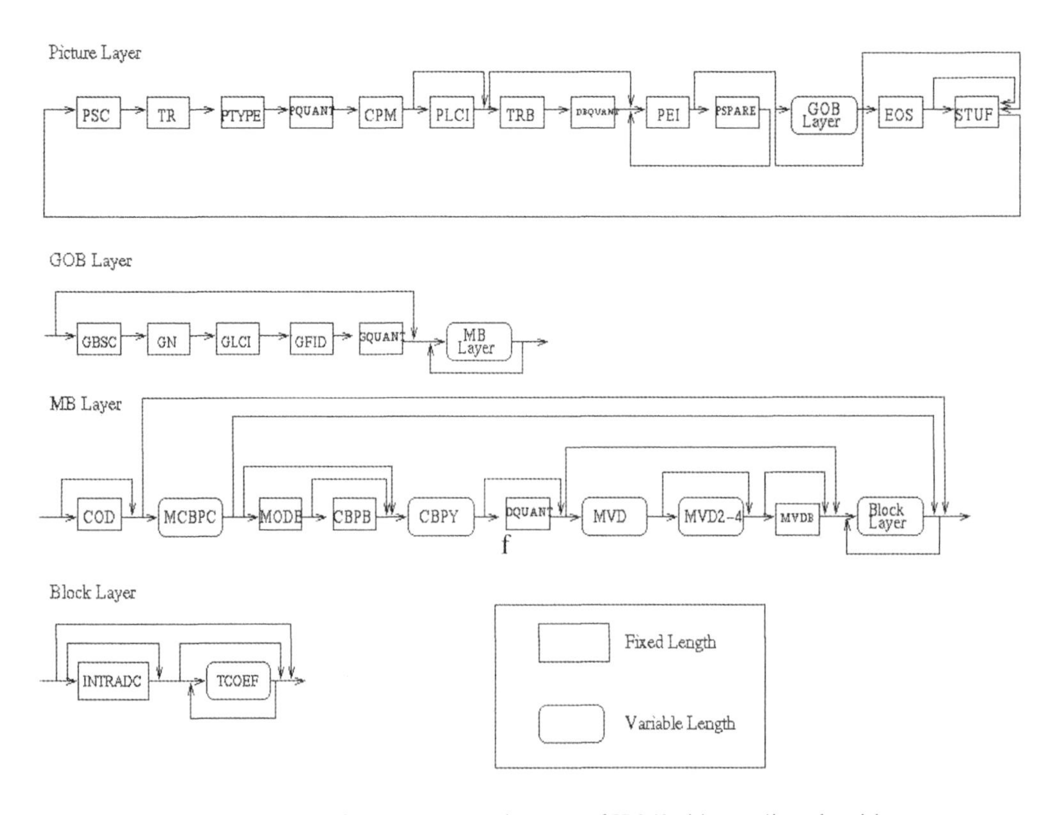

Figure A.1 Layering structure and syntax of H.263 video coding algorithm

A.1.3 Picture type information (PTYPE) (13 bits)

Information about the complete picture

Bit 1: always 0, for distinction with H.261

Bit 2: always 1, in order to avoid start code emulation

Bit 3: Split screen indicator, 0 off, 1 on; split screen indicator is a flag that indicates that the upper and lower half of the decoded picture could be displayed side by side. It has no direct effect on the encoding or decoding of the picture

Bit 4: Document camera indicator, 0 off, 1 on

Bit 5: Freeze picture release, 0 off, 1 on; freeze picture release is a signal from an encoder which responds to a request for packet retransmission (if not acknowledged) or fast update request, and allows a decoder to exit from its freeze picture mode and display the decoded picture in the normal manner.

Bits 6–8: Source format, 000 sub-QCIF, 001 QCIF, 010 CIF, 011 4CIF, 100 16CIF, 101 reserved, 110 reserved, 111 reserved

Bit 9: Picture coding type, 0 INTRA, 1 INTER

Bits 10–13 refer to the optional modes that are only used after negotiation between encoder and decoder. If bit 12 is set to 1, bit 10 shall be set to 1 as well as described earlier in the advanced prediction mode

Bit 10: Optional unrestricted motion vector mode, 0 off, 1 on

Bit 11: Optional syntax-based arithmetic coding mode, 0 off, 1 on

Bit 12: Optional advanced prediction mode, 0 off, 1 on

Bit 13: Optional PB-frame mode, 0 normal picture, 1 PB-frame

A.1.4 Continuous presence multipoint (CPM) (1 bit)

A one-bit codeword that signals the presence of multiple destinations. In a multipoint connection, a multipoint control unit (MCU) can assemble two to four bit streams into one video bit stream, so that at the receiver up to four different video signals can be displayed at the same time in a sort of quad-screen.

A.1.5 Picture logical channel indicator (PLCI) (2 bits)

A fixed length codeword of two bits that is only present if CPM mode is indicated. A maximum of four channels can therefore be indicated to the decoder. The codeword is a binary representation of the logical channel number of the picture header and all following information until the next picture or GOB start code.

A.1.6 Quantiser information (PQUANT) (5 bits)

A fixed length codeword of five bits which indicates the quantisation parameter (Qp) to be used for the picture until updated by any subsequent GQUANT or DQUANT. This codeword is then the binary representation of the values of QUANT which range from 1 to 31.

A.1.7 Temporal reference for B-frames (TRB) (3 bits)

TRB is present if PTYPE indicates PB-frame and indicates the number of non-transmitted pictures (at 29.97 Hz) since the last P- or I-picture and before the

B-picture. The codeword is then the binary representation of the number of non-transmitted pictures plus one.

A.1.8 Quantisation information for B-pictures (DBQUANT) (2 bits)

DBQUANT is present if PTYPE indicates the presence of PB-frame mode. In the decoding process, a quantisation parameter QUANT is obtained for each MB. However, with PB-frame on, QUANT is used for the P-block, whereas for the B-block a different quantisation parameter, namely BQUANT, is used. DBQUANT indicates the relation between QUANT and BQUANT as defined in Table A.1, where / means division by truncation. BQUANT ranges from 1 to 31; if the value of BQUANT resulting from Table A.1 is less than 1 or greater than 31, it is clipped to 1 and 31, respectively.

A.1.9 Extra insertion information (PEI) (1 bit)

A one-bit flag which, when set to 1, signals the presence of the following optional data field.

A.1.10 Spare information (PSPARE) (0/8/16... bits)

If PEI is set to 1, then 9 bits follow consisting of 8 bits of data (PSPARE) and then another PEI bit to indicate if a further 9 bits follow, and so on. This segment of bits is left for future use by ITU in case bit additions are required. IF PSPARE is followed by PEI = 0, PSPARE = xx000000 is prohibited in order to avoid start code emulation. Therefore, 4 out of 256 values would be prohibited in this case.

A.1.11 End of sequence (EOS) (22 bits +0–7 stuffing bits)

This is a codeword whose first 17 bits are similar to those at the beginning of a

Table A.1 DBQUANT and relation between QUANT and BQUANT

DBQUANT	BQUANT
00	(5 × QUANT)/4
01	(6 × QUANT)/4
10	(7 × QUANT)/4
11	(8 × QUANT)/4

PSC. Its value is 0000 0000 0000 0000 1 11111. Its insertion in the bit stream indicates to the decoder the end of a sequence.

A.1.12 Stuffing (STUF) (variable length)

A variable-length codeword which consists of 0 bits. Encoders insert this codeword directly after the last bit of a frame so that the video bit stream including STUF is a multiple of 8 bits from the first bit in the bit stream. Decoders are designed to discard STUF.

A.2 Group of Blocks Layer

Each GOB consists of a GOB header followed by data for MBs. Each GOB contains one or more rows of MBs. For the first GOB (with number 0), no GOB header shall be transmitted. For all other GOBs, the GOB header may be empty, depending on the encoder strategy.

A.2.1 Group of block start code (GBSC) (17 bits +0–7 stuffing bits)

A word of 17 bits and value 0000 0000 0000 0000 1. GOB start codes are made byte-aligned by inserting less than 8 zero-bits before the start code such that the first bit of the start code is the first bit of a byte.

A.2.2 Group number (GN) (5 bits)

A fixed length codeword of 5 bits used to represent the number of a GOB within a frame. For the GOB with number 0, the GOB header consisting of GBSC, GN, GFID and GQUANT is empty. Group number 0 is used in PSC, as mentioned earlier. Group number 31 (11111) is used in EOS and the values from 18 to 30 are reserved for future use by the recommendation.

A.2.3 GOB logical channel indicator (GLCI) (2 bits)

A two-bit codeword that is present only if CPM mode is indicated. It represents the logical channel number for the GOB header and all following information until the next picture or GOB start code.

A.2.4 GOB frame ID (GFID) (2 bits)

This is a fixed-length two-bit codeword. GFID will have the same value in every GOB header of a given picture. Moreover, if the PTYPE indicated in the picture header is the same as for the previous transmitted picture, GFID will have the same value as in that previous picture.

A.2.5 Quantiser information (GQUANT) (5 bits)

This is a five-bit codeword used to indicate the quantiser QUANT to be used for that GOB until updated by any subsequent DQUANT. Since it represents a value for QUANT, this word can take a value from 1 to 31.

A.3 MB Layer

Each MB consists of an MB header followed by data for blocks. COD is only present in pictures for which PTYPE indicates INTER. MCBPC is present when indicated by COD or when PTYPE indicates INTRA. MODB is present if PTYPE indicates a PB-frame. CBPY, DQUANT, MVD and MVD2-4 are present when indicated by MCBPC. CBPB and MVDB are only present if indicated by MODB. Block data is present when indicated by MCBPC and CBPY.

A.3.1 Coded macroblock indication (COD) (1 bit)

A one-bit flag which when set to 0 signals that the MB is coded. If set to 1, the MB is not coded and the remaining part of the MB layer is empty.

A.3.2 Macroblock type & coded block pattern for chrominance (variable length)

The MB type gives information about the MB and which data elements are present in it. The coded block pattern for chrominance signifies Cb and Cr blocks when at least one non-INTRADC coefficient is transmitted. CBPCN = 1 if any non-INTRADC coefficient is present for block N.

A.3.3 Macroblock mode for B-blocks (variable length)

MODB is present for MB-type 0–4 if PTYPE indicates PB-frame mode and is a variable-length codeword indicating whether B-coefficients and/or B-vectors are transmitted for this MB.

A.3.4 Coded block pattern for B-blocks (CBPB) (6 bits)

CBPB is present only if indicated by MODB. CBPBN = 1 if any coefficient is present for B-block N, else 0, for each bit CBPBN in the coded block pattern.

A.3.5 Coded block pattern for luminance Y (CBPY) (variable length)

A variable-length codeword giving a pattern signifying the Y-blocks for which at least one non-INTRADC transform coefficient is transmitted. CBPYN = 1 if any non-INTRADC coefficient is present for block N, else 0.

A.3.6 Quantiser information (DQUANT) (2 bits)

A two-bit code tailored to define the change in the quantiser value. The differential value is transmitted.

A.3.7 Motion vector data (MVD) (variable length)

MVD is included for all INTER MBs (in PB-frame mode also) and consists of a variable length codeword for the horizontal component followed by a variable length codeword for the vertical component.

A.3.8 Motion vector data (MVD2–4) (variable length)

The three codewords MVD2–4 are included if indicated by PTYPE (Advanced Prediction Mode) and by MCBPC. Each of them consists of a variable length codeword for the horizontal component, followed by a variable length codeword for the vertical component.

A.3.9 Motion vector data for B-macroblock (MVDB) (variable length)

MVDB is only present if indicated by MODB, and it consists of a variable length codeword for the horizontal component, followed by a variable length codeword for the vertical component of each vector.

A.4 Block Layer

A MB comprises four luminance blocks and one of each of the two colour blocks. INTRADC is present for a block when indicated by MCBPC. TCOEF is present if indicated by MCBPC or CBPY.

A.4.1 DC coefficient for INTRA blocks (INTRADC) (8 bits)

An eight-bit codeword representing the DC coefficient of an INTRA block. The codes 0000 0000 and 1000 0000 are not used.

A.4.2 Transform coefficient (TCOEFF) (variable length)

The codewords represent the output of the run-length coder. They encode the 63 remaining coefficients of a block (excluding the first INTRADC coefficient) by a set of EVENTs. An EVENT is a combination of a last non-zero coefficient indication (LAST is assigned a 0 when there are more non-zero coefficients in the block and a 1 when this is the last non-zero coefficient in the block), the number of successive zeros preceding the coded coefficient (RUN), and the non-zero value of the coded coefficient (LEVEL). The remaining combinations of (LAST, RUN, LEVEL) are coded with a 22-bit word consisting of 7 bits ESCAPE, 1 bit LAST, 6 bits RUN and 8 bits LEVEL.

Appendix B

Description of the video clips on the supplementary CD

All the video clips on the supplementary CD have been compressed using the yuv2avi Microsoft Video 1 coder with 25 per cent quality loss in order to enable their near-real-time playback on limited processing-speed PC computers. The AVI files are further compressed using the windows media eight video encoding utility, namely wmeutil8, to produce the compressed WMV video files. The WMV video clips can be played using the Microsoft Windows Media Player tool. The sequences are all QCIF-size and encoded at 25 f/s without any frame-skip applied. The original Suzie sequence consists of 150 frames, while Foreman and Claire sequences have 200 frames each. The error-resilience tools used for MPEG-4, wherever explicitly mentioned, include resynchronisation, data-partitioning and two-way decoding.

1. claire-orig.wmv*
   Claire original sequence.

2. claire_h263.wmv*
   Claire sequence encoded with H.263 at 64 kbit/s (error-free).

3. claire_h263_erred.wmv*
   Claire sequence encoded with H.263 at 64 kbit/s and subject to random errors with BER = 0.0001. No error-resilience mechanism is used before transmission.

4. claire_h263_erres.wmv*
   Claire sequence encoded with H.263/M (refer to Chapter 4 for details) at 64 kbit/s and subject to random errors with BER = 0.0001.

5. fman_10q1_20q2_errfree.wmv*
   Foreman sequence encoded with MPEG-4 at 87 kbit/s (Qp1 = 10), and transcoded to 47 kbit/s (Qp2 = 20) using the closed-loop transcoding scheme with MV refinement, where the MV refinement window size = ± 5 pels (error free).

6. fman_cs2_12db_prone.wmv*
   Foreman sequence encoded with MPEG-4 at 87 kbit/s (Qp1 = 10), and trans-

coded to 47 kbit/s (Qp2 = 20) using the closed-loop transcoding scheme with MV refinement, where the MV refinement window size = ±5 pels. The transcoded bit stream is then transmitted over a GPRS channel of C/I = 12 dB, using 4 timeslots and CS-2. No error-resilience mechanism is used at the homogeneous transcoder.

7. fman_cs2_12db_resil.wmv*
   Foreman sequence encoded with MPEG-4 at 87 kbit/s (Qp1 = 10), and transcoded to 47 kbit/s (Qp2 = 20) using the closed-loop transcoding scheme with MV refinement, where the MV refinement window size = ±5 pels. The transcoded bit stream is then transmitted over a GPRS channel of C/I = 12 dB, using 4 timeslots and CS-2. AIR is used at the homogeneous transcoder while preserving the same transcoder bit rate of 47 kbit/s.

8. for_h2h.wmv*
   Foreman sequence encoded/decoded with H.263 at 100 kbit/s, then re-encoded/decoded with H.263 at 100 kbit/s (cascaded).

9. for_h2m.wmv*
   Foreman sequence encoded/decoded with H.263 at 100 kbit/s, then encoded/decoded with MPEG-4 at 100 kbit/s.

10. for_htm.wmv*
   Foreman sequence encoded with H.263 at 100 kbit/s, transcoded to MPEG-4 and then decoded with MPEG-4.

11. for_m2h.wmv*
   Foreman sequence encoded/decoded with MPEG-4 at 100 kbit/s, then encoded/decoded with H.263 at 100 kbit/s.

12. for_m2m.wmv*
   Foreman sequence encoded/decoded with MPEG-4 at 100 kbit/s, then re-encoded/decoded with MPEG-4 at 100 kbit/s (cascaded).

13. for_mth.wmv*
   Foreman sequence encoded with MPEG-4 at 100 kbit/s, transcoded to H.263, then decoded with H.263.

14. foreman-orig.wmv*
   Foreman original sequence.

15. suz_18kbs_000343.wmv*
   Suzie sequence coded with MPEG-4 at 18 kbit/s, with error-resilience tools enabled, and transmitted over a GPRS channel with C/I = 15 dB using CS-2. INTRA frame refresh is used once in the sequence.

16. suz_32kbs_001229.wmv*
   Suzie sequence coded with MPEG-4 at 32 kbit/s, with error-resilience tools enabled, and transmitted over a GPRS channel with C/I = 15 dB using CS-2.

INTRA frame refresh is used once in the sequence.

17. suz_64kbs_000732.wmv*
Suzie sequence coded with MPEG-4 at 64 kbit/s, with error-resilience tools enabled, and transmitted over a GPRS channel with C/I = 15 dB using CS-2. INTRA frame refresh is used once in the sequence.

18. suz_h2h.wmv*
Suzie sequence encoded/decoded with H.263 at 56 kbit/s, then re-encoded/decoded with H.263 at 56 kbit/s (cascaded).

19. suz_h2m.wmv*
Suzie sequence encoded/decoded with H.263 at 56 kbit/s, then re-encoded/decoded with MPEG-4 at 56 kbit/s.

20. suz_h_128.wmv*
Suzie sequence encoded/decoded with H.263 at 128 kbit/s (error-free).

21. suz_h_16.wmv*
Suzie sequence encoded/decoded with H.263 at 16 kbit/s (error-free).

22. suz_h_32.wmv*
Suzie sequence encoded/decoded with H.263 at 32 kbit/s (error-free).

23. suz_h_64.wmv*
Suzie sequence encoded/decoded with H.263 at 64 kbit/s (error-free).

24. suz_htm.wmv*
Suzie sequence encoded with H.263 at 56 kbit/s, transcoded to MPEG-4, then decoded with MPEG-4.

25. suz_m2h.wmv*
Suzie sequence encoded/decoded with MPEG-4 at 56 kbit/s, then re-encoded/decoded with H.263 at 56 kbit/s.

26. suz_m2m.wmv*
Suzie sequence encoded/decoded with MPEG-4 at 56 kbit/s, then re-encoded/decoded with MPEG-4 at 56 kbit/s (cascaded).

27. suz_m_128.wmv*
Suzie sequence encoded/decoded with MPEG-4 at 128 kbit/s.

28. suz_m_16.wmv*
Suzie sequence encoded/decoded with MPEG-4 at 16 kbit/s.

29. suz_m_32.wmv*
Suzie sequence encoded/decoded with MPEG-4 at 32 kbit/s.

30. suz_m_64.wmv*
Suzie sequence encoded/decoded with MPEG-4 at 64 kbit/s.

31. suz_mth.wmv*
Suzie sequence encoded with MPEG-4 at 56 kbit/s, transcoded to H.263, then decoded with H.263.

32. suzair_18kbs.wmv*
Suzie sequence coded with MPEG-4, with AIR enabled, at 18 kbit/s (error-free).

33. suzair_18kbs_000660.wmv*
Suzie sequence coded with MPEG-4 at 18 kbit/s, with error-resilience tools enabled, and transmitted over a GPRS channel with C/I = 15 dB using CS-2. AIR is also used to mitigate the effects of transmission errors.

34. suzair_32kbs.wmv*
Suzie sequence coded with MPEG-4, with AIR enabled, at 32 kbit/s (error-free).

35. suzair_32kbs_001135.wmv*
Suzie sequence coded with MPEG-4 at 32 kbit/s, with error-resilience tools enabled, and transmitted over a GPRS channel with C/I = 15 dB using CS-2. AIR is also used to mitigate the effects of transmission errors.

36. suzair_64kbs.wmv*
Suzie sequence coded with MPEG-4, with AIR enabled, at 64 kbit/s (error-free).

37. suzair_64kbs_000782.wmv*
Suzie sequence coded with MPEG-4 at 64 kbit/s, with error-resilience tools enabled, and transmitted over a GPRS channel with C/I = 15 dB using CS-2. AIR is also used to mitigate the effects of transmission errors.

38. suzie_14q1_direct_encdec_only.wmv*
Suzie sequence encoded/decoded with MPEG-4 at 32 kbit/s for Qp = 14.

39. suzie_6q1_12q2_mvrefine4_errorfree.wmv*
Suzie sequence encoded with MPEG-4 at 95 kbit/s and transcoded to 39 kbit/s using the closed-loop transcoding scheme with MV refinement, where the MV refinement window size = ± 4 pels (error-free case).

40. suzie_6q1_12q2_mvrefine4_errorprone_seed21500.wmv*
Suzie sequence encoded with MPEG-4 at 95 kbit/s and transcoded to 39 kbit/s using the closed-loop transcoding scheme with MV refinement, where the MV refinement window size = ± 4 pels. Then the transcoded bit stream is subject to random errors at BER = 0.001, while no error-resilience is incorporated in the homogeneous transcoder.

41. suzie_6q1_12q2_mvrefine4_errorresil_seed21500.wmv*
Suzie sequence encoded with MPEG-4 at 95 kbit/s and transcoded to 39 kbit/s using the closed-loop transcoding scheme with MV refinement, where the MV refinement window size = ±4 pels. Then the transcoded bit stream is subject to random errors at BER = 0.001, while AIR (Adaptive INTRA Refresh) is used at the homogeneous transcoder for error resilient transmissions. The transcoded bit rate of 39 kbit/s is maintained during the AIR operation.

42. suzie_8q1_14q2_cascaded_encdec.wmv*
Suzie sequence encoded/decoded with MPEG-4 at 64 kbit/s, then re-encoded/decoded with MPEG-4 at 32 kbit/s (cascaded enc/dec).

43. suzie_8q1_14q2_clsdtrnscd.wmv*
Suzie sequence encoded with MPEG-4 at 64 kbit/s and transcoded to 32 kbit/s using the closed-loop transcoding scheme with MV re-use.

44. suzie_8q1_14q2_mvrefine4.wmv*
Suzie sequence encoded with MPEG-4 at 64 kbit/s and transcoded to 32 kbit/s using the closed-loop transcoding scheme with MV refinement, where the MV refinement window size = ±4 pels.

45. suzie_8q1_14q2_mvrestimate.wmv*
Suzie sequence encoded with MPEG-4 at 64 kbit/s and transcoded to 32 kbit/s using the closed-loop transcoding scheme with MV re-estimation.

46. suzie_8q1_14q2_opentrscd.wmv*
Suzie sequence encoded with MPEG-4 at 64 kbit/s and transcoded to 32 kbit/s using the open-loop transcoding scheme.

47. suzie_8q1_direct_encdec_only.wmv*
Suzie sequence encoded/decoded with MPEG-4 at 64 kbit/s for Qp = 8.

48. suzie_cs1cs3_12db_28.wmv
Suzie sequence coded with MPEG-4 and transmitted over GPRS of C/I = 12 dB, using 2 channel bearers with stream prioritisation (refer to Chapter 5). The low-priority texture stream is transmitted using CS-3 while the high-priority header and motion stream is transmitted using CS-1.

49. suzie_cs2_12db_25.wmv
Suzie sequence coded with MPEG-4 and transmitted over GPRS of C/I = 12 dB, using CS-2 over a single channel bearer.

50. suzie_orig.wmv*
Original suzie sequence.

Glossary of Terms

2D	Two-dimensional
2.5G	Evolved second-generation mobile networks
3D	Three-dimensional
3G	Third-generation mobile networks
ACK	Acknowledgement
AIR	Adaptive INTRA Refresh
ANSI	American National Standards Institute
APM	Advanced Prediction Mode
ARQ	Automatic repeat request
ASO	Arbitrary Slice Ordering
ATM	Asynchronous Transfer Mode
AVT	Audio Video Transport
AWGN	Automatic White Gaussian Noise
BER	Bit Error Ratio
B-Frame	Bi-directional Frame
bit/s	bits per second
BL	Base Layer
BM	Block Matching
BSS	Base Station Subsystem
CBPY	Coding Block Pattern for Y (luminance)
CCITT	International Committee for Communications by Telephony and Telegraphy
CDMA	Code Division Multiple Access
C/I	Carrier-to-Interference ratio
CIF	Common Intermediate Format
CIR	Cyclic INTRA Refresh
CRC	Cyclic Redundancy Check
CS	Coding Scheme
DCT	Discrete Cosine Transform
DQUANT	Differential Quantiser
DSP	Digital Signal Processor
DVB	Digital Video Broadcasting

DVD	Digital Versatile Disc
EDGE	Enhanced Data rate GSM Evolution
EEP	Equal Error Protection
EGPRS	Enhanced GPRS
EL	Enhancement Layer
EREC	Error Resilient Entropy Coding
ERPS	Enhanced RPS
ETSI	European Telecommunications Standards Institute
EZW	Embedded Zero-tree Wavelet
FEC	Forward Error Correction
FGS	Fine Granular Scaleability
FGST	FGS-Temporal
FLC	Fixed Length Coding
FLOP	Floating Point Operations
f/s	frames per second
FTP	File Transfer Protocol
FTYPE	Frame Type
GBSC	GOB Start Code
GGSN	Gateway GSN
GN	GOB Number
GOB	Group Of Blocks
GOP	Group of Pictures
GPRS	General Packet Radio Service
GQUANT	GOB Qp
GSM	Global Mobile System
GSN	GPRS Support Node
HEC	Header Extension Code
HDTV	High-Definition Television
HSCSD	High Speed Circuit Switched Data
HTTP	Hyper Text Transport Protocol
HVS	Human Visual System
IDCT	Inverse DCT
IETF	Internet Engineering Task Force
I-Frame	INTRA Frame
I-f/s	INTRA frames per second
ILB	Intelligent Buffer
I-MB	INTRA MB
INTRAQ	INTRA MB Qp
IP	Internet Protocol
ISDN	Integrated Services Digital Network
ISO	International Standard Organisation
ITU	International Telecommunication Union

JPEG	Joint Picture Experts Group
kbit/s	kilo bits per second
LBG	Linde-Buzo-Gray
MAC	Medium Access Control
MB	Macroblock
MCBPC	MB Mode and Coding Block Pattern for Chrominance
MCU	Multipoint Control Unit
ML	Multi-Layer
MoMuSys	Mobile Multimedia Systems
MPEG	Motion Picture Experts Group
MS	Mobile Station
MSE	Mean Square Error
MV	Motion Vector
MVD	Motion Vector Difference
MVDB	Motion Vector Difference for B-MB
NACK	Negative ACK
NV	NetVideo
OMBC	Overlapped Block Motion Compensation
PB-Frame	Predicted and Bi-directional frame mode
PDTCH	Packet Data Traffic Channel
PDU	Packet Datagram Unit
P-Frame	Predicted frame (INTER)
PLR	Packet Loss Rate
PQUANT	P-frame Qp
PSC	Picture Start Code
PSNR	Peak-to-peak SNR
PSTN	Public Switching Telephone Networks
PTYPE	Picture Type (FTYPE)
QCIF	Quarter CIF
QoS	Quality of Service
Qp	Quantisation parameter
RCPC	Rate Compatible Punctured Code
RLC	Radio Link Control
ROI	Region Of Interest
RPS	Reference Picture Selection
RRC	Reduced Resolution rate Control
RS	Reed-Solomon
RTP	Real-time Transport Protocol
RVLC	Reversible VLC
SAD	Sum of Absolute Differences
SGSN	Serving GSN
SL	Single Layer
SNDC	Sub-Network Dependent Convergence

SNR	Signal-to-Noise Ratio
SRC	Scaleable Rate Control
TCP	Transport Control Protocol
TCOEFF	DCT Transformed Coefficients
TDMA	Time Division Multiple Access
TM5	Test Model 5
TMN	Test Model Near-term
TR	Temporal Reference
TRB	TR for B-frame
TS	Timeslot
UDP	User Datagram Protocol
UEP	Unequal Error Protection
UMTS	Universal Mobile Telecommunication Systems
UTRAN	UMTS Terrestrial Radio Access Network
VLC	Variable Length Coding
VLSI	Very Large Scale Integration
VoD	Videoon Demand
VOP	Visual Object Plane
VRC	Video Redundancy Coding
WCDMA	Wireless CDMA
Y-PSNR	PSNR for Luminance (Y) component
YUV	1 component for luminance (Y), 2 components for chrominance (U and V)

Index

Printed and bound in the UK by
CPI Antony Rowe, Eastbourne